T0143259

Learned Patriots

Learned Patriots

*Debating Science, State, and Society in
the Nineteenth-Century Ottoman Empire*

M. ALPER YALÇINKAYA

The University of Chicago Press   Chicago and London

M. ALPER YALÇINKAYA is assistant professor in the Department
of Sociology/Anthropology at Ohio Wesleyan University.

The University of Chicago Press, Chicago 60637
The University of Chicago Press, Ltd., London
© 2015 by The University of Chicago Press
All rights reserved. Published 2015.
Printed in the United States of America

24 23 22 21 20 19 18 17 16 15 14   1 2 3 4 5

ISBN-13: 978-0-226-18420-3 (cloth)
ISBN-13: 978-0-226-18434-0 (e-book)
DOI: 10.7208/chicago/9780226184340.001.0001

Library of Congress Cataloging-in-Publication Data

Yalçınkaya, M. Alper, author.
  Learned patriots: debating science, state, and society in the nineteenth-
century Ottoman Empire / M. Alper Yalçınkaya.
      pages; cm
  Includes bibliographical references and index.
  ISBN 978-0-226-18420-3 (cloth: alk. paper) — ISBN 978-0-226-18434-0
(e-book) 1. Science—Social aspects—Turkey—History—19th century.
2. Science and state—Turkey—History—19th century. 3. Islam and
science—Turkey—History—19th century. 4. Science—Moral and ethical
aspects—Turkey—History—19th century. 5. Science—Turkey—Public
opinion—History—19th century. I. Title.
  Q175.52.T9Y35 2015
  338.956'0609034—dc23   2014017518

♾ This paper meets the requirements of ANSI/NISO Z39.48-1992
(Permanence of Paper).

Contents

Introduction

Import the science of the West, import its arts;
And carry out your task with utmost pace.
For no longer is it possible to live without them;
For art and science alone belong to no nation.

These lines are from a celebrated poem by the Turkish poet Mehmet Akif Ersoy (1873–1936), a household name in his country and the creator of some of the most well-known stanzas in Turkish, including the one above.[1] The author of the national anthem of Turkey, Ersoy is also a hero of Turkish Islamic conservatism. This is due particularly to his determinedly critical attitude toward the speed and form of the Westernizing reforms of Kemal Atatürk and his comrades in the Turkish Republic of the 1920s and the 1930s—an attitude that, allegedly, led to his self-imposed exile in Egypt.[2] Yet how is it, then, that a man with Islamist proclivities and a critic of the Westernization of Turkey espoused the "science of the West" so unequivocally? Can a person be truly "pro-science" if he or she is "anti-Western"?

In a way, the answer to these questions is also provided in the same stanza: according to the poet, science could perhaps be found in its most perfect form in the West, but it wasn't inherently Western. Indeed, in its essence, science was not a product or representative of any one nation and could not, in itself, be a threat against what was "native."

But Ersoy was also quick to attach a warning to this description elsewhere in the same poem: while science was neutral, it did not, so to speak, travel alone. As a result, it was imperative for Turks in particular and Muslims in general to be very cautious when allowing science to pass through their borders. And in this endeavor, they had a model to emulate: the Japanese. Of all that the civilization of the West had to export, only the contemporary sciences had been able to enter Japan, and even that had been possible only thanks to the "permission" of the judicious Japanese people.[3]

Thousands with foresight and faith stood by the shores,
And kicked out of the gates depravities of all sorts![4]
Western goods will proceed only if truly precious,
Evil that arrives as fashion will rot at customs.[5]

Thus, the picture is more complicated than at first appears: science itself is inert, but it is to be imported from the West, where it happens to reside among depravities. Therefore, it is the duty of the "foresighted" and "faithful" "non-Westerner" to separate science from depravity at the border. Or perhaps it is the ability to distinguish between "beneficial science" and "fashionable perversion" that proves that one has faith and foresight. In either case, Ersoy's depictions of science that are intended to encourage its rapid importation are simultaneously suggestions about the type of person that the "non-West" needs. Indeed, his collection of poems *Safahat* (Phases), from which the above stanzas come, contains many similar examples: an argument about the merits of science is, almost as a rule, followed by a warning about remaining true to the moral character of the community.

Most Turks are familiar with this way of talking about science. Ersoy's poems are but one example among myriad similar texts that suggest that while sciences should be imported from the West, "our moral values" should be kept intact; importing science and imitating the Europeans are separate things, and while the former is desirable, the latter is extremely perilous. After quoting Ersoy's verses, Turkey's prime minister R. Tayyip Erdoğan himself made the following remark on January 23, 2008: "We did not import the sciences and arts of the West. Unfortunately, we imported its immoralities that contradict our values. We should have raced to import its arts and sciences [instead]."[6] What makes Erdoğan's remark particularly significant is its context, as the prime minister made this statement while addressing a group of students who were awarded government scholarships for graduate studies abroad. Thus, his argument about importing the "sciences of the West" is at the same time, and perhaps more importantly, an argument about the kind of individual that these students, or the young people of Turkey in general, must strive to become. What we have before us, then, is a way of talking about science that makes it a matter tightly connected to the definition of good citizenship. Discussions about science are discussions about much more than science.

But how and why did this way of talking about science become so dominant in Turkey? What are the historical roots and functions of this discourse? These are the questions that this book attempts to answer. For this purpose, I look at what it meant for educated, Turkish-speaking Muslims to talk about science in an era of drastic transformation in the last of the Islamic empires: the multi-ethnic, multi-religious Ottoman Empire in the nineteenth century.

A word of caution is in order at the outset. It is a common and commonly criticized tendency seen especially in the works of Turkish-speaking researchers to refer to "Ottomans" even when the focus is only on Turkish-speaking Muslims within the empire. It is true that an intricate version of the Turkish language was the language of the Ottoman state, and the legitimizing ideology of the Ottoman state was rooted in Turkish-Islamic traditions. However, Ottoman society was composed of a great variety of ethnic and religious groups, and this was a reality that members of these groups took for granted. Hence, while the specific ethnoreligious group that this study focuses on had a special position within Ottoman society, it was not the only one, and the extent to which the findings I discuss in this book reflect the experiences of other groups would be the topic of another study. Nevertheless, in order to avoid tedious repetition, I do refer to "Ottomans" or "Muslim Ottomans" throughout the text, rather than "Turkish-speaking Muslim Ottomans." I should also state that as the political, cultural, and intellectual capital of the Ottoman Empire, Istanbul was the location in which the major participants of the debate resided. Due to this, as well as my emphasis on the "official discourse" on science, the sources I use are almost exclusively those published in Istanbul. The way discourses emerged and evolved in the provinces is an equally important issue but is outside the scope of this study.

I analyze what these, mostly young, Muslim men[7] talked about when they talked about science, and I study the emergence and development of a discourse that rendered science and morality two concepts that were inseparable from one another.

Analyzing Muslim Ottomans' characterizations of the relation between science and morality is an important aspect of this study, but not the only one. At a broader level, I am interested in the observation mentioned above, namely that a reference to science is, almost as a knee-jerk reaction, followed by a reference to morality in this way of speaking. What this study finds is that from the outset, and for *all* participants of the Ottoman debate on science, what truly mattered were the qualities of the present and future subjects of the sultan, or the citizens of the Ottoman state. The debate, consequently, was more on the characteristics of the "man of science" than on science itself. Even arguments that were specifically about whether science was in harmony with or neutral toward "our values" can be seen in a similar light as, in this fashion, the participants of the Ottoman debate assured that talking about science was always about what "our values" were or, even more fundamentally, who "we" were. The ultimate issue was social order and the key question "Who are we and who do we want to be?"—a question that, as we will see, could not be answered without also talking about who we aren't and *shouldn't* be.

Why Talk about Science?

A brief skimming of the Ottoman newspapers, literary works, and official documents from the period could be sufficient to conclude that the Ottomans did indeed talk a lot about science in the nineteenth century. Two words that were commonly used to refer to science, namely, *ilm* and *fen*, regularly appear on the pages of countless texts. But this observation would probably not be too surprising to many readers, and a well-established narrative about the nineteenth-century Ottoman Empire, if not all Muslims in this period, is precisely shaped around this common-sense notion. After all, this was a period when not only the Ottomans but many societies all over the globe had to come to terms with the seemingly unstoppable advance of the major European powers. In the face of economic, political, and military domination, it was simply "natural" that Ottoman elites talked about the key reason behind European ascent: advances in science and technology.[8]

It is not the purpose of this study to discuss the extent to which the "success" of colonialist powers can be attributed to their development and deployment of science and technology.[9] Instead, I focus on a perhaps more subtle, yet equally problematic, presupposition of the above narrative: that "science" is one and only one "thing," and one knows it when one sees it. Just as the contemporary analyst is somehow able to locate the exact place of science in the past, waiting to be imported, as it were, it is assumed that should have been the case for nineteenth-century Muslims as well. Furthermore, it was, or should somehow have been, apparent to anyone that this "thing" called science was beneficial—an assumption that implies that the motives of individuals that made a case for science do not need to be scrutinized: they were simply "doing the right thing."[10]

In a way, such views stem from a broader and rather popular assumption, namely that science has always been a part of human experience and can be identified in any period or society. This, however, is essentially a presentist outlook that has long been challenged effectively, particularly by historians of science. After all, in the analysis of any type of patterned social interaction—and scientific activity is no exception in this respect—we need to pay attention to the cultural meanings that social actors attribute to their actions, the socially acceptable ways of justifying and classifying actions, the types of actors that are involved, and the place of the specific activity within a broader structural and institutional context. Briefly put, rather than imposing our assumptions on nineteenth-century Muslim Ottomans, we should ask questions such as: What were *they* seeing? How were *they* interpreting what they saw? *Whose* perceptions and interpretations mattered in the end?

When approached in this manner, it appears as a kind of symbolic violence to imply an equivalence between twentieth- or twenty-first-century science and the institutions and perspectives of other times and places.[11] Indeed, in the nineteenth-century Europe where Ottomans "met" science, the category did not have the same connotations for all, and the meaning and institutional location of science had not yet stabilized.[12]

In addition, recent work in science and technology studies has reminded us of yet another basic but crucial fact: "science" is ultimately a cultural category and, like all cultural categories, what it means and what it does or does not include are always potentially a matter of debate and struggle. Many sociologists as well as students of science now use a cartographic metaphor to make better sense of such matters. We navigate through social space thanks to the "maps" cultures provide; we classify and hierarchize people, objects, and phenomena, follow particular lines of action rather than others in specific contexts, and have "appropriate" expectations from people and situations thanks to these maps.[13]

But as this makes clear, cultural cartography has unique consequences. Classifications of different types of human activity and the accompanying principles about their respective worth have significant implications for groups engaged in these activities. Hence, groups embark on what the sociologist Thomas Gieryn famously referred to as "boundary work"—the attribution of selected characteristics to a specific activity or institution in order to construct a boundary between it and other activities or institutions.[14] In contemporary societies where the label "scientific" is commonly associated with prestige and credibility, defining, challenging, and defending specific boundaries around "science" becomes a pressing issue for its practitioners and aspirants. Employing this approach, many sociologists of science have studied how scientists define science (or their particular field within it) in varying ways in order primarily to establish, expand, or protect their authority and autonomy.[15]

Clearly, however, the definition of boundaries and categorization is a process that achieves a lot more in social relations in general. The category a social actor or an activity is placed in can have momentous, potentially irreversible consequences for the actor and the activity; hence, classification struggles and boundary disputes constitute a fundamental form of social struggle.[16] Moreover, by implication, these struggles are of a relational nature. As a variety of sociological studies have demonstrated, these struggles involve claims for group membership and non-membership; boundaries are set for including "us" and excluding "others."[17] Categories ranging from "immoral" or "criminal" to "vulgar" or "amateur" separate not only deeds from one another but distinguish "us" from their doers.

What Muslim Ottoman elites did in the nineteenth century involved a lot of cultural boundary-work. In their attempt to make sense of and provide solutions for the hardships the empire was going through, they grappled with a diverse set of cultural quandaries and engaged in numerous projects of cultural cartography. They invented ways of interpreting the social conditions of the Europeans and the Ottomans as well as the reasons behind these conditions, and they not only constructed new cultural categories but reinterpreted the meaning and boundaries of the existing ones. And their discussions on science, which I focus on in the following chapters, were prime examples of these endeavors.

But all this did not happen in an ethereal world of ideas, as some versions of intellectual history seem to suggest. These elites were particular types of people, products of particular settings and experiences, and had particular dispositions, cultural repertoires, and, ultimately, interests.[18] As a result, when they talked about the meaning of science, they were also consciously or unconsciously talking about "themselves" and "others." In short, they were drawing social boundaries. The Ottoman debate on science was at the same time a debate on how to classify people in a changing Ottoman Empire.

Ways of Talking and Dominating

As I underscored above, what we see in Ottoman documents of the nineteenth century are specific *ways* of talking about science—ways that connect science to morality in a general sense. Hence, my emphasis is less on specific arguments than on discourses about science. Discourse, in its general usage in social analysis, signifies a particular way of not only representing but constituting the world. We talk about and ascribe meaning to otherwise ambiguous, if not inherently meaningless, social experiences and phenomena. Discourses shape perception, establish criteria for judging the validity and appropriateness of statements about reality, and, indeed, outline the positions subjects can occupy. They do not determine individual statements but provide a *range* of concepts, images, metaphors, and rules with which it becomes possible to make "proper" statements.[19] Commonly, social reality is characterized by the coexistence of competing discourses, rather than the absolute domination of a single one.

Now it is important to emphasize that boundaries of categories such as science are potentially always open to debate, and that discourses establish a "range of acceptability" rather than determining all statements in a coherent fashion. But it is also important to reiterate the unique consequences of establishing—albeit temporarily—specific definitions and specific sets of boundaries on social phenomena for specific social groups. Hence, an analysis of "ways of talking" is incomplete unless it takes into account the unequal distribution of different types of resources in society and its impact on the outcomes of discursive struggles about classifications. Cultural maps *are* flexible, discourses *are* multiple, and consequently there is always potential for the emergence of centrifugal forces; but these potentials always coexist with centripetal forces that are directed toward establishing uniformity. Therefore, our studies on cultural cartography will be flawed if we fail to pay attention to the strategies of powerful social groups and institutions, particularly the state. In the words of the social theorist William Sewell Jr., who adopts the cartographic metaphor as well, what we also need to look at are "official cultural maps."[20] This is particularly important in the analysis of the Ottoman case, as the state was almost the ultimate point of reference for the actors we will focus on.

The establishment of the boundaries of symbolic categories and social groups is clearly a quintessential state issue. As the sociologist Pierre Bourdieu indicated in his famous characterization, the state claims the legitimate use of not only physical but symbolic violence within a territory.[21] This entails both the construction and the imposition of specific definitions and categories, and the distribution of privileges based on them. In the meantime, alternative ways of dissecting reality may be assigned labels such as "marginal," "illegitimate," or simply "wrong." Institutions will be established around official categories, labels, and definitions, which are inculcated in schools and diffused by the media.[22]

Two nineteenth-century developments in the Ottoman Empire assume particular importance in this context. First, the impact of the gradual integration of the Ottoman Empire into the world capitalist system on not only the Ottoman economy but also on its political and social structure is difficult to overestimate. Sectors within the non-Muslim communities were able to reap considerable benefits from this economic transformation, and members of these communities acquired the characteristics of a commercial bourgeoisie in the Ottoman Empire.[23] In the meantime, positions within the Ottoman bureaucracy appeared as the most viable employment opportunity for educated Muslim Ottomans.

A well-established approach in studies on the nineteenth-century Ottoman Empire suggests that a key aspect of the social transformation that occurred in this period was the bifurcation of the Ottoman bourgeoisie into commercial and bureaucratic segments along ethnoreligious lines.[24] Theorists provide a variety of reasons for this bifurcation, such as the higher likelihood of non-Muslim Ottomans to be able to join the economic and social networks of Europeans with which they had religious and cultural ties, the Islamic Ottoman tradition that ascribed more prestige to administrative work than to commercial activity, the patronage European powers provided for non-Muslim Ottomans in the nineteenth century primarily for political purposes, and the Islamic-Ottoman administrative principle of legal pluralism that made it easier for non-Muslims to operate under the more commercially beneficial Western law within the empire.[25]

It is important to note, however, that recent studies suggest that the transformation in question was more complex than a model of smooth and simple bifurcation could imply. First, it was by no means the case that all members of all non-Muslim communities benefited from this process. Second, the degree to which different religious communities participated in sectors of the economy varied from region to region. And third, this bifurcation, to the extent that it existed, did not involve a strict segregation; not only were many non-Muslims employed in the Ottoman bureaucracy, but many Muslims could be found in the sphere of commerce as well as in the new working class. The main point is the disproportionately low participation of Muslim Ottomans in the emerging capitalist economic structure and, in particular, in its higher echelons.[26] For the purposes of this study, it is most relevant to note that, in such a context of transformation, many members of the Muslim community perceived the Ottoman state as the most reliable provider of employment opportunities, most visibly in Istanbul. This, in turn, gave rise to a different type of apparent divide—one within the Muslim middle class itself. In the city of Istanbul, where this study focuses, higher-status Muslim bureaucrats who were products of a new type of education had much in common with members of the non-Muslim middle class, particularly in terms of cultural interests and tastes. This helps explain why a sense of split and fragmentation emerged within the Muslim middle class, as the following chapters will illustrate. Also worth adding is that, in addition to the transformation of the bureaucracy into the most appealing field of employment for many Muslim Ottomans, in this era the power of the bureaucracy grew steadily at the expense of the authority of the sultan, and bureaucrats increasingly identified themselves with the state.[27] In this respect, the apparent split also involved questions about what type of bureaucrat should represent the state, and what types of knowledge and taste the Ottoman state should endorse.

Second, the degree of control that the Ottoman state enjoyed over the fields of both education and the press needs to be underlined. This was particularly the case for the institutions that catered primarily to Muslim Ottomans—the group that constitutes the subject of the present study. Non-Muslims were not entirely exempt from these pressures, as especially after the 1850s the official Ottoman policy aimed at the construction of an "Ottoman" identity that would transcend ethnoreligious identities. This entailed the establishment of educational institutions where Muslim and non-Muslim Ottomans would study together. But as in the case of the economy, in the field of education Muslims had fewer options than non-Muslims, who also had access to communal schools, some of which were of a much higher quality than state schools. It is true that there did exist private schools run by Muslims in the Ottoman Empire, especially after the 1870s, and privately owned newspapers achieved significant sales figures. But the top-level schools were state run, and the state exercised considerable control over all schools and the press. School curricula, textbooks, and publications of all sorts were monitored effectively, and the press was subjected to strict censorship.

As a result of these factors, struggles within the domain of the state itself became central to struggles regarding the meaning and significance of science in the nineteenth-century Ottoman Empire. All ways of talking are not equal, and the significance of the official discourse cannot be neglected in any discursive context. But especially if we consider their above-mentioned circumstances, the importance of the struggle for constructing, transforming, and challenging the official discourse on any topic for educated Muslim Ottomans in the nineteenth century becomes more apparent. Not surprisingly, thus, the specter of the state hovers over the entire debate that I analyze in this book.

Power Struggles and Science

There is not a monolithic, autonomous entity that is "the state."[28] In-stead, we can see the state as an arena of struggle within which social ac-tors with different types and amounts of resources interact. The limited sense in which I use the term "the state" in this book refers to a snapshot—the specific outcome of these interactions at a given moment.[29]

What are the resources that actors bring into the struggle? And what exactly is the struggle for? Clearly, actors engaging in this struggle are individuals with talents, credentials, know-how, inclinations, manners, aspirations, and expectations ("cultural capital") as well as economic re-sources ("economic capital"). They are also members of groups and have institutionalized, reliable relations with others that they can benefit from ("social capital").[30] But it is also true that these types of capital are effective particularly if they are accepted by other actors to be legitimate, relevant, and worthy of respect, and it is only then that the individual also acquires "symbolic capital."[31]

Now what these social actors endowed with varying amounts of vari-ous types of capital struggle for is what Bourdieu calls "statist capital"—a "meta-capital" the acquisition of which enables social actors to "exercise power over the different fields and over the different particular species of capital" themselves.[32] With statist capital it becomes possible to tinker with the relative worth of specific types of capital and the ways in which they can be converted into one another. For instance, through state pol-icy, a particular kind of cultural capital, say, possessing a diploma from a certain type of school, can be rendered compulsory for becoming a legiti-mate member of a specific profession, which could, in turn, translate into the accrual of symbolic and economic capital. Similarly, having access to statist capital can enable groups to transform the type of cultural capi-tal they enjoy into one with high amounts of symbolic capital. Thanks to the state's ability to exert considerable influence over education, law, and cultural production, holders of statist capital can impose specific symbolic boundaries and classifications as the superior, if not the only legitimate, ones.[33]

During the nineteenth century, the state domain went through sig-nificant changes in the Ottoman Empire. Among the most consequential of these transformations were the steady rise in size and influence of the Ottoman bureaucracy, and the change in the makeup of this class—two developments that had already started in the eighteenth century. In the traditional imagination of the Ottoman ruling class, "men of the pen" (*kalemiye*, i.e., scribes) constituted a "relatively modest sector" in comparison to "men of knowledge" (*ilmiye*, i.e., the religious elite; doctors of Islamic law) and "men of the sword" (*seyfiye*).[34] But due to the militarily weakened empire's need for diplomacy and for cadres that could handle the relations with European powers effectively, the scribes gradually emerged as the group with the appropriate know-how. Similarly, new schools established in order to reorganize the Ottoman military along European models started to produce young men with a new kind of cultural capital—types of knowledge and skill inculcated by no other institution within the empire. The role that these groups played within the Ottoman state mechanism gradually increased, and the state domain witnessed a remarkable struggle regarding the respective values of particular types of cultural capital.[35]

As a new group without (yet) considerable prestige and credibility that they could take for granted, this rising class was composed mostly of young men who needed to establish their worth and distinguish themselves from other sectors of the ruling elite. Ultimately, it was these young men's cultural capital itself that had to be deemed significant by their competitors for them to truly acquire symbolic capital. Consequently, the discourse that they constructed about themselves and the state contained many references to the type of knowledge to which they had more access than their challengers did, that is, knowledge about "how things truly worked in the real world." And a key component of the new bureaucrats' alleged expertise about the new world that the European powers were constructing was their knowledge concerning "the new sciences of the Europeans."

Finding a Place for Science

Scientists, as most approaches to "boundary work" suggest, define the boundaries of science in a way that will help them maintain if not increase their authority and credibility. What is not so convincing, however, is to imply that instances of boundary work can be explained most effectively by science- and scientist-centered analyses.[36] If we employ the cartographic metaphor in studying struggles regarding the boundaries of science, or indeed, any cultural category, it is obvious that these struggles are simultaneously about the boundaries of other categories.[37] Put simply, one cannot make an argument about where science starts and, say, religion ends, without implying what the boundaries of the category "religion" are, or should be. Similarly, a struggle that involves attributing specific characteristics to the activities of scientists in order to keep contenders "out" cannot but attribute specific characteristics to the contenders as well. Defining science and describing scientists is always at the same time defining non-science and non-scientists. Precisely for this reason, those other categories and actors are equally central to any debate on science and scientists.[38] And it is a more interesting and ultimately more rewarding approach, thus, to ask which actors and which categories become relevant in debates on science in different social and historical contexts, and why.

Furthermore, debates and struggles do not take place in the abstract, and participants work with strategies (rhetorical and political) and tools (linguistic and conceptual) that are available to them. Attention to culture in general as well as the specific social locations of the actors involved is important because only in this way can we understand what it means to have a debate about science in the first place. It is only then that we can make sense, for instance, of a lecture on science that transforms into a lecture on language, an article on science that focuses on the differences between Arabs and Turks, or an essay on theater that first turns into an essay on science, then becomes a treatise on ethics—some examples from the Ottoman texts that this book analyzes. How and why do such connections become relevant? Why do certain actors employ discursive strategies that link specific concepts to one another in specific ways?

In light of these observations and questions, the ultimate question I ask in this study is not "How were the boundaries of science defined in the Ottoman Empire?" but the more naive-sounding "What were the Ottomans talking about when they talked about science?" Hence, this book is an exploration of what it meant for literate Turkish-speaking Muslim Ottomans themselves to debate science in the nineteenth century.

When we pose the question in this fashion, the concepts, categories, and rhetorical tools that were available to the participants of the debate become particularly significant. In fact, in a sense, the people who constructed the official and alternative discourses that I discuss in the following chapters were all engaging in an act of translation.[39] Granted, translation, especially in the sense of "cultural translation," is a murky topic. And the task at hand is a particularly complex one, as not only were the Ottoman actors I focus on translating ideas into specific cultural maps, but this book translates their interpretations into English, from a specific perspective, and for an audience located in different social and historical contexts. Cultural translation involves, as the anthropologist Talal Asad once noted, not "an abstract matching of two sentences, but [a] social practice rooted in modes of life."[40] The meaning of a word, or an idea, is never fixed and effortlessly shared by all "natives," and for social actors "understand-ing" is possible only within a historical and discursive context, and from a particular perspective. So we need much more than a dictionary in order to translate, and to understand an act of translation; it is the social settings and the webs of social interaction within which concepts operate that we should focus on.

Even at the lexical level, the issue is more complicated than could be as-sumed, however. Educated Muslim Ottomans found primarily two catego-ries in the dominant cultural lexicon that they could refer to in order to translate *la science*: *ilm* and *fen*. Of the two, *ilm* was indubitably the more prestigious and significant. *Ilm* (pl. *ulûm*) is an Arabic word that essen-tially denotes "knowledge," but has additional, less mundane uses, and is the word commonly used as the equivalent of "science." In the Islamic tradition, branches of learning like Qur'anic exegesis and jurisprudence as well as mathematics and medicine were referred to as *ilm*. A distinction Muslim scholars commonly drew between branches of knowledge was be-tween "intellectual *ilm*s" (*ilm-i aklî*), such as astronomy and medicine, and "transmitted *ilm*s" (*ilm-i naklî*), which included sciences that were directly about the teachings, the prophet, and the holy book of Islam. Note that while the distinction appears as one between secular and religious sci-ences, the distinction itself is religious, as it is constructed within a com-prehensive Islamic perspective. Furthermore, the sciences, as a whole, in-dicate a unity and are, ultimately, inseparable from the knowledge of God itself. Indeed, in Islamic philosophy, *ilm* in the singular also denotes *the* knowledge: knowledge possessed by God.[41] The word *âlim*, derived from *ilm*, means "one who knows," and is used to describe both a scholar, and once again, God. Religious scholars, or the doctors of Islamic law, are re-ferred to as the "ulema," which is the plural form of *âlim*, and in the Otto-man Empire, the class comprising the ulema was referred to as *ilmiyye*: the class of knowers. Similarly, students of the *medrese*s—institutions that in the Ottoman Empire of the nineteenth century were essentially devoted to religious education—were called *talebe-i ulûm*, students (literally, seek-ers) of *ilm*s. Therefore, despite the use of the term also for what would be called secular sciences today, it is crucial to note the religious significance of the term *ilm*.[42]

Fen (pl. *fünûn*), on the other hand, is a word primarily meaning "branch," and in traditional Ottoman usage indicated those types of knowledge with a more overtly practical component. Hence, we see ref-erences to the expertise of scribes or civil servants as *fen*s. Similarly, sur-gery, military arts, and architecture were also commonly referred to as *fen*s. In this sense, then, the connotations of *fen* are closer to that of "art" than "science." Yet, as we shall see, this word was also frequently used in nineteenth-century Ottoman texts on the sciences of the Europeans. Indeed, in his seminal work, the sociologist Niyazi Berkes noted that the new sciences had been called *fen* in the Ottoman Empire in order not to attract the derision of religious scholars who monopolized the concept *ilm* for their own expertise, and this argument has been repeated numerous times over the years.[43]

But a detailed analysis of Ottoman texts casts doubts on this interpretation. Ottoman speakers sometimes used these words interchangeably, sometimes as complementary to one another, but not particularly consistently. Not only were the words used in a variety of ways, but Ottoman authors themselves continued to debate which term should be used in which context well into the twentieth century.[44] Moreover, as Şükrü Hanioğlu notes, the many instances in which the word *ilm* was used for the "new sciences" are of particular importance, as this strategy implied that these disciplines were in the same category as the ones taught in the *medrese*s, and made calls for their importation more palatable.[45] In a sense, the word operated, at least to a limited extent, as what science studies scholars refer to as a "boundary object"—abstract or concrete objects that bridge different social worlds and enable different groups to communicate while commonly not having exactly the same signification or function for each group.[46]

Yet to this we should add that the way words were used had further, potentially even more significant implications, as they had to do not simply with communication but with domination. The concepts represented social hierarchies; *ilm* and *fen* were represented by specific groups, and it was not words but actual social actors with varying degrees of power that came into contact. The unique religious and moral connotations of the *possession* of *ilm* were particularly important for the new Ottoman elite, as these associations were vital to their justification of the status and social esteem they sought. Put bluntly, they were not interested in being merely "men of know-how," they wanted to be "men of knowledge."[47] Hence, what Turkish-speaking educated young Muslim men engaged in was not simply an effort to find a place for *la science* in their cultural lexicon; they were carving out a niche for themselves in a social hierarchy.

Science, "Men of Science," and Virtue

Public discussions about science are discussions about people—people who represent, speak and act in the name of, praise, condemn, manage, fund, are exposed to, or have to somehow deal with science.[48] How the persona of the scientist is constructed, how science and scientists (and nonscience and lay people) are perceived and represented are thus key questions to dwell on in order to acquire insight into both the way science operates in any society and how debates on science take shape. As Steven Shapin's work has shown, the idea of the scientist as a reliable, trustworthy individual was central to the making of modern science, and relations of interpersonal trust among practitioners of science maintain their importance.[49] Evaluations of other people, moral assessments, and references to virtues play an important part in science—a part that has consistently been neglected until recently. Furthermore, perceptions of science take place within social and institutional contexts, and the way social actors perceive the "representative of science" is likely to have a defining impact on how they imagine and deal with science.[50] And not only are allusions to specific definitions and types of "moral values" common in the public understandings of science, scientists themselves are able to benefit from maintaining an image that portrays them as virtuous people.[51]

In the Ottoman case that I analyze, it becomes apparent that the construction of a discourse on the merits (or dangers) of science implies the construction of arguments about the virtues (or vices) of the representatives of science. The emerging elite class of the early nineteenth century represented science as a particularly, almost fascinatingly, beneficial type of knowledge. Yet they also appropriated European orientalists' narratives on the early Islamic scholars' contributions to science and constructed a discourse that, while emphasizing the significance of the new sciences, also linked them to the Islamic tradition. Additionally, they frequently referred to the sciences that they represented as *ilm*, with all its epistemological, religio-ethical, and social hierarchical implications. As the representatives of a special type of knowledge, they portrayed themselves as not only knowledgeable but virtuous and, consequently, worthy of respect and admiration. Finally, and relatedly, this portrayal implied that scientific knowledge provided an incontrovertible account of how things worked and, thus, learning "some science" would render the learner "able to understand" (and hence obey) the rulers. Familiarity with scientific knowledge made the ruler virtuous and the ruled obedient.[52]

Characterizations of this sort—and not a supposedly unavoidable battle like "science versus religion" or "modernity versus tradition"—were at the root of the conflict that the Ottoman debate on science embodied. The alternative discourse on science that challenged this emerging official discourse was constructed primarily by the disillusioned members of the rising class themselves, and their criticism was directed not against "science" in the abstract but against the Muslim representatives of science in the Ottoman Empire, particularly due to their "Europeanized" lifestyles and consumption patterns. In a historical juncture where Muslim Ottomans from different walks of life were increasingly concerned about their location within the transforming social order, as well as the growing ability of European powers to influence Ottoman domestic policy, the characteristics of the person who represented the "new knowledge" became a critical issue. While the emerging official discourse portrayed the members of the new elite as selfless, learned "saviors," the alternative discourse characterized them "imitators," "arrogant fops," and, ultimately, "misguided materialists."[53] Obedience was still a key issue, but it was the representative of science himself who needed to prove his deference to the "authentic" members of the nation. What we see at the end of the nineteenth century is a merging of these discourses leading to the particular way of talking about science that I delineated at the beginning of this chapter: science is needed, uniquely useful knowledge, but a "man of science" should be, and be able to prove that he is, one of "us."

Outline of the Book

Chapter 1 introduces some key themes concerning the cultural transformation of the Ottoman Empire from the late eighteenth through the early nineteenth centuries, and describes the early stages of the emergence of a new class of "knowers." Following a brief discussion on the transformation of the empire in this period, I discuss the early attempts to establish institutions with more emphasis on the new sciences in their curricula. The students of these new Ottoman schools, young men sent to Europe for education, and the new diplomats of the empire constitute the members of the emerging new "learned class." Using their writings and data from official documents, I show how, at the beginning of the nineteenth century, the portrayal of European science as "new knowledge" gradually emerged. I illustrate how members of the emerging class started to portray this new knowledge as their distinctive property, making it grounds for redefining power and prestige in the empire. Yet this portrayal was far from established, and the question of ownership of this new knowledge, and the political implications of possessing it, remained unresolved in this period.

In chapter 2, I focus on the period following the Tanzimat (Reorganization) Decree of 1839—a period characterized by numerous political, social, and cultural reforms, and by the rise to power of the new class of bureaucrats. I discuss the lifestyles and dispositions of the members of this class, analyze the way they talked about science, and illustrate their fanciful representations of this "new knowledge" produced by the Europeans. Science is commonly referred to as the opposite of "ignorance" in these texts. These styles of talking about science are strategies for claiming distinction, and they indicate the new bureaucrats' struggle to define themselves as the new "knowing class." In this analysis, I also trace the birth of one of the key discourses on science in the Ottoman Empire of the nineteenth century: science as the route to patriotism. Science, as knowledge that the holders of state power possessed, was identified with the state in this discourse, and scientific knowledge was presented as knowledge that made one respect the state. The important conclusion is that right at the outset the debate on science was fundamentally about social order and the characteristics of virtuous men, be they high-ranking officials or humble subjects.

Chapter 3 focuses on the 1860s—a particularly significant decade that witnessed both the first attempt to establish the Ottoman University and the birth of the first successful periodical in Ottoman Turkish, the *Journal of Sciences*. I analyze these two developments in detail and present them as instances illustrating the consolidation of the official discourse that was gradually constructed in the previous decades. I focus particularly on the articles published in the *Journal of Sciences* and analyze the rhetorical strategies used to associate scientific knowledge with moral virtue. In this context I also draw analogies between the British useful knowledge movement and the Ottoman elites who praised science and men of science in reference to the concept of usefulness. Finally I show the links between the new political order that granted equal status to Ottomans of all religious groups and the idea of science as a kind of knowledge embodied within the state that all religious groups could unite around.

In chapter 4, I analyze the birth of the reaction to the new elites and its implications for the debate on science. This period also started in the 1860s and was characterized by the Young Ottoman movement, which was a reaction against the monopolization of political power by a small faction. The populistic political discourse of the Young Ottomans also entailed an alternative discourse on science. While expressing the frustrations of the Muslim community in general, they also inserted direct references to Islam in their arguments on science. Emphasizing the scientific contributions of early Muslim scholars and the prestige of traditional Islamic disciplines, the Young Ottomans made science an issue that could hardly be discussed without an overt reference to tradition and community. Moreover, they invented the most popular stereotype used in Ottoman literature to criticize overly Westernized elites: the fop. Depicted as a person who incessantly praised European countries and the merits of European science without a sound knowledge of either, the fop would become the figure against which any advocate of European science would be evaluated. Ultimately what the Young Ottoman reaction made clear was that the Ottoman debate was not about science but rather the qualities of the man of science.

I devote chapter 5 to demonstrating the richness of the debate in the 1870s and to analyzing the key themes of the emergent alternative discourse on science. For the litterateurs of the 1870s, just as it had been for the bureaucrats of the Tanzimat, debating science was a matter of establishing the proper qualities of an Ottoman subject in general and of a Muslim Ottoman man of science in particular. The key differences were that for the late-century litterateurs, an awareness of the scientific knowledge of Europe did not necessarily make one virtuous or respectable; those who spoke in the name of science also bore the burden of proving their loyalty to the community. The tone of such criticisms got increasingly harsh in this period, and the portrayal of the superficial science enthusiast as a treasonous buffoon became common. However, I show that this emphasis on community had an unintended consequence, as it required defining this community. Whether the scientific works of early Muslim scholars still mattered and, more significantly, if they could truly constitute a legacy for *Turkish*-speaking Ottoman Muslims if they were written in Arabic became hotly debated issues in this context. Thus, while the alternative discourse firmly established moral virtue and loyalty to the community as the criteria for evaluating a man of science, it also irreversibly rendered the debate on science as also a debate on the identity of the community.

Chapter 6 is the first of two dedicated to the analysis of the debate during the reign of Abdülhamid II (1876–1909). I argue that in this period a new official discourse on science emerged, synthesizing the official and alternative discourses discussed in the previous chapters. Science was officially endorsed knowledge in the new official discourse, but that the man of science should be a loyal Muslim subject was also central to it. I show that three legacies of the 1860s and 1870s dominated the debate in this period: science as an issue regarding the identity of the community, science as useful knowledge, and the use of cautionary figures to discipline the young science enthusiast. This period was characterized by an increasing emphasis on the Islamic identity of the empire, yet the debate on science continued to problematize this definition, thus rendering its participants potentially "dangerous." The prestigious Western-style Ottoman schools continued to produce ambitious young men, many of whom exhibited a similarly dangerous self-assurance. Hence, a new figure emerged in literature to accompany the fop: the "confused materialist." This stereotype, which was used effectively to discipline young men of science, became a permanent component of the debate. Yet I also show that in order to fend off such attacks, science enthusiasts made efforts to once again define themselves as the virtuous ones and, in turn, directed their criticisms against poets. The debate on science was once again a debate on moral virtue, as the topic became whether scientists or poets were truly virtuous and useful.

In chapter 7, I analyze the arguments of the so-called confused materialists. I show that while these young men were commonly portrayed as atheists, enemies of tradition, and naive Europhiles, in their texts on science they repeatedly included references to Islam and morality. Emphasis on moral virtue was already at the heart of the pro-science discourse of the mid-nineteenth century, so this was not an entirely new development. But in the 1880s and 1890s, such references were made with much more overt references to Islam and the authority of the sultan. The "harmony" between Islam and science was almost taken for granted in these texts. Moreover, portrayals of the man of science as an obedient, harmless figure who was eternally grateful to the sultan became uniquely prevalent. I use evidence from textbooks, newspaper articles, literary works, and texts written specifically for children to provide an analysis of this portrayal. I also show that just as scientific texts were filled with references to morality, books on morality contained many references to science, often presenting obedience as a key quality of the good man of science. Consequently, I argue that at the end of the nineteenth century, science and moral virtue were concepts that were inseparably connected in the Ottoman debate on science. Materialistic and scientistic conceptions *were* espoused by many young critics of the Hamidian regime, but not only was theirs not the dominant discourse, an emphasis on moral virtue was integral to it as well.

ONE

A New Type of Knowledge for a New Social Group

Introduction

Helmuth van Moltke, the military strategist and legendary chief of staff of the Prussian army in the second half of the nineteenth century, had resided in the Ottoman Empire between 1835 and 1839 when he was a young captain, and where he had been employed as a military adviser to Sultan Mahmud II (reigned 1808–39). In one of the letters he wrote during this period, he recounts an incident that he witnessed while he was a consultant to Hafız Ahmed Pasha, the general in command of the Ottoman troops fighting against the forces of the rebellious governor of Egypt, Mehmed Ali Pasha. During a council, a religious dignitary who often advised the Ottoman general asserted that the empire still maintained its formidable power, and suggested in a proud manner that ten thousand Ottoman soldiers could get on horseback and, "trusting in Allah and in the strength of their sabres," enter Moscow. "Why not," an officer replied, "if their passports have been properly visaed at the Russian Embassy?" The ironic comment of the young officer Reşid Bey, a man educated in Paris, was incomprehensible to the audience, however, as it had been made in French.[1]

Moltke's narrative may have been colored by the affinity he probably felt toward the European-trained Ottoman officer. Similarly, in its superficial portrayal of an antagonism, the anecdote conjures up one of the most popular—and most criticized—themes of twentieth-century political discourse about not only Turkey but the entire Muslim world: the educated, modernizing military officer vs. the ignorant religious demagogue. Yet despite these easy observations regarding potential bias and simplism, the anecdote remains significant. For one thing, it is not an isolated example, as we shall see below. Second, while it would indeed be too simplistic to interpret it as an example of the "inevitable" clash between religion and modern knowledge, the incident is certainly indicative of the attitudes of a new group of individuals in the early nineteenth-century Ottoman Empire—a new group composed of young men who claimed to possess a superior knowledge of how things "actually" worked in the contemporary world. That the comment had been made in French is not insignificant, either, as French was, in a sense, to become the official language of this new group. The young officer undoubtedly knew that his comments would not be comprehensible; the language he was able to speak was a mark of his distinction from the rest of the council. It is this attitude and this particular group that I will focus on in this chapter and the next.

Reşid Bey, known as Reşid "the Spectacled," was a product of an early Ottoman initiative to send students to European schools. The man behind the initiative was a vizier of Mahmud II, Hüsrev Pasha (?–1855), a champion of military reform and commander-in-chief of the new model Ottoman army created after the decimation of the Janissaries in 1826. Hüsrev was strictly against reform outside the military, yet in 1831 he was able to have four young members of his household be the first group of Ottoman students sent abroad, to the preparatory school of Jean-François Barbet in Paris to receive a European-style education.[2] This apparent paradox is probably due to Hüsrev's attempt to preserve his own position by making sure some top-level officials of the empire in the future would be "his men" as well—a strategic move within a state mechanism where legal-rational procedures remained much less effective and reliable than personal networks and loyalties.[3] Hüsrev's move demonstrates his foresight, as he must have predicted that the future would belong to those educated in the "European way." Indeed, in a letter he sent to his protégés in 1832, he wrote:

When I picked you to be educated in France out of all the youths I raised before my eyes, I effectively entrusted with you all the hopes regarding the education of Muslim youth. Our state dignitaries will look at you and decide whether to follow my example and entrust the future of their children to the knowledge of Europe.[4]

His protégés would not fail him: of the first four students sent to Paris, one later became a grand vizier—the future Ibrahim Edhem Pasha, who, after Barbet's school, attended the École des mines and is considered the founder of geology in Turkey. Another became a colonel and a third, an artillery general.[5] But these young men were ultimately members of a truly new group, a group with a heightened sense of self-esteem and high status expectations. They would embark on a struggle to redefine what "knowledge" and "ignorance" were to denote and connote to Ottoman Muslims.[6]

A. Defining "New Knowledge," Determining "the Ignorant"

The changes that made people like Reşid possible started in the eighteenth century in the Ottoman Empire. Key in this context were efforts to reorganize the Ottoman military with the assistance of European consultants. This process involved the establishment of new schools, such as the short-lived *Hendesehane* (School of Military Engineering) in 1734, as well as the more influential *Mühendishane-i Bahri-i Hümayun* (Imperial School of Naval Engineering) in 1773 and the *Mühendishane-i Berri-i Hümayun* (Imperial School of Military Engineering) in 1795.[7] Similarly, the first Ottoman ambassadors were sent to European capitals in this century. These significant attempts were sporadic, however, and the true institutionalization of these efforts took place in the nineteenth century.

It was also in the nineteenth century that "ignorance" became one of the most frequently used words in Ottoman texts on the state of the empire. Appearing regularly in the official documents of the early 1800s, "the perils of ignorance" became one of the leitmotifs of the official discourse on the empire's problems, particularly in the second half of the century. But as Reşid Bey's remark exemplifies, not only did references to ignorance involve direct or indirect attributions of ignorance and knowledgeability to different groups, but they were also built on particular assumptions about what constituted knowledge itself. As knowledge and what was expected of those who possessed it came to be defined differently, ignorance and the characteristics of the ignorant were also described in varying manners.

In the early to mid-nineteenth century, a new group emerged in the Ottoman Empire comprising individuals who had been to Europe, or who spoke a European language, or who had been educated in Europe or at a European-style school in the Ottoman Empire, or some combination of the above. These experiences and skills were crucial particularly for the new generation of Ottoman bureaucrats whose role was much more vital at a time when the military might of the empire proved incomparably less effective against European powers than in the past. Using Bourdieu's terminology, we can argue that this new type of cultural capital enabled these individuals to gradually acquire statist capital as well, and, using their newly acquired status as "men of the state," they propagated new definitions of knowledge and ignorance. Defining the knowledge they possessed as useful and true knowledge, they sought to legitimize their power. This effort not only had implications for the status of the representatives of other forms of knowledge, but entailed new imaginations of social and political hierarchy within the Ottoman Empire.

1. Early Characterizations

A rather early example of the way in which knowledge and ignorance came to be redefined under the impact of European-style education is the famous treatise of Seyyid Mustafa, one of the first graduates of, and later a teacher at, the Imperial School of Military Engineering opened in 1795, during the reign of the reformist sultan Selim III (reigned 1789–1807).[8] Written in French and published in Istanbul in 1803 under the title *Diatribe de l'Ingénieur Séid Moustapha sur l'état actuel de l'art militaire, du génie et des sciences à Constantinople*, this is a work that contains in a nutshell many key themes of the future debates regarding science.[9] Similarly, as it was a work that was also sent to Ottoman embassies in Europe, it can be read as promotional material, depicting the way the empire wished to present itself to European powers at the time of Selim III.

In the autobiographical introduction of the *Diatribe*,[10] Mustafa argues that although he was tremendously interested in scientific knowledge even as a child, he was not satisfied by what Turkish masters could teach him.[11] He learned French, "the most universal language"—a comment that the editor of the French publication emphasizes with a footnote[12]— and read at a young age the works of European scientists that he was able to acquire. He was particularly impressed with the impact mathematics had had on the development of military tactics and architecture in Europe. Luckily for him, it was Selim III's reign in the Ottoman Empire— a sultan who was convinced that "welcoming the sciences and the arts" would be the most intelligent deed for a ruler and would bring the most benefits for his people.[13] Hence, the European-style school of engineering was opened in Istanbul, and Mustafa became one of its first students.

Mustafa's experiences in this new institution evidently heightened his sense of distinction from those who were unaware of the sciences he held so dear. When doing fieldwork in public, he and his fellow students found themselves surrounded by "the voice of incompetence and ignorance" coming from every corner. They were "molested, almost persecuted" by the people around them, who were screaming, "Why do you draw these lines on these papers? What is their use? Warfare cannot be conducted with a compass and a ruler."[14] The actions of the people disheartened the students, and they felt it impossible that the people could be disabused, but it was once again the benevolence of the sultan that helped them: Selim III followed their progress carefully and gave them opportunities to demonstrate to people of all classes "the great benefits of mathematical sciences applied to the art of war and to fortification."[15] In other words, when the ignorant public ridiculed and disillusioned them, their patron the sultan restored their hope and self-esteem. It is unclear how audiences were made able to perceive what Mustafa and his colleagues demonstrated, but the sultan's authority is without doubt what the students, lacking an authority of their own, sought refuge in. "New knowledge" and its representatives were under state protection, and the authority of new knowledge rested on the legitimacy of the holders of state power.

Did such a legitimacy exist, then? As future events would demonstrate, the answer was hardly positive. For Mustafa, too, it was clear that the sultan himself had faced problems. Old glories had led the Ottomans into lethargy, and "the class of the idiots and the superstitious" were fooling the simple-minded into believing that any innovation based on imitation was an offense.[16] But with his zeal as well as composure, Selim III silenced the "cowardly reproach," and "shut the mouth of ignorance and forced all classes of people to follow his example."[17] Hence, Mustafa states optimistically in conclusion, his country is now how he had always wanted it to be: "enlightened more each day by the torch of sciences and arts."[18]

With these comments, Seyyid Mustafa makes clear what he regards as ignorance: lacking the new knowledge produced by the Europeans, or, more fundamentally, knowledge about the uses and significance of science, especially those sciences that had recently been developed in Europe. In order to make a case for the new military school, Mustafa refers to the saying of the prophet that permits Muslims to use the weapons of the enemies of Islam when fighting them—the principle of "due reciprocity" that was used commonly in this period to justify the importation of the arts and sciences of the Europeans.[19] But it is clearly not the military sciences alone that Mustafa endorses, as his constant emphasis on "sciences and arts" indicates; in a way reminiscent of Reşid's comment cited at the beginning of this chapter, Mustafa's emphasis is on the practical. What matters is utilizable knowledge about how things "really" work in the new world: one *can* know a lot in this new world, but can at the same time remain "ignorant" if the knowledge in question is "old knowledge."

Mustafa's presentation of a particular image deserves constant attention as well, as it would continue to appear in a variety of forms in future works on the new sciences of the Europeans: the sultan as the protector of science and those who possess it. The "benign" sovereign is portrayed as a wise ruler in full support of the new students, both by seeing to their every need, and by "silencing the mouths of the ignorant" when necessary. Therefore, learning the new sciences and applying them is, in a sense, the *duty* of the students toward their protector, the sultan. Seyyid Mustafa and his fellow students had been presented with a gift from the sultan that necessitated reciprocation.

Ironically, Mustafa's patron Selim III—the sultan who had "shut the mouth of ignorance"—would be killed during the revolts of 1807–8 along with a number of other representatives of the "New Order" that he had endorsed.[20] A commonly made argument regarding these revolts portrays the main actors behind them as the reactionary alliance of the Janissaries with factions within the ulema. Focusing on such an alliance of the groups designated as "ignorant" by possessors of "new knowledge" could encourage facile and anachronistic portrayals of the revolts as an episode in the "wars" between tradition and modernity or religion and science. As students of the period argue, however, a more nuanced reading would be that the revolts were primarily a result of a conspiracy carried out by a clique aggravated by the favoritism of Selim III.[21] That the New Order posed a significant threat to the status of a number of groups representing the Old Order (such as the Janissaries) is undeniable, but it is also important to see how Mustafa's *Diatribe* itself makes clear this identification of state power with a small group endorsed by the sultan.

Ultimately, Seyyid Mustafa is but one representative of a broader group composed of people who identified themselves not only with "the new"— defined in a variety of ways—but also with state authority. The particular subset of which Mustafa was a member contained young men trained in Europe, or in European-style schools within the empire, and who were learned about the new sciences that were not taught in the traditional Ottoman institutions of higher education for Muslims, the *medreses*. This subset was remarkably small at the time of Mustafa but would grow consistently throughout the nineteenth century. These men, who perceived and portrayed themselves as representatives of new knowledge, were the major contributors to the debate on the meanings of knowledge and ignorance throughout the nineteenth century. Consequently, that they so tightly associated the types of expertise they acquired with state authority would prove fundamental to the way Ottomans debated science.

It should be noted at this point that it is not particularly reasonable to imagine these schools as immaculate institutions where the new types of knowledge were studied in a state-of-the-art manner. The schools in question came in many different types and with varying levels of sophistication, and, consequently, it would be erroneous to portray these students as masters of the new sciences. Indeed, due to the absence of a comprehensive educational system designed around these new institutions, many schools that were intended to provide higher education ended up having to teach literacy first. Nevertheless, this is not a particularly relevant matter if we want to understand the attitudes of the members of this new group. Ultimately, what matters is the very fact that they did receive some training in fields that were not taught in traditional Islamic institutions such as the *medrese*s, rather than how competent the students actually were, as the perception of distinction is not a function of actual competence and experience.

Besides these students, an equally important subset of this group consisted of those who had somehow visited or actually been employed in Europe and had a chance to personally observe the new curiosities of the Europeans. Most typically, these were Ottoman diplomats—a group that emerged in the eighteenth century and played an ever-growing role in Ottoman politics and thought.

One such person, a contemporary of Seyyid Mustafa and a member of the same circle, was Mahmud Raif Efendi (1760/1–1807)—a man known as Mahmud "the English" due to the post he had occupied as the secretary to the first Ottoman ambassador to London in 1793. Mahmud Raif discussed his experiences in England in his report, which is a brief, plainly descriptive account that he wrote, once again, in French—a first in Ottoman diplomatic history.[22] In this report Raif at times praises England and its capital rather enthusiastically: London is full of beautiful buildings, great schools and hospitals; commerce in England is "très considérable," lands "extrêmment bien cultivé," products "bien bonnes." The people of England are ordinary, like the people of all countries, but they are better educated than the people of other nations—a comment that obviously applies to the Ottomans as well. Indeed, for Raif, there is "beaucoup de science et d'instruction" in England.[23]

While in London, Raif also composed a geographical treatise in French that was published in Istanbul in 1804.[24] Like Mustafa the engineer, Raif the diplomat started his work by praising Selim III, who had revived the "mathematical sciences [that] had been abandoned and neglected in the Islamic countries simply due to love of idleness and indolence." The celebrated works of the old masters were written "in the way of the ancients" and because they were so detailed, they appealed only to the elite. Furthermore, unlike earlier works that were manuscripts, Raif's book was printed, thanks to which it was easily accessible for "enthusiasts of science and knowledge."[25]

Mahmud "the English" is also the author of *Tableau des nouveaux règlements de l'Empire ottoman,* a noteworthy book on the reforms of Selim III, who also endorsed it as a publicity move, as he had Seyyid's *Diatribe.*[26] This work is also characterized by an emphasis on the merits of "the new" as opposed to "the old," and it paints a particularly rosy picture of the future if not the current state of the empire.[27] Interestingly, the book contains an illustration depicting the Académie Royale des Sciences of the Ottoman Empire. There was no such institution within the empire in this period, and, indeed, the illustration is actually that of the Imperial School of Military Engineering. The caption could perhaps be read as a hint of the models Ottoman elites had in mind, or, and more certainly, as a sign of how they wished to publicize the fruits of their efforts to European audiences. But it is still telling that it was possible for the emerging Ottoman elite of the early nineteenth century to represent the Imperial School of Military Engineering as the Ottoman Academy of Sciences; the institution associated with science was the school, and science entailed primarily a type of knowledge to be learned, not produced. As we shall see, these associations would remain dominant throughout the century.

In these texts, neither Raif's nor Mustafa's arguments are simply on the relative merits of old and new types of knowledge; the contrast is more comprehensive. The old world and its representatives are associated with lethargy, esotericism, elitism, and unawareness, while the representatives of the new—that is, the holders of new knowledge—are portrayed as hardworking, truly beneficial subjects of the sultan, and true enlighteners of the people. Possessing new knowledge almost necessarily transforms one into a new kind of subject—one who is productive, industrious, and, ultimately, good. But while these representations would dominate the Ottoman debate particularly after the 1850s, for this particular generation they appear somewhat ironic: like his patron Selim III, Mahmud Raif would be murdered by the Janissaries during the revolts while serving as the superintendent of a division of the new army.

2. Mahmud II: Ignorance as a State Issue

While revolts cost many their lives, reform would continue under Sultan Mahmud II, who ascended to the throne fourteen months after the death of Selim III. After strengthening his power base via an initial alliance with the local notables and taking advantage of the acquiescent attitudes of the higher ulema, Mahmud II carried out a number of reforms intended to centralize the government and reorganize it in ways similar to the contemporary European states. The chief objective was the elimination of all alternative centers of power and the construction of a loyal and reliable administrative cadre. Cognizant of the central government's desperate need for cash and threatened by local notables' political power, which had increased significantly after the introduction of the lifetime tax farming in 1685, Mahmud II eliminated many of the notables and strove to centralize the administration of taxation by "combining negotiations, ruse and force."[28] Related to the same aim of centralizing and increasing state income, the revenues of all endowments, including religious ones (*waqf*s), were taken under the roof of the newly established Imperial Ministry of Endowments in 1826, which turned the ulema into paid officials who were economically dependent on the central government.[29] The chief of the ulema, the Şeyhülislam himself, was now a minister in the newly established cabinet rather than the head of a class with some autonomy. Furthermore, and most significantly, the Janissaries were literally exterminated in the same year, thus destroying one of the oldest Ottoman institutions and a chief obstacle against military reform.

Efforts for centralization were accompanied by cultural innovations. Clearly linked with his political objectives, Mahmud II ordered his portrait to be hung in government offices, invited the Italian composer Giuseppe Donizetti to train the new, European-style military band that would replace the Janissary band, and introduced European-style clothing as well as the fez to replace turbans among all sectors of society excluding the ulema.

Most significantly, he started in 1831 the publication of the official gazette, the *Takvim-i Vekayi* (Calendar of Events)—the first Ottoman newspaper in the Turkish language. The leading article in the first issue argued that the newspaper was not necessarily based on an entirely new idea, as its mission was similar to that of the works of court chroniclers. However, it was essential to keep the people aware of the daily actions of the government to avoid misunderstandings and unfair reactions, as "[it is] human nature to object to and criticize the things the truth and essence of which one does not know." Hence, the government would now explain and legitimize its actions to its subjects; put differently, the intention was to *make* new subjects by disseminating information about the "true nature" of things. In addition to governmental matters, the article stated, it would also be beneficial to convey information on "sciences, fine arts and trade."[30]

Indeed, while the bulk of the newspaper was devoted to sections entitled Internal Affairs, Military Affairs, and Foreign Affairs, also included, albeit irregularly, was a section entitled Sciences/Arts (*Fünûn*). This section was devoted mostly to the presentation of brief information on books of all types published at the Imperial Press, but it also occasionally included news about recent inventions and developments in agriculture and industry as well as some statistics. The very fact that such a section was envisioned even at the outset, and then published within the Ottoman Official Gazette, is a strong indication of what the government deemed beneficial knowledge and endorsed during the reign of Mahmud II. It is also noteworthy that the Foreign Affairs section was dedicated almost exclusively to news about Europe, fostering a more Eurocentric approach than most European newspapers,[31] and news about European inventions and manufactures sometimes appeared in this section as well. The idea of Europe as the "sole origin of science and industry" that was a common theme of nineteenth-century debates was clearly conveyed by the gazette, and this fact is all the more striking as the director as well as the staff of the gazette came from the ranks of the ulema—an issue that will be taken up later in this chapter.[32]

Yet it should also be noted that the newspaper's actual audience proved not to be the general public. Rather than seeking subscribers, "a list was made of all state officials, people of learning, and notables, both in the capital and in the provinces, as well as foreign ambassadors and ministers, and mostly all of the five thousand copies printed were distributed according to that list."[33] The low rates of literacy throughout the empire and the underdeveloped state of the newly organized postal service can account for this decision, but the basic consequence was that it was the state elite and the notables who were exposed to this new knowledge regarding science as well as politics.

In such a period of reform, the tone of the 1824 decree of Mahmud II on the importance of basic education may appear surprising. In this decree, the sultan complained that most parents tended to end their children's education at the age of five or six so that the children could start apprenticing and making money right away. Instead, the decree maintained that "it is necessary to prioritize the learning of the fundamentals of religion to all worldly affairs."[34] Ignorance was becoming a serious problem, the decree stated, as these children lost interest in learning entirely. The consequence was that the majority of the people were ignorant about the basic tenets of their religion—a state that was the "sole reason behind the absence of divine aid" and would result in punishment both in this world and in the afterlife. Hence, the decree required that all children should remain in school until puberty and learn to read, particularly the Qur'an and catechism.

Once again, it would be anachronistic to note this emphasis on religion as a setback for the spread of "modern" knowledge. Nevertheless, the overtly religious tone of the decree and the political strategies of Mahmud II mentioned above may justify considering it a maneuver on Mahmud II's part to appease the ulema before the extensive reforms of the following years. Moreover, as the historian Selçuk Akşin Somel argues, the decree of 1824 can also be regarded as an antecedent of the educational reforms that would take place in the following decades. Not only does the decree encourage literacy—a skill that goes well beyond religious use— but it emphasizes the worldly punishments to which an ignorant people may be subjected, thus implying that literacy (in the guise of religious knowledge) is essential for the Ottoman Empire to recover from its weakened state.[35] In any event, while it is evident that the decree unequivocally emphasizes the importance of religious training, it certainly indicates an attempt to make education an issue of urgency, and a field that needs to be standardized and closely supervised by the central authority. Furthermore, even though the promotion of literacy per se is somewhat shadowed by the strong emphasis on religious education, the theme of ignorance as the reason behind the calamities striking the empire (in the absence of divine aid) was certainly established by the decree of Mahmud II. The ignorance in question was not identical to the one defined by Seyyid Mustafa, but Mustafa's conceptualization of ignorance (lack of appreciation for/ knowledge of science) would prevail in the following decades, taking the place of religion in the formulation brought forth by Mahmud II (religious ignorance as the reason behind the empire's decline).

3. Knowledge: Old and New

In the meantime, and in a way foreshadowed by the writings of Seyyid Mustafa and Mahmud Raif, "new" and "old" became increasingly common ways of classifying knowledge. Indeed, even though the revolts of 1807–8 had dealt a blow to the "New Order" that Selim III had initiated, "new" continued to be an appealing adjective. Importantly, "new knowledge" was ever more frequently associated with European languages in this period; one needed to learn a European language—most prestigiously French—to be proficient in the new sciences. But this emerging "common sense" entailed an increasingly heavy burden for the state, the provider of public education. Similarly, it rendered the task of the would-be "master of knowledge" exceedingly difficult; there was, in a sense, too much knowledge, and too many languages, to learn.

We see an example highlighting this heavy burden in the writings of one such master: Mustafa Behçet Efendi (1774–1834), the chief physician of Mahmud II, who commented in 1826 on the need for a new school of military medicine to serve the new army that was replacing the Janissaries. Behçet, who would indeed play a key role in the founding of the Imperial Schools of Medicine and Surgery, complained that "most Muslim physicians' practice is founded on the methods of old medicine, and they are not equally familiar with the methods of new medicine."[36] Importantly, for Behçet Efendi, a true physician needed to be able to utilize the methods of *both* the old and the new medicine as appropriate. But as it was obvious that "acquiring . . . the specialization based on the new method . . . is absolutely dependent on learning foreign languages," the students had to learn medicine in the French language.[37] Yet they also needed to know the Arabic and Turkish names of plants, substances, and diseases, in addition, of course, to the fundamentals of Islam. These had to be taught by a Muslim, but for the teaching of French and medicine, foreigners could be employed.

Mustafa Behçet Efendi, like his grandfather and father, was a member of the ulema. Coming from an aristocratic family, he was well educated, and in addition to his *medrese* training, he also learned European languages—a particularly rare phenomenon among the ulema. He specialized in medicine, but he also held many religious and legal appointments including the chief judge of Anatolia and Rumelia. In the words of a contemporary observer, Behçet "embodie[d] in his single person the various attributes of law, physic [*sic*] and theology."[38] In a sense, Behçet Efendi's elevated expectations from the new school were based on his own upbringing: knowledgeability brought about by simultaneous expertise in both the old and the new, not only Arabic and Turkish but also French.

Behçet's works have constituted a puzzle for contemporary readers. On the one hand, he is a scholar who translated sections from Buffon's *Histoire naturelle* as well as wrote and translated texts on new approaches to pharmacy, smallpox, and syphilis. His treatise on cholera based on a contemporary Austrian manual was widely circulated within the empire and used in territories as far from the capital as Tunisia.[39] Yet Behçet, and after him his brother Abdülhak Molla, also worked on a compilation of traditional and popular beliefs on medical cures that were published posthumously under the title *Hezar Esrar* (A Thousand Mysteries). That a man like Behçet Efendi, who was so interested in new medicine, could also take "superstition" seriously has been referred to commonly as an oddity in the few existing studies on him; at best, he has been described as a transitional figure.[40] But the compilation in question starts with an introduction that presents the contents as beliefs and arguments worth examining and studying experientially, if possible, rather than as indisputable truths.[41] The discussion of the "mysteries," too, is characterized by an assured approach that presents as worthy of reevaluation not only popular practices and beliefs but also the arguments of "past masters" such as Galen and the Muslim physician Ibn al-Baitar (1197–1248). Notably, the introduction also alludes to magnetism as a phenomenon comparable to the mysteries discussed in the compilation, with references to concepts like qualities and sympathy. While these concepts were particularly significant in Aristotelian and Galenic as well as Islamic traditions, they had hardly been abandoned in eighteenth- and early nineteenth-century European medicine.[42] Indeed, the idea of sympathy as an invisible force of attraction between objects had acquired renewed popularity in the forms of mesmerism and animal magnetism in France and England that lasted well into the late nineteenth century.[43] Hence, Behçet's position can be better understood as a confident attempt by an Ottoman scholar to contribute to the emerging methods and views in European biology and medicine, with arguments and examples from traditions with which he—and many contemporary European men of science themselves—were familiar.

Behçet Efendi's approach is also one of the numerous examples of ulema participation in the transformation of the Ottoman Empire. As historian Uriel Heyd demonstrated, the reforms of Selim III and Mahmud II found supporters especially among the upper echelons of the *ilmiyye* class.[44] Indeed, Behçet's main rival and the author of a five-volume compendium of medical science (*Hamse*, The Pentalogy), Şanizade Ataullah Efendi (1769?–1826), was also a member of the ulema. But in addition to his *medrese* education and training in medicine, Şanizade had also studied at the Imperial School of Military Engineering, thus acquiring knowledge from many different types of educational institution that existed in the empire. It is known that, as a result, he was able to read in French, Italian, and Greek—at least to some degree—as well as Arabic and Persian. In the introduction to one of the volumes of his compendium entitled the *Miyârü'l-Etıbbâ* (The Standard for Physicians, the translation of an Austrian manual[45]), he underlines how important it is for a good physician of his time to have a good education, work hard, and be "aware of the curiosities of the sciences."[46] He also notes that practitioners of medicine who only have practical skills cannot be considered true physicians: it is the medical *knowledge* that one possesses that makes one a "complete physician."[47] And the knowledge in question is now a "new" type of knowledge, built on a variety of different sources, and presented to the community of practitioners by Şanizade, with his extraordinarily vast learning that enables him to speak about "the new" from within "the old."

The ease with which Şanizade handled "the new" was also seen in the first volume of the compendium the *Miratü'l-Ebdan*, which was the first book by an Ottoman Muslim that included European-style engravings of body parts. This section in particular acquired some popularity in Europe as well. Indeed, as in the case of Seyyid Mustafa, Şanizade's work received praise from French readers, also with an indication clarifying the context of the Ottoman interest in the new sciences: In the fervently positive review of Şanizade's work that he published in the *Revue encyclopédique*, the French orientalist Thomas Xavier Bianchi argued it was remarkable that, in a Muslim society where people "blindly" followed the religious elite who had always opposed innovation, such a work came from a member of the ulema itself. Very significant, though, was that Şanizade's work was based on *French* sources; noting the importance of French influence on the Ottoman Empire, Bianchi wrote that it was "honorable" for France to have contributed in this manner to the welfare of the Ottoman society.[48] Based on Bianchi's arguments, the *Oriental Herald* made the following remark:

Emulation, that main-spring of human action, and without which man would never have emerged from his first state of barbarism, has at last manifested its power over the indolent Turks; at least upon one of them, who, feeling ashamed of the inferiority of his countrymen in [these] matters, compiled this work on anatomy. . . . Whether this sudden infraction upon ancient prejudice and habit may prove the harbinger of civilisation among the Turks, we will not venture to predict; but . . . we may hazard the opinion that either they must very speedily become zealous proselytes of European civilization, or within a few years they will be blotted out from the map of nations.[49]

As I will further discuss below, such assertions make it clear that what Ottoman authors and politicians commonly regarded as the importation of the "new sciences" took place within a context shaped thoroughly by colonialism. This meant that the Ottomans had not only to "translate" these new sciences but also to grapple with an ideology about the meaning and significance of them. While men like Behçet and Şanizade could approach these simply as new sciences that they could confidently comment on and integrate with the old, their successors would increasingly adopt the language of European observers, and "the old" would come to be seen as "ancient prejudice." Moreover, as Bianchi's arguments also clarify, to their European addressees, the Ottoman interest in the "new sciences" was a matter of influence and patronage; that the Ottomans were learning "the new" from the *French* was at least as important as their interest in learning them. Bianchi, a member of the Société Asiatique and at the time interpreter at the French Embassy in Istanbul, made the point even more clearly in the introduction to the French-Turkish lexicon he published in 1831. Stating that many French works on mathematics, geography, navigation, and the military arts had already been translated into Turkish, Bianchi argued that the French language appeared to be steadily acquiring additional influence and popularity in the Ottoman Empire. This, he noted, should lead to a strong French influence, "of which our policy and industry can take great advantage one day."[50] This conflict between the Ottoman and European characterizations of the process of "translation" would play a great role in the shaping of alternative discourses on science in later parts of the nineteenth century.

The participation of such members of the ulema as Behçet and Şanizade in the dissemination of new sorts of knowledge should not be interpreted as an indication of a swift change. First, it is worth noting that the ulema who supported the changes in question were educated in medical sciences themselves, albeit in a different form. The ulema may have perceived the change as one within which they could still maintain their authority, as medicine had always been seen as a respectable practice related to theology in the Islamic tradition, and Ottoman interpreters could comment on the changes in a confident fashion. Furthermore, it is also a fact that foreign or non-Muslim physicians had come to dominate the practice in the nineteenth century due to the shortage of *medrese*-trained physicians. Hence, the ulema may have regarded the opening of the new school and the publishing of new books as an opportunity for more Muslim physicians to enter the field. Indeed, the new medical school was initially to admit Muslim students alone. Thus, in many respects, the contributions of men like Şanizade and Behçet can be regarded as indications of the competition within the field of medicine as well as the ulema in general— their emphases on the mastery of both the old and the new were about raising the stakes in the fields of which they were members.[51]

As a general trend, most of the support for Mahmud II's reforms came from the higher ulema. Nepotism and the competition for the few top positions within the *İlmiyye* body led the higher-ranking ulema to compete for the sultan's approval rather than pass judgment on his actions. Furthermore, aristocratic ulema families had started to emerge due to networks of patronage, and their duties within the government caused top-ranking members of the ulema to have a closer experience of the workings of the government, which, in turn, led to a more nuanced outlook along with a primary consideration for the *raison d'état*.[52] It is in that sense hardly surprising that Şanizade was the son of the chief judge of Mecca, whereas Mustafa Behçet's grandfather was a grand vizier and his father a chief clerk of the Imperial Council.

The resultant acquiescent, if not celebratory, attitudes of these high-ranking members of the ulema aggravated the low-ranking ulema as well as the students of the *medreses* throughout the empire, and it was these groups indeed that would voice the most serious criticisms of the changes they witnessed, including those concerning the definitions of knowledge and ignorance.[53] We should also emphasize that the attitudes of the high ulema drew attention to the association between new knowledge and *the state*, rather than only the new bureaucrats who gradually became its most vocal advocates. Knowledge endorsed by the state was knowledge endorsed by those who identified themselves with the state.

Concerns about low rates of attendance and success at the Imperial School of Military Engineering, on the other hand, enabled another interesting figure to come into prominence. In a memorandum he wrote to the sultan in 1830, chief of staff Hüsrev Pasha stated that the required courses could not be taught in an orderly and appropriate way due to the incompetence of the teachers. It would be advisable to appoint European engineers as well as a new Muslim principal who was "familiar with the needed sciences and arts."[54] Very importantly, this new classification of knowledge openly defines the sciences taught at the Imperial School of Military Engineering as the ones that the empire "needs," simultaneously defining the individuals educated in them—military engineers—as those truly needed by the empire. "The new" and "the needed" were increasingly one and the same, with unmistakable implications regarding the prestige of those representing "the new."

Hüsrev Pasha's recommendation for the post of the principal was Ishak Efendi (1774?–1836), a Jewish convert from Ionnina, former chief translator to the Imperial Council and a graduate of the school himself. Ishak did become the principal, and during his tenure, he published his four-volume magnum opus, the *Mecmua-i Ulûm-ı Riyaziye* (Compendium of Mathematical Sciences, published between 1830 and 1834)—a compilation from European sources and a pioneering work in the Ottoman Empire that contained detailed information on mathematics, chemistry, physics, astronomy, biology, botany, zoology, and mineralogy. Mahmud II expressed his appreciation to Ishak with 250 gold pieces.[55]

The short introduction to the *Mecmua* explained precisely what Hüsrev Pasha's comment had stated briefly: the sciences taught at the Imperial School of Military Engineering and included in the *Mecmua* were the "needed" sciences. It was these sciences that needed to be learned in order to be faithful to the exalted order for holy war in the contemporary world. This was because the organization of the soldiers was now based on arithmetic and algebra, measurements depended on trigonometry, the manufacture of weapons and warships required knowledge of mechanics, the actual movement of ships depended on understanding astronomy and physics, and so on.[56]

Interestingly, Ishak also stated that "past masters ha[d] specific treatises on each of these, and some of these ha[d] Turkish translations," yet because they were all separate pieces on specific areas, he intended to compose an encompassing work that would make instruction easier.[57] Nevertheless, the fact is that many sections in his work, such as the one on chemistry, are novel for the Ottoman audience. Ishak appears to have understated their novelty, perhaps aiming to present his arguments in a less alienating way,[58] but his emphasis on the unique utility of the "mathematical sciences" for holy war "in these times" is a rather palpable insinuation of novelty. What is also important to underline is that Ishak's praise for the sciences is built exclusively on the practical uses of the new sciences in general and their positive impact on the military in particular. The reference to "holy war" clearly indicates a religious type of motivation for learning the new sciences, yet what it ultimately attributes to the new sciences is essentially limited to practical benefits.

Indeed, unlike the representatives of up-and-coming groups like Seyyid Mustafa the engineer and Mahmud Raif the diplomat, the "masters" who represented both the old and the new—like Şanizade, Behçet, and Ishak—played down, if not entirely disallowed, the potential impact of the new sciences on the character of their students. As we have seen, the narratives of Mahmud Raif and especially of Seyyid Mustafa imply that acquaintance with, if not expertise in, the new types of knowledge render one a "superior subject"—a new type of Ottoman individual whose services to the "Sublime State" would far surpass that of any other. Both Raif and Mustafa highlight their sense of distinction even in the brief autobiographic sections of their works: both *knew* even at a very young age that times had changed, new types of knowledge had arisen in Europe, and one had to learn European languages to be able to acquire them. They had followed this path, and thus their "fresh" minds would enable them to provide to the empire unique benefits—benefits that truly mattered. As we will see in the following sections, this novel attempt to associate new knowledge with good subjecthood—of which we see but a few examples in the early nineteenth century—would become much more dominant in subsequent periods, rendering the Ottoman debate on science essentially a debate on civic virtue and vice.

One final point worth mentioning is that reading Ishak's book more or less served as *the* instruction at the Imperial School of Military Engineering. According to James De Kay's observations, classes were essentially dedicated to dictation in its most basic sense, with inattentive students being asked to repeat the last sentence of the professor.[59] Furthermore, when asked what textbooks they used, the future military engineers of the Ottoman Empire replied "that when they had faithfully gone through [Ishak's] volumes, they would have acquired all the knowledge in the world."[60] If we are to trust the author, the new schools were not fundamentally different than the traditional *medreses*: education was regarded primarily as learning by heart the contents of a book, under the supervision of a master. The new sciences were, essentially, a new set of information to be learned, a compartment of the "wealth of knowledge"—a theme that remained prominent throughout the century.

B. The Context of Reaction: The World of the Representatives of "New Knowledge"

The so-called Ottoman encounter with science is not an encounter that took place in the world of ideas; it was an encounter that involved real people with specific cultural repertoires, dispositions, expectations, and interests. Ottoman diplomats, the students and graduates of the new Ottoman schools, and Ottoman elites educated in Europe perceived and represented the new, "beneficial" knowledges produced by the Europeans within specific social and political contexts. And, ultimately, it was these contexts and the characteristics of the people involved in these encounters that would shape the perception of science in the Ottoman Empire of the nineteenth century. As future decades would make increasingly clear, observations about the types of people who came to represent new knowledge in the Ottoman Empire, and the conditions within which they interacted with Europeans, constituted the core of the Ottoman debate on science. This is hardly surprising, as the adoption of these new types of knowledge by the Ottomans was no less a political and social process than an intellectual one, for both the Ottomans and their European counterparts.

A relevant example in this context involves the establishment of the School of Military Sciences (*Mekteb-i Ulûm-ı Harbiye*) in 1834—an institution that would gradually evolve into the most prominent symbol of new knowledge in the empire. The chief responsibility in the founding of the new school was assigned to a young but very experienced military bureaucrat, Mehmed Namık Pasha (1804–1892). More like Seyyid Mustafa and Mahmud Raif than Şanizade and Mustafa Behçet, Namık Pasha was a graduate of the Imperial School of Military Engineering. He had also studied at the École militaire in Paris and been sent as a military attaché to St. Petersburg[61] and as an ambassador plenipotentiary to London. The schools from which he graduated and the duties he performed afterward clearly render Namık a perfect example of the new Ottoman elite. Fluent in French and English, Namık had also visited military schools and factories in England, and he was particularly eager to acquire knowledge about steam power.[62] Intending to establish a school similar to the French Military Academy of Saint-Cyr, Namık Pasha made sure the school had a large library, new maps and tools all brought from England, and its own printing press and hospital.

However, unsure of the prospects and worth of the new school—an indication that "new knowledge" still had insufficient cultural and symbolic value—wealthy families failed to send their children there. As a result, the government ended up recruiting stray children as the first students. Similarly, Namık Pasha's introduction of the European school desk—putting an end to the tradition of sitting on the floor with the books placed on bookrests (rahle)—raised concern.[63] According to a contemporary European visitor, the exquisite mosque attached to the school had been opened and the students compelled to pray due to the concerns of the parents who had misgivings about the school.[64]

The concerns in question—concerns that accompanied the entire process of the appropriation of the new sciences—make sense particularly if the influence of the world-political context on the opening and organization of the school is taken into consideration, rendering it vulnerable to criticism. As has been established firmly by recent historical studies, it is a flawed approach to analyze the Ottoman reforms of the nineteenth century simply as impositions of European powers. Domestic factors not only motivated but significantly shaped the form of these efforts. However, it is still important for our purposes to note that the introduction of European-style schools and new knowledges into the Ottoman Empire was an issue directly connected to the dynamics of international politics. The Great Powers of Europe, if they themselves were not interested in playing this role, were at least wary of the interests of their rivals in becoming the educational and intellectual guides of the Ottoman Empire.

What the English historian and traveler Julia Pardoe (1806–1862) notes about the state and future of the new military school is particularly significant in this context, as her comments illustrate what lay behind the calls for furthering progress in the Ottoman Empire. Calling the military school "a body without a soul," Pardoe laments the apparent inadequacy of the teaching cadres in terms of experience and talent, rendering all the enthusiasm fruitless: "Could sentiment be deepened into science, and inclination wrought into ability, the Military College would take high ground, . . . but where the means are limited, the effects must be comparatively inconsequent."[65] Yet the real fault lay with some Europeans who praised these new Ottoman institutions beyond their real worth:

And thus, flattered into a belief of their own sufficiency on the one hand, and misled by misstatements on the other, the influential individuals connected with the unhappy College have abandoned it to the ruin which must ultimately, and at no distant period, overtake it; from the hopeless incapacity of a set of men, who, familiar with the name of every science under Heaven, are most of them profoundly ignorant of all save the first rudiments of each; and who are, consequently, ill calculated to work that great moral change so ardently desired by all the true friends of Turkey.[66]

Turkey needed to acquire the intellectual level of Europe by "train[ing] up her youth to habits of reflection and scientific research," yet Russia, fearful of the possible effects of such schools, managed to have Ottoman reformers believe that the school needed no further development and the Ottomans already had the knowledge they needed. Pardoe's conclusion was simple: "England must resolve [this] question."[67]

Comments made by the French author Alphonse Royer indicate a similar approach, but this time from a French point of view. Describing Namık Pasha as a perfect statesman, Royer praised him as someone who "contributed by his own example to the spread of the desire to study *our* language and *our* sciences" among Ottoman youth.[68] This understanding that the Ottoman interest in the new sciences and faith in the guidance of French sources in this endeavor should be seen as great achievements by France had also been promoted by the orientalist Thomas Bianchi, as stated earlier. Also worth remembering is that the works of Seyyid Mustafa and Mahmud Raif were intended primarily for European audiences with the aim of making a statement about the direction that the Ottoman Empire had taken.[69] Additionally, we have a report presented to the sultan in 1837 by Reşid Pasha, the influential statesman of the reform period, which discusses in detail the conflicts among France, Britain, and Russia over the employment of European experts and teachers by the Ottoman Empire.[70] Finally, the French policy of using the *grandes écoles* model to help shape the new Ottoman elite class for a great part of the nineteenth century is well documented.[71]

Thus, a central role of the new bureaucrats and the advocates of new knowledge involved negotiating with the Great Powers the terms under which "their knowledge" would be imported into the Ottoman Empire. In these negotiations, where they were not necessarily able to set the terms, they strove to appease the various parties involved, and, within a broad process of drastic economic and legal transformation, these maneuvers contributed to their perception as imitators whose primary concern was not the interests of the Ottoman people. An example of the position of the "Europeanized" bureaucrats is a comment by Namık Pasha himself, who is reported to have said during his visit to Manchester: "Ours is not a manufacturing country, and we have no pretension to compete with the science and capital of England. But our fertile territory and happy climate enable us to furnish you with many of the materials which you require."[72]

The parents who hesitated to send their children to the new schools (or were happy to see that a mosque was built) are unlikely to have been aware of all such transactions and comments. Moreover, as the recent literature on the nineteenth-century Ottoman Empire as well as the following chapters illustrate, the Ottomans were by no means clueless copiers of European ways or helpless followers of the instructions of European states. Nevertheless, we cannot ignore the context within which most Muslim Ottomans perceived these new schools and bureaucrats, and interpreted such instances as skirmishes between "European ways" and Islam, or as the surrender of Ottoman bureaucrats to the Great Powers. With yet little credibility and symbolic capital, and perceived as men who simply complied with the demands of European powers, the new bureaucrats could not render prestigious the types of knowledge that they represented. Expressions of hesitation and doubt that we observe in such cases indicate social, rather than epistemological, conflicts.

Conclusion

What we observe in the early nineteenth-century Ottoman Empire is the early stages of the emergence of a new social group and a new discourse on knowledge or, rather, a discourse on "new knowledge." Now it is a fact that "new" had already started to become a popular adjective in the eighteenth-century Ottoman Empire,[73] and erudite scholars like Şanizade and Mustafa Behçet were able to approach "new" and "old" types of knowledge in an integrative, rather than antagonistic, way. But as the bloody end of the New Order of Selim III itself indicates, an emphasis on novelty is not always palatable for all, especially if it involves social and political changes of a potentially threatening nature for specific cliques or groups.

Furthermore, the emerging discourse on the benefits of the new, as illustrated by the writings of Seyyid Mustafa and Mahmud Raif, was not simply one that separated one type of knowledge from another; it established a hierarchy between both them and their possessors. In other words, it classified types of knowledge and contrasted types of people. However, as this chapter has also illustrated, these examples from the beginning of the century remained isolated attempts. In fact, the authors themselves found out the hard way that the empire they lived in was not exactly as they represented it in their publicity pamphlets. The close circle around Selim III, the enlightened patron of the new elites, did not survive the revolts of 1807–8, and what Seyyid Mustafa and Mahmud Raif referred to as "ignorance" and "indolence"—that is, of course, the groups to which they attributed these characteristics—carried the day.

But the basic arguments and the tone of their writings achieved prevalence in later decades, as the number of individuals similar to them grew. The new generation of Ottoman bureaucrats would include more men like Namık Pasha: speaking European languages, interacting frequently with European politicians and notables both within the Ottoman Empire and abroad, these bureaucrats, more than any other group or even the sultan himself, would influence the policies that the empire would follow in what is known as the Tanzimat (Reorganization) Era (1839–76). Indeed, by destroying the Janissaries in 1826, pacifying the local notables, and limiting the influence of the ulema in order to strengthen central authority, Mahmud II had laid the groundwork for a regime in which Istanbul bureaucrats would hold the reins of government without significant checks and balances. These bureaucrats, as we shall see, constructed an official discourse that unequivocally glorified the new—including new knowledge and its representatives.

TWO

Speakers, Institutions, Discourses of Science in a New Regime

Introduction

The "new sciences of the Europeans" had few devoted spokes-people in the Ottoman Empire at the turn of the nineteenth century. Institutions and experiences that produced individuals who made enthusiastic cases for these new sciences—and, as an inevitable implication, for those who were aware of the new types of knowledge—were rare. Note, for instance, that the Ottoman Embassy in London where Mahmud Raif worked was established in 1793 as the first permanent Ottoman Embassy. Similarly, the schools where the new sciences were taught to some degree were products of the late eighteenth and early nineteenth centuries. But especially after the 1830s, not only the number but also the political power of such individuals increased significantly. As the new diplomats, bureaucrats, and graduates of the European-style schools played an ever-growing role in Ottoman politics and culture, their characterizations of knowledge and ignorance gradually shaped what we can call an official discourse on science—a discourse that established certain interpretations of science as "matters of fact" by constantly repeating them in official texts and reproducing them in educational institutions.[1] And the way these individuals described science was tightly connected to how they perceived themselves vis-à-vis other elites and "commoners."

In this chapter, I analyze this connection and focus in more detail on both the key components of the emerging official discourse on science and on the characteristics of the social actors who constructed it. Even though what follows touches on the biographies of specific individuals, the personalities themselves are less important for our purposes than the types of social groups and dispositions that they represent. Similarly, while I refer to several developments particularly in education policy, what matters more is the representations of knowledge ("old," "new," "useful," "useless," etc.) in the texts that accompanied them, as it is these representations that enable us to identify the links between specific portrayals and the groups that produced them. What such an analysis reveals, as I will show below, is that, especially after the 1830s, science was discussed not simply as a matter of knowledge or in relation to the question of "saving the empire," but as a matter of personal and civic virtue and vice. In other words, it was the characteristics, merits, and demerits of the person who was or was not familiar with the new types of knowledge that formed the core of the official discourse on science in the Ottoman Empire. And this was due primarily to the fact that an alleged awareness of the "unique" properties of scientific knowledge constituted a marker of distinction for the members of a new social group.

What follows first is an exploration into the context within which this new type of social actor acquired increasing authority and into the qualities of these actors. Then I will focus on the official discourse on science that they constructed after the 1830s.

A. The Speakers of Science

1. Men of the Tanzimat

While the early nineteenth century witnessed the slow but steady production of military bureaucrats at the new schools of the empire, many other influential elites of the period were primarily autodidacts, usually possessing those skills still rarely found within the empire, especially among the Muslim community: literacy and some familiarity with a European language. Typically after elementary religious education, the teenager would start working as an apprentice secretary in one of the government offices. He would continue learning on the job and hope to grab the attention of a higher-ranking bureaucrat who could become his patron.

Wishing to create a more reliable, efficient, and loyal cadre, Mahmud II attempted to standardize the training of bureaucrats as well. An official document from 1838 complains that those secretaries employed in government offices so far tended to have only some training on the Qur'an, and "perhaps have never heard even the names of the mathematical sciences and geography, the instruction of which is most important and most needed for clerks to be employed both at home and abroad."[2] Hence, the school founded the same year to produce Mahmud II's civil servants (*Mekteb-i Maarif-i Adliyye*, The School for Learning), had a curriculum that included, in addition to courses on grammar, Arabic, and Persian, courses on mathematics, French, and geography.

However, the setting that produced the most prominent bureaucrats of the Ottoman Empire around the mid-nineteenth century was another institution founded by Mahmud II: the Translation Bureau (*Tercüme Odası*) where clerks (both Muslim and non-Muslim, along with the occasionally employed foreigners) not only translated European documents into Ottoman Turkish but also received basic training on subjects similar to those at the school for public servants. Most crucial, however, was the teaching of French. Opened in 1821 but a full-fledged department only after 1833, the Translation Bureau raised not only many of the leading statesmen of the Tanzimat (Reorganization) Era but also their critics.[3] Many "graduates" of the bureau also had the opportunity to work in Europe, typically in Ottoman embassies, where some of them followed courses in universities and occasionally made the acquaintance of European intellectuals and scholars.

It is the accomplishment of these bureaucrats—who gained significant power at the expense of the ulema as well as the sultan himself—that specific representations of knowledge, ignorance, and their virtues and vices gradually became official after the 1830s. These are, after all, the men who penned many of the key legislative and administrative texts in this period. Among the basic features of the representations that these texts contained were the identification of the sciences of the Europeans with "needed" and "true" knowledge, and the possessors of this type of knowledge as *the* knowledgeable group within the empire. At the same time, the representation of this new type of knowledge was based on the way the new bureaucrats related to it: this was a type of knowledge that was to be "possessed," "known," not produced.

But, once again, we should note that the emergence and consolidation of this discourse did not take place within a vacuum. In fact, the years 1838 and 1839 witnessed two events that defined the context within which the Ottoman Empire would experience the rest of the nineteenth century: the Ottoman-English trade agreement of 1838, and the Imperial Decree of Gülhane in 1839 marking the official beginning of the Reorganization Era. The former was an economic turning point in that it turned the empire into a free-trade zone for English merchants.[4] The agreement was signed by the Ottomans in return for much-needed English aid in the Ottoman military campaign against the rebellion of the governor of Egypt, Mehmed Ali Pasha; but similar agreements would have to be signed with the other Great Powers soon afterward. The result was an Ottoman market filled with cheap European imports that dealt a blow to Ottoman manufactures, the effects of which lasted until the 1870s.[5] Added to the resultant social disruption were the apparent associations between class and ethnoreligious community membership: numerous members of the non-Muslim communities (Greek, Armenian, and, to a lesser extent, Jewish) were able to seize the role of middleman between Ottoman products and European merchants, thus assuming the shape of a commercial bourgeoisie.[6] While there was no rigid split between Muslims and non-Muslims particularly in the lower echelons of society, the lifestyles and consumption patterns of this new class were alienating to some members of the Muslim community. More specifically, the growing differences among Ottoman bureaucrats themselves resembled a bifurcation within the Muslim middle class of Istanbul.

The 1839 Imperial Decree of Gülhane, on the other hand, involved the declaration that the Ottoman state would undertake a series of administrative reforms and that rights to life and property were guaranteed by the state, tax farming would be abolished, the system of conscription would be made fair, and, crucially, the new laws would apply to all communities within the empire, regardless of their ethnic or religious identity.[7] This so-called Tanzimat Decree, which gave its name to the period of reform that it initiated, was not necessarily entirely welcome by the non-Muslim communities at first, but one of its most conspicuous impacts was the disillusionment of members of the Muslim community, which saw themselves as losing their privileged position within the Ottoman Empire.[8] The new bureaucrats' attempts to construct a common Ottoman identity that would transcend all religious identities and ideally help keep the empire intact (commonly referred to as the policy of Ottomanism) were not particularly harmonious with the Islamic-Ottoman conceptions of social order.

In sum, a significant portion of the Muslim population perceived the Tanzimat as an era of submission to European powers—an era in which non-Muslim Ottomans were favored at their expense by the estranged bureaucrats of Istanbul. Further proof was provided by the everyday lives of the new bureaucrats themselves: the lifestyles of top-level bureaucrats increasingly resembled that of the non-Muslim bourgeoisie and the Europeans.

How did these bureaucrats perceive and represent themselves? How were these representations perceived by the discontented groups? Focusing on several specific examples can help to explore these questions.

Mustafa Reşid Pasha (1800–1858), the chief architect of the Tanzimat Decree of 1839, was a diplomat who had been the Ottoman ambassador to France in 1835 and then to England in 1836. On his return to Istanbul in late 1837, he became foreign minister. The report he presented to Mahmud II on his return can give the reader an idea about the way in which new diplomats like Reşid operated. In this text, which is primarily a list of European expectations from the Ottoman Empire, Reşid mentions, for instance, that the completion of the new building of the Military School would not only bring about numerous benefits but also "attract the attention of the Europeans." He recommends the termination of tax-farming "to which all Europeans object," as well as the adoption of the quarantine system, the absence of which causes many problems in transportation and commerce about which "all Frenchmen complain."[9]

This indeed was the basis of the prestige of the new top bureaucrats of the empire who had been in Europe: being able to transmit and evaluate the opinions of the Great Powers at a time when the Ottoman Empire could no longer survive by its own means. As Carter Findley puts it:

Where Mahmud's diplomats really produced their impact was not so much as representatives of the Ottoman Empire in the states of Europe as in their unprecedented ability to absorb and respond to their experiences abroad, and in their role in mediating the demands of the major powers to their own people. Thus, in representing the West to the Ottomans, more than the other way around, they quickly acquired an influence that extended in Ottoman official circles far beyond the field of foreign affairs as narrowly defined.[10]

The bureaucrats not only deliberated on the expectations of the Europeans, but as long as they remained in their posts, they were able to act on those expectations in ways they saw fit. Indeed, all the suggestions in Reşid's memorandum would be realized within the period of reform that started with the Imperial Decree of 1839.[11]

Security of life and property in addition to the growing centralization of political power transformed the new bureaucracy into an entirely new class that no longer resembled the scribes of past centuries. Making the most of their skills, knowledge, and relations, the higher-ranking members of this new class exhibited an unprecedented degree of self-confidence. For instance, Reşid Pasha was able to tell the British foreign secretary Lord Palmerston that Sultan Mahmud II "had no knowledge whatsoever of the skills needed in administrating the affairs."[12] A contemporary observer argued that the sultan's power was more than balanced by that of the "Machiavellian" new bureaucrats who, thanks to their relatively superior erudition and practical experience, had usurped the state, turning imperial authority into "but a phantom."[13]

As Ottoman statesmen faced the reality that the survival of the empire depended on navigating and exploiting the balances of power in Europe, the foreign ministry became a most prestigious post and, ultimately, a stepping-stone to the prime ministry. The bureaucrats who occupied these posts in the Tanzimat Era—most importantly Reşid, Âli, and Fuad Pashas—had diplomatic experience, and they frequently interacted with European diplomats and visitors in Istanbul in European fashion, impressing them immensely. The American author Edwin de Leon "praised" Reşid Pasha, stating that "both in intellect and character [he] looked less like an Oriental than any Eastern man I have ever seen. . . . No more prepossessing man, no more subtle statesman, no more accomplished diplomat could be found in the ranks of the *corps diplomatique* than this representative Turk."[14] In the obituary published for Fuad Pasha on February 16, 1869, the *London Times* made a similar comment: "People could hardly believe that the elegant and cultivated person who spoke so well, who told such good stories and uttered witticisms that Talleyrand would not have disowned, and whose manners were so polished, could be a Turk."[15]

Safvet Pasha, another top bureaucrat who had worked at the Translation Bureau and occupied several ministries during his long career, was praised by the American Oriental Society, of which he was an honorary member, as "an enlightened and scholarly Turkish gentleman."[16] At a later age, Safvet stated in a letter to his son that even the most awful city in Europe was superior to Istanbul and it would take centuries to turn the Ottoman capital into a Vienna. He wrote: "I am utterly regretful that I wasn't able to spend some twenty years of my life . . . in Europe. If I had been able to do that, now I would at least be cherishing the memories of the things I would have been able to see during that time."[17]

Another example is Halil Şerif Pasha (1822–1879), who was also educated in Paris. He was let down when he was appointed Ottoman ambassador to Athens, rather than Paris, as he had expected. Âli Pasha wrote him a letter teasingly comparing Athens to Paris where he stated he could not "believe that there could be on earth a more moving and more seductive song than the *Marseillaise*."[18] The allure of Paris for young Ottoman bureaucrats is clarified by another letter Âli sent to Halil Şerif, this time when he was appointed to St. Petersburg: "It is better that you go there [Paris] later, because you will then be less young and you will arouse . . . less envy among those who remain here to labor far from the charms with which you would be surrounded in that fairy capital."[19] While in St. Petersburg, Halil Şerif would become one of the founders of the Ottoman Society of Science, which I will focus on in chapter 3.

These high-ranking bureaucrats relied on the rather high salaries paid by the state, instead of fluctuating land revenues like the officials of the past. Their consumption patterns, too, differed widely from their predecessors, as well as that of the other lower-class, particularly Muslim, groups within the empire.[20] They represented new tastes: they were increasingly more interested in European goods, particularly those that had a strong symbolic value in terms of "Europeanness," such as pianos, and they liked to frequent the quarters of Istanbul where non-Muslims and Europeans lived.[21]

Therefore, the builders of the official discourse on science can be seen as members of a group who appeared increasingly more alien to significant portions of the Muslim community of Istanbul. Yet it is also important to avoid making hasty generalizations at this point. First, we should underline that this group was a small minority who not only lacked legitimacy in the eyes of the public but also embodied significant competition within it, making patronage relations highly consequential. As a result, reversals of fortune were rather common for the members. Second, while the divergence between the ruling elite and the public certainly led to the popular view that the new bureaucrats were irreligious admirers of the infidels, the official discourse on science, knowledge, and ignorance was much more complex than what this view would imply. The "men of the Tanzimat" did construct a discourse that praised the sciences of the Europeans, but at the same time they linked the possession of scientific knowledge to moral duties and responsibilities. Science was not simply related to economic and military might; it was a moral issue.

2. A Manifesto for Science: Mustafa Sami

The first of the caveats above is best illustrated by a Tanzimat bureaucrat who is the author of one of the most passionate Ottoman paeans to the sciences of the Europeans. Mustafa Sami, the author of the famous *Avrupa Risalesi* (A Treatise on Europe), worked as a scribe in various civil offices and became a senior clerk in 1833.[22] After gaining some initial familiarity with European affairs during his employment as the secretary of the Ottoman embassy in Vienna, he was sent to Paris as the chief secretary of the ambassador in 1838.[23] On his return in 1839, he became the minister of postal services and soon afterward published his treatise.

The treatise starts with Sami's statement of purpose: serving the nation by making it aware of things of which it is ignorant. Cognizant of his own "inadequacy" and "insignificance," yet aspiring to be of service to his nation, Sami shares his observations and opinions in order to encourage "greater scholars" to express their own views, thus enabling hitherto unshared knowledges to be revealed.[24] The author's expression of humility is a well-established convention in Ottoman texts, but the emphasis on "serving the nation by spreading knowledge" is significant, as this was an approach used commonly by the members of the new "knowing class."[25] The "true patriots" of the new era would be the ones who possessed useful knowledge and shared it, almost as a *mission civilisatrice*. Indeed, this emphasis on "sharing" is typical of the writings of the new elite (as we also observed in the case of Mahmud Raif in chapter 1). This rhetorical strategy presents the new bureaucrats as selfless enlighteners, in contradistinction to the representatives of "old knowledge" who are associated with esotericism. In a sense, the new elite attempt to turn the tables on their critics by portraying them as the truly arrogant ones who have neglected the people.

Sami describes Parisians as epicures but also as morally upright and patriotic individuals—a view that was not necessarily congruent with established wisdom in the early nineteenth-century Ottoman Empire. Yet what is most impressive about the people of Paris is their interest in learning, which Sami swiftly generalizes to all Europeans: everyone in Europe, even an ordinary porter or a shepherd, is literate; even the blind and the handicapped can study and make a living on their own.[26] "Thanks to learning and accomplishment," Europeans discovered the true nature of all things, which enables them to organize their lives well and maintain their health. Literacy allows them to keep their accounts in order; their knowledge of mathematics and chemistry helps them to improve their crafts.[27] They publish books on all sciences and even on subjects like pest management, as a result of which they benefit and assist even the people of other lands. Their expertise in geometry and mechanics enables them to build wide, smooth roads and to plan their cities better.[28]

What, then, is the reason behind all this order and might? Sami's answer is simple: science. "Europeans realized and admitted that the greatest embarrassment and disgrace in the world is ignorance," and made spreading education their primary goal.[29] As a result,

just as in our lands poetry became the basis of *belles lettres*,[30] in theirs . . . the progress of geometry enabled the improvement of algebra, making possible the invention of steam engines, thanks to which goods that would take one year to produce are manufactured in one day; similarly, due to the progress of the science of chemistry, the science of lithography was discovered. . . . No country in the continent of Europe, save Italy, has an agreeable climate or fertile soil. They have stepped forward thanks only to science and knowledge.[31]

Note that the opposite of ignorance is defined as the spread of *scientific* knowledge in this section—an approach that we will observe in other texts as well. It is also worth underlining that Sami does not simply link science to material progress; his rather daring praises that extend to the character of Europeans in general are also related to the spread of science. Acquiring scientific knowledge, Sami suggests, made Europeans more productive, self-reliant, caring, patriotic, and, in sum, "better" people.

But is science a European invention? In other words, are the sciences of the Europeans "*European* sciences"? Mustafa Sami's answer is a resolute no. Far from being related to European customs or Christianity, the contemporary sciences of the Europeans are based on the sciences developed by Muslim Arabs. So science is one, and it is part and parcel of the Islamic heritage.

This approach was not uncommon in Sami's time. Societies experiencing similar challenges and going through similar social and cultural transformations as the Ottoman Empire in the Europe-dominated world of the nineteenth century produced discourses that characterized the new in terms of the existing: they praised the new sciences and tradition (which they redefined in the meantime) simultaneously. In his travelogue published in 1834 after his visit to Paris, the Egyptian polymath Rifa'a al-Tahtawi had made one of the earliest and strongest cases regarding the impact of early Islamic scholars on early modern European natural philosophers—a vibrant theme in the orientalist works of the period.[32] Similarly, Chinese scholars and bureaucrats commonly emphasized that the classical branches of Chinese learning were the sources of Western science during most of the nineteenth century.[33]

To Sami, once science was disseminated throughout Muslim lands *as in the past*, there would be no more need for European products. Yet, and very significantly, this is not all: thanks to the revival of science in the Ottoman Empire, the people will "learn to appreciate the value of their fatherland and nation," Sami asserted.[34] As a result, the poor will be protected, hospitals for the needy will be opened, *medreses* and dervish lodges will be built, and ruined mosques and bridges will be repaired, thus serving the afterlife as well.

All this may be seen simplistically as the coming to grips of an Ottoman bureaucrat with the realities of the new world. Yet how exactly did

Sami reach these conclusions about science? How did he make these observations? Strikingly, as he confesses at the end of his work, Mustafa Sami did not speak any European languages. His impressions appear to be based on what he had been told during his time in Europe, not on rigorous personal investigation. But this fact makes his arguments even more representative: Sami's descriptions of science as knowledge that enables one to appreciate one's own allegiances, as built on the legacy of early Muslim scholars, and as the sole reason behind European supremacy are ultimately the clichés of the emerging official Ottoman discourse on science. Seeing them as but truisms also explains the flaws in Sami's reasoning, such as the supposed link between the progress of algebra and the invention of the steam engine, the highly exaggerated descriptions of the level of education of commoners in Europe, and the absence of any sound explanation regarding the reasons behind the progress of science. Mustafa Sami, as a member of the new class of bureaucrats who remained unsure of their legitimacy, "markets" science and its benefits, and this assumes the form of a mystification, not an analysis. And this mystification often involves presumptions about the impact of science not only on the economy and the military but also on the qualities of individuals; science is not simply a matter of knowledge and practical benefits, it is a matter of personal and civic virtue.

Their effort to transform their practical know-how into prestigious cultural capital led bureaucrats like Sami to present science as the knowledge that the empire needed more than anything. That these new, aspiring Ottoman elites possessed, or at least appreciated, this knowledge was then used to portray these men as true patriots and *useful* subjects. Sami's brief digression after his discussion on the invention of lithography is telling in this context: praising European states for rewarding and respecting authors who publish books thanks to which nobody's effort goes wasted, Mustafa Sami implies what he deserves after his useful treatise.[35]

Finally, we should remember that while Sami has nothing but flattery for Europeans (which comes very close to implying that they are *morally* superior to Ottomans), at the end of his treatise he attempts to make his arguments more palatable: science does not belong to Europeans, it is the true legacy of Muslims. Furthermore, as science produces both good people and generates wealth, religious buildings will be maintained better—note, however, that Sami mentions this along with the saving of the poor and the building of hospitals for the needy, which adds a hint of condescension to his remarks.

These arguments proved insufficient, however, as the fact remains that new elites like Mustafa Sami still lacked legitimacy, and their efforts to convert their cultural capital into symbolic capital more often than not led to censure.[36] The late nineteenth-century court chronicler Ahmed Lütfi writes that Sami attracted "derision"; yet "derision" would still be too light a word to describe the reaction against Sami's work if we examine the satirical poems written on him. The young poet Üsküdarlı Hakkı Bey (1822–1895) referred to Sami as the "Devil-faced dissolute," "the leader of the confounded," "the collaborator of the Zoroastrian and the Christian," and "a gypsy in European clothes" who would face ruin in both this world and the afterlife.[37] Hakkı Bey's poem was full of condemnations of Mustafa Sami that were rooted in the Islamic tradition, and he openly called him an infidel. Praise for Hakkı Bey's attack came from older poets like Safvet (1794–1866) and Lebib Efendis (1785–1867). While Safvet commended Hakkı for being so truthful (*hak-gû*, a play on the author's name), Lebib wrote: "You turned into hell all sides of the foe of the Prophet / Those fiery verses hit the enemy right in the heart."[38]

It is important to note here that Sami's critics were not prominent ulema but other bureaucrats—a case exemplifying the divide within the bureaucratic middle class. An examination of the particular positions the protagonists of the story occupied at the time can shed some light on the roots of this antagonism. Lebib Efendi had been appointed minister of quarantine in January 1840, but soon afterward his office was put under the control of the minister of commerce, who, at the time, was none other than Mustafa Sami's patron, Ahmed Fethi Pasha.[39] Sami's *Avrupa Risalesi* was published during this period, in July 1840. Safvet, on the other hand, had been Lebib's protégé since 1822.[40] In sum, then, Sami's work appears to have been an excuse for the manifestation of the rivalry between two patrons and their protégés.

But it is also the fact that the two groups had a crucial difference: Lebib had not been educated or employed in Europe, and bureaucrats like him, and their protégés, were less likely to occupy prestigious posts in this period. Granted, neither Ahmed Fethi nor Mustafa Sami spoke European languages, and they were not well educated in the new sciences.[41] It appears that their past employment in Europe and their connections were the sole basis of their prestige on their return, and the feeling of injustice this may have caused among other bureaucrats can be considered as the foundation of the hostility. Sami's treatise illustrates rather well the sense of distinction and entitlement he felt.

Of Sami's other critics, the poet Ibrahim Hakkı was eighteen when he wrote his poem. He worked in the ministry of endowments most of his life, but without rising to higher ranks in the bureaucracy. He wrote several poems in which he complained about his poverty and disillusionment, and he suffered a severe mental breakdown from which it would take him almost twenty years to recover.[42] Similarly, Safvet's most famous poem, *Beranjer*, is based on a comparison between his absolute poverty and the esteem with which poets were regarded in France.[43]

Their poetic imagery aside, the works of Lebib, Safvet, and Hakkı appear to represent a common perception; they indicate that men like Sami still constituted a minority with limited influence. Sami's patron, Ahmed Fethi Pasha, collaborated with the chief architect of the Tanzimat edict, Mustafa Reşid Pasha, and Mustafa Sami was appointed director of the Imperial Press. Yet, according to the chronicler Ahmed Lütfi, Sami became the object of utter contempt during these years as he "denounced and deplored established ways and customs, and talked about European customs heart and soul to anybody he saw."[44] This seems to have been quite costly for Mustafa Sami, as his career afterward is characterized by a series of posts he held rather briefly before he was removed from office, such as his ambassadorships in Berlin and Tehran. Like his critic Hakkı, Sami was suffering from severe mental disorders at the time of his death in 1855.[45] The fields of culture and state were changing rapidly in the Ottoman Empire, leaving behind many disillusioned casualties. And references to science and its implications constitute a unique indicator of these changes.

3. Science and Morality: Sadık Rifat Pasha

While Sami's treatise is a brief but enthusiastic exposé by a lesser bureaucrat, the example of Sadık Rifat Pasha (1807–1856) enables us to see the approach of a more influential ideologue of the Tanzimat Era. A collaborator of Reşid, Rifat had become the Ottoman ambassador to Austria in 1837, and even though he also experienced a volatile career due to intra-elite competition, Rifat remained a high-ranking bureaucrat until his death.

Key among his published contributions are his observations about Europe and, in particular, European civil servants. Seeing a striking contrast between European bureaucrats, whose many rights were recognized and duties well-defined, and their Ottoman counterparts, who did not even enjoy the right to life in the traditional Ottoman system, Rifat wrote some of the most important texts expressing the aspirations and concerns of the new bureaucratic class.[46]

In his discussions on knowledge and ignorance—discussions that had become central to any political treatise in this period—Rifat praised Austrian schools, the science classes in their curricula, and their combination of the theoretical and the practical.[47] In another text on education in the Ottoman Empire, however, he wrote the following:

It is necessary to strive for . . . the elimination of the ignorance of the people and their acquisition of the needed sciences.[48] . . . Ignorance is the true source of all evils and improprieties . . . [so it is] required to educate the people with respect to the science of ethics as much as possible. . . . Knowing everything, that is, some subtleties that do not concern them may, among common people, give rise to hazards like a certain licentiousness, and in the end, disobedience. Hence, those types do not need to know a lot of things, and it is sufficient if they are taught to read and write.[49]

Nevertheless, it was crucial, according to Sadık Rifat, to provide a comprehensive education for those who would be in state service.

Sadık Rifat Pasha, an admirer of Metternich and his policies for retaining social order,[50] was not the only bureaucrat of the period concerned with the dangers of social upheaval and disobedience. Indeed, Ottoman bureaucrats were terrified by the social unrest in Europe that they had heard about or personally witnessed. Mustafa Sami, the author of the *Avrupa Risalesi*, for instance, wanted to resign from office at the Berlin Embassy due to the uprisings of 1848.[51] During a conversation on slavery while he was in Paris, Ahmed Fethi Pasha (Mustafa Sami's patron) proclaimed, "It is better for the happiness of everyone that everyone stays in his place, this is the surest way not to die of hunger and not to arouse evil passions."[52] Âli Pasha, too, is known to have said, "The Lord has entrusted the well-being of the state to five or six people. These should govern the fate of the state."[53]

The new bureaucrats were enamored by the "civilization" they had observed in Europe.[54] The curiosities they witnessed (which they broadly labeled as the products of the new sciences) fascinated them, and they regarded the teaching of these new sciences essential in the Ottoman Empire. But it was precisely their experiences in and testaments about Europe and its marvels that their authority and distinction were founded on. The new bureaucrats *knew* how things worked in Europe, they were able to communicate with the Europeans and manage the transfer of European ideas and goods into the empire. Hence, the spread of the types of knowledge they monopolized was a double-edged sword for them: while, on the one hand, it could facilitate the implementation of the schemes of the central government, it would also help produce new competitors for the power they wielded and potentially disrupt social order in ways they had observed in Europe. The outcome was the representation of new knowledge but with an emphasis on its certainty, which was simply to be learned from books, just as religious sciences were learned in the *medreses*. The question of social order did not matter less for the new bureaucrats than for the ulema, which led to the emergence of the new synthesis: a sufficient amount of properly understood new knowledge, coupled with an understanding of one's place that should not be challenged, that is, the combination of science and morality.

It is in this respect not surprising that Sadık Rifat, the devotee of *sivilizasyon*, was also the author of a textbook on morality, the *Risale-i Ahlak*, first published in 1847 and required in both Qur'an schools and the new elementary schools until 1876.[55] Written as a series of brief discussions on desirable and undesirable traits, the text does not cite the classics of Islamic ethics, and while God is referred to as the ultimate judge of actions, the book explains that virtuous acts are stipulated by both religion and reason.[56] The authority of the teachings is thus rendered undefeatable.

The very first topic discussed in this textbook on ethics is the acquisition of knowledge. Ignorance, the author contends, is lacking the knowledge that is essential for being human, and the ignorant are always derided by their peers. Yet knowledge should not be acquired in order to be able to call others ignorant; the learned should hope to educate and be useful to the people and, thus, to be properly respected.[57]

Even in this elementary text, it is possible to trace the reasoning of the enlightened bureaucrat: he is to acquire knowledge, educate others, and earn their respect. Indeed, the book contains many examples about earning respect by acting properly. But Sadık Rifat was even clearer in the supplement he published in 1857. His *Zeyl-i Risale-i Ahlâk* starts with a blunt proclamation that obedience to religious commands and the sultan are the prerequisites of being considered a moral individual.[58] This volume also contains a section on knowledge where Rifat commends the new developments in the sciences and arts, and, as a consequence of these changes, the imposition of "beneficial laws and useful regulations."[59] The powerful steamboats, vast factories, and all the new inventions are further products of science, and science progresses more rapidly each day as it even has a language devoted to it: French. But ultimately the source of all science is the intelligence God bestowed on all men, and, as a result, science does not belong to a nation, it is the common property of humanity.[60] Hence, scientific knowledge should be respected and utilized by all; it *must* be imported by the Ottomans. Yet it appears that it is those who do speak French who truly understand the nature of science and its benefits, according to Sadık Rifat.

Rifat's use of the word *ilm* is of particular importance. In the section devoted to the sciences, Rifat states: "As '*ilm*' means 'to know,' and to learn and know the better of everything is the most esteemed privilege of being human, everyone should strive to learn what he does not know."[61] By identifying *ilm* with the sciences he discusses so enthusiastically, Rifat simultaneously imagines ignorance essentially as the lack of knowledge pertaining to those sciences. Furthermore, when the concept associated with religiously significant knowledge is appropriated for science, the virtues associated with being knowledgeable can be ascribed to those who possess scientific knowledge.[62]

Sadık Rifat also notes that sciences and arts enable individuals to make a living without demanding aid from the state.[63] Hence, "everyone should provide the education for their children that will allow them to be good subjects of their Sultan and subsist without being a burden on the state and the nation."[64] Not everybody should be a civil servant; one could earn his life as a locksmith as well. Indeed, since Rifat states at the outset that he wrote his book exclusively for those who would become civil servants,[65] it appears that those on state service should be allowed to know the intricacies of science, but the rest should get just the "right dose."

Hence, Sadik Rifat's writings illustrate the ambiguity that the new elites' characterizations of science embodied. On the one hand, for reasons discussed in the preceding sections, familiarity with the new types of knowledge was to be praised, and this could even take the form of equating them with religiously endorsed knowledge. It was implied that this familiarity was one key reason why the new elite constituted the new class of "knowers." But what about the impact of new knowledge on "common people"—people whose participation in government was not envisaged? In this case, scientific knowledge becomes knowledge that should teach individuals the proper order of things and provide them with skills that will render them hardworking and productive. Consequently, scientific knowledge makes the ruling elite fit to rule and transforms the ruled into disciplined and deferential servants. Particularly the latter part of this formula can be seen in a number of documents prepared by the Tanzimat bureaucrats, a significant one of which I analyze below.

B. Science: The Route to Patriotism?

Perhaps the most striking and consequential contribution of the 1830s to the Ottoman debate on science in this context is the emergence of a distinctive theme: familiarity with science as a prerequisite for being a good, patriotic Ottoman subject. This theme was introduced discreetly in a document that is also very significant in its portrayal of the new official meanings of knowledge and ignorance: a memorandum on the state of education within the empire prepared in 1838 by the newly founded Council of Public Works.[66]

The memorandum stated at the outset that it was undeniable that "education and the sciences" (*maarif ve ulûm*) were the basis of power and glory, as well as all the arts and industries that generate wealth. Furthermore, just as religious sciences lead to salvation in the afterlife, the memorandum proclaimed, other sciences (*fünûn-ı sâire*) bring about the perfection of the conditions of mankind on earth. There was no ambiguity about what these other sciences were: astronomy, by facilitating maritime transport, helped stimulate trade; mathematical sciences both helped better organize military forces and enabled the emergence of "many useful and curious things that amazed the philosophers of the past, such as steam power."[67] These changes had made ignorance particularly detrimental, as it could lead to impediments to trade and decline in industry.

But this was not all, as the ultimate question in the nineteenth-century Ottoman Empire was not the characteristics and consequences of knowledge itself, but the characteristics and attitudes of the people who possessed or lacked knowledge. Hence, the memorandum made the remarkable statement that learning was essential, since the ignorant could not "[truly] know the state of whose auspices they exist under, and what love for the fatherland means." According to the memorandum, the Ottoman state had established many schools in the past to promote knowledge, and many remarkable books had been written by the early Ottoman scholars. "Certain affairs and disturbances" had stalled this process until the time of Mahmud II, and some problems persisted during his reign as well. These problems had to be eradicated, the memorandum stated, as ignorant individuals, due to their lack of appreciation for what their state provided them with, ended up being useless both to themselves as well as to their nation. Hence, learning was important in that it enabled one not only to understand and appreciate the state (that is, wielders of state power), but to become a good servant of it. It is worth remembering at this point that in its first issue the Ottoman official gazette also noted the need for the people to avoid attacking the state and to truly understand its actions. The "training" of the citizenry through tools like the press and education was clearly a priority for the new state elite who sought to establish legitimacy in the eyes of the Ottoman public.

Now all this does not mean that ignorance is defined only as the lack of scientific knowledge in this and similar documents. Indeed, some of the most powerful sections of the text are about the need for ameliorating elementary, that is, primarily religious, education.[68] The proposals do not extend so far as to encourage teaching the sciences (which the memorandum praises passionately) in elementary schools, either. Hence, the document can be read as an initial attempt to combine the authorities of the emerging new knowledge and religion in order to create the desired type of subject. The two aspects of the report can be seen as the components of an unarticulated dual structure, one (pro-science) demonstrating the will for modernization, the other (religious) indicating the intent of "social disciplining."[69] Yet the way the memorandum discusses the benefits of the "other sciences" demonstrates forcefully that the praise for science is also closely related to its effect on the subject vis-à-vis the state: learning the knowledge produced by the new sciences not only enables the individual subject to generate more wealth for himself and his state, but also allows him to appreciate what his state provides for him, making him more patriotic. Science is no less essential than religion for generating productive *and* deferential subjects.

We should also note that the official discourse that emerged in this period did not simply link scientific knowledge to patriotism by emphasizing that learning enabled one to appreciate one's state. As Seyyid Mustafa's *Diatribe* had already demonstrated in 1803, and as the tone of this and similar other reports on education, as well as the forewords of many if not all books on science published throughout the nineteenth century suggest, Ottoman sultans were increasingly keen to assume the title "patron of education and science." In this characterization, learning science becomes one's duty toward the sultan, the fulfillment of which would make one a good subject: after all, the sultan (and the state elite) demanded reciprocity for the "gift of knowledge" they gave the subjects. We will see in the following chapters further examples of the popularity of this portrayal and of the ways it would be challenged.

C. New Institutions for New Knowledges

1. Schools, Students, Virtues

In a period where ignorance and knowledge were redefined by an emerging status group, it is not surprising that the question of institutionalization also came to the fore. The new sciences were taught in a small number of prestigious schools in the Ottoman Empire of the 1840s, yet due to the absence of a well-designed educational ladder, these schools had to admit students who were hardly prepared for such studies. There also remained a shortage of qualified educators and teaching materials, primarily textbooks. Hence, and possibly with the encouragement of top bureaucrats, in January 1845 Sultan Abdülmecid issued an edict complaining that despite his strongest will, the condition of his subjects and his lands had not been substantially ameliorated save for the improvement of the military.[70] Expressing his disappointment, he described the most urgent problem of the empire as the "elimination of the ignorance of the subjects in all issues, religious and worldly." His edict then asserted that new schools should be established—schools that were "the origin of knowledge and science, and the source of arts based on learning."[71]

Abdülmecid's blunt distinction between religious and worldly knowledge is striking. It appears that the principal tenets of the 1838 report of the Council of Public Works were now taken for granted: religious and worldly knowledges were different. The emphasis on knowledge-based arts also indicates that instruction in the schools he envisaged would not have to do with religion. It would be reasonable to conjecture that Abdülmecid also recommended a sequence based on elementary religious training to later be followed by "worldly knowledge" and vocational education.

The Provisional Council of Education that was formed to deliberate on the issue of education soon after this edict comprised one chair, six members, and one secretary. Four members, including the chair, were from the ulema, three were bureaucrats, and one was a military officer: Mehmed Emin Pasha, an engineer educated in Paris. In addition to Fuad Pasha and the secretary, Recai Efendi, two other civil servants with extensive bureaucratic experience constituted the "men of the pen" contingent. The representatives of the ulema, significantly, were some of the best-educated religious dignitaries of the time.[72]

In an 1846 report based on the council's recommendations, we observe a similar logic to the one indicated by Abdülmecid: Schools of elementary levels should be devoted to the religious sciences that are essential for everybody. Second, a *Darülfünûn* (House of Sciences) should be opened in Istanbul for "those who desire to learn and acquire all the sciences and knowledges in order to achieve human perfection and those who wish to be employed in an office of the Sultan." No science would be neglected by this institution, and its students would "strive to achieve maturity under the enlightening auspices of [the sultan]."[73] These measures were of utter importance as education was defined as the basis of prosperity. Moreover, and very importantly for our purposes, the ignorant were *dangerous*: "Those who are devoid of sciences and knowledges know neither patriotism, nor divine or human law and remain in the state of animals, and their natures, due to ignorance, would be inclined to pick up all kinds of evils."[74]

What is conspicuously absent from this document is the status of the *medreses*, even though it does address the issue of higher education. As it was specifically mentioned that the proposed *Darülfünûn* would be the path to follow for those pursuing government jobs, it was implied that *medrese* graduates would likely not be able to get offices outside of the strictly religious realm. Additionally, the document refers to the future students of this new institution as the *talebe-i ulûm*, literally, students of knowledge (*ilm*), which was the term traditionally used for *medrese* students. And very significantly, the document presents another example of the crucial formula of the official discourse: learning the sciences makes one virtuous. Ignorant people not only are unaware of the meaning of patriotism—a phrase that had obviously turned into a truism by then—but they are prone to vices. In a sense, they betray the qualities instilled in them by God, and in their animal state, they constitute a danger to the nation. What this characterization of the ignorant implies about the moral virtues of those who do possess the "sciences and knowledges" does not need elaboration. Indeed, students of the future House of Sciences are described as those who would thus achieve *kemâlât*: a concept that entails both knowledgeability and morality.[75]

Once again, what we observe is the transfer of a concept more commonly related to the acquisition of religiously significant knowledge to "all sciences" that the document promises will be taught at this new establishment. It is true that the report does not specifically exclude religious sciences, but it is precisely this generality of the way "knowledges and sciences" are referred to that, in a sense, disenchants the idea of *kemâlât*. Virtuousness, or human perfection, is very tightly linked to knowledge and science in general, rather than to a particular type of knowledge, and those who acquire knowledge and good morals within the *Darülfünûn* are portrayed as those who would be appreciated both by God and the state.

2. The Academy

One of the main obstacles before the establishment of an institution of higher education as the *Darülfünûn* was the lack of textbooks. Hence, the Council of Education concluded that an academy of sciences should be established in order to prepare the textbooks. In yet another memorandum of great significance, the council asserted on February 11, 1851 (9 Rebiüla-hir 1267), that the attainment of prosperity and civilization was founded exclusively on the spread and growth of the various sciences, and this depended entirely on the support of the state.[76] Indeed, according to the report, history proved that those states that strove for the spread of knowledge had always provided their subjects with prosperity and dominated other states. In other words, science had to be under state patronage and protection, as it, in turn, rendered the state more powerful. The Ottoman past was an illustration of this fact: the advent of the Ottoman state had enabled the "sun of sciences and arts to shed its light over eastern lands," and many books on the "needed sciences of the day" had been written by Ottoman authors.[77] However, the authors had ignored Turkish and preferred to compose their works in Arabic or Persian, "in order to demonstrate their brilliance." Furthermore, they had devoted most of their effort to poetry and belles lettres and ignored other disciplines.

Even though Muslims had always known that in each era different sciences and arts were current and treasured and, accordingly, written and translated works on those particular sciences, there had come a state of negligence after a while, "due to some reason." In addition, people of authority were either deprived of knowledge and talent or of the disposition to work for the sake of the state and the fatherland. Thanks to the enthronement of Sultan Abdülmecid, however, the wave had turned, and unprecedented progress had been observed in the realm of education. Due to this speedy development, no time should be wasted until the opening of the *Darülfünûn*, and the production of the required textbooks should be the responsibility of a new Ottoman Learned Academy, the *Encümen-i Dâniş*, in the meantime. The members of this academy were to be competent in at least one field of science and able to compose books or translate works from Arabic, Persian, or foreign languages.[78]

We observe in this memorandum further very clear statements about the imagination of science as "state property." Those who engaged in science would be under state protection, as their works would strengthen the state and help it improve the conditions of its subjects; state sponsorship was the sole alternative. Furthermore, the clearly articulated will to spread the sciences within the empire had made it obvious that books written in languages other than Turkish were, ultimately, of no use. Indeed, within one document we observe references both to the Muslim world in general and the Turkish-speaking people in particular as the audience of these comments and suggestions. In this respect, the council's report serves as an invaluable indication of how the discussion on science inevitably faced the question of which of the sultan's subjects were its true intended addressees.[79]

While the memorandum refers to poetry and belles lettres along with other sciences, it also implies they are useless and outdated. The works of the old masters are praised, but it is also noted that they are not necessarily useful in the new era, as each period has its own sciences and arts. This of course indicates the internalization of a concept of linear progress by the bureaucrats who composed the report and their attempt to rewrite Ottoman history: the old is not necessarily valid or respectable anymore. The empire needs, as it were, new masters, possessing new knowledges and new skills. It is also worth noting how the discussion uses the metaphor of the sun that enlightens the people—paving the way to the perception of the nation in terms of an antagonism between "the enlightened" and those "in the dark." The issue is clearly not just about the material benefits of the new knowledges, it is about the characteristics of people that ultimately matter and, thus, talking about science is simultaneously talking about virtue.

Şeyhülislam Arif Hikmet Efendi, a former member of the Provisional Council, declared his office's positive opinion about the new institution, and the Ottoman Learned Academy was opened on July 18, 1851. In his brief opening speech, Mustafa Reşid Pasha expressed his gratitude to the sultan who had made so much effort to disseminate the sciences and knowledges that "teach men their humanity, and lead everyone toward happiness and well-being in both this world and the afterlife."[80] Note once again the connection between knowledge, humanity, and virtuousness.

Cevdet Pasha, the author of the opening speech representing the members of the academy, was a *medrese* graduate who had also learned French and chosen to leave the ranks of the ulema class to join the "men of the pen." In his speech, read by Hayrullah Efendi, the vice president of the academy, Cevdet stated that arts and sciences were the sole bases of prosperity, order, the well-being of both the elite and the commoner, safety, as well as all the curiosities that were witnessed all around. Cevdet also specified that the survival and the fulfillment of the physical needs of man, as well as his achievement of the civilization to which he is naturally inclined, depended on natural and mathematical sciences. His spiritual side, on the other hand, leaned toward metaphysics and found pleasure in poetry and belles lettres.

As he was speaking about an academy whose key task would be to produce new books on new types of knowledge, Cevdet also proclaimed that languages did not acquire distinction unless literary and scientific books were written in them.[81] Hence, the importation of science was simultaneously a means through which the Ottoman language would be reconstructed. Indeed, its official regulations stated that the chief objective of the academy was to serve "the generation of the needed books on various sciences in the Turkish language and to serve the progress of the language."[82] The books to be published by the academy would be in *âdi Türkçe*, or plain Turkish, so that everybody could understand them. In this respect, the new sciences were declared not only the useful ones but also the ones that common people would understand, as opposed to the old sciences that remained esoteric, as they were commonly in Arabic.

Now it is a fact that the dominant political ideology of the era was Ottomanism—an ideology that entailed the construction and dissemination of a supra-ethnic, supra-religious Ottoman identity. Yet as this early example suggests, an unintended consequence of the Ottoman debate on the new sciences would be the gradual emergence of an association between the Turkish language and the new sciences. This association would become more fully formed and explicit in the following decades, thus adding a new dimension to debates on science and making them at the same time debates on who the Ottomans were.

The composition of the academy indicates the Ottomanist agenda of the period and its aim to resemble a European-style learned academy in which the nationality or religion of the members would not matter. It is true that of the seventy-three academy members only sixteen belonged to the ulema, thus indicating the continuation of the trend to lessen the presence and influence of the ulema in institutions of knowledge production. But it is also important that, in addition to the twelve non-Muslim Ottomans, three Europeans were also among the founding members of the academy: the orientalists James Redhouse, Thomas Bianchi, and Joseph van Hammer. Later, the American orientalists Edward E. Salisbury and Charles Johnson joined as well, and the Smithsonian Institute, under the directorship of Joseph Henry, sent eleven books as a gift in return for a book on the church of Hagia Sophia donated by the Ottoman academy.[83] In 1850, one year before the establishment of the academy, moreover, the names of two of its most prominent members, Fuad and Safvet, had appeared among the honorary members of the American Oriental Society.[84]

Despite the initial enthusiasm and its cosmopolitan attitude, the Ottoman Learned Academy was rather short-lived—its name disappeared from official almanacs after 1862.[85] The books that had been presented to the academy within this period included a number of translations and original works on European and general history, in addition to a translation of Buffon's *Histoire naturelle*.[86] The more celebrated works associated with the academy, however, are Cevdet's *History*, commissioned by the academy but completed only in 1892, and Cevdet and Fuad's coauthored work on Ottoman grammar (*Kavaid-i Osmaniye*).[87]

Another work that stands out among the products of the academy is the first book on geology in Turkish: *Ilm-i Tabakatü'l-Arz*, based on translations of sections from Elie de Beaumont's works and published in 1853. The translator, Mehmed Ali Fethi, a member of the academy, was from the ulema, and he translated the work from an Arabic edition. As a clear demonstration of the backing behind this work, the first pages of the volume were dedicated to the appreciative comments of prominent state officials and ulema who were also members of the academy. Âli Pasha, for instance, wrote that the "noble science"[88] this new book contained had not been discovered in the "land of Turkish language" as yet, resembling a gem left unnoticed under the earth. Hence, Fethi's translation was full of benefits.[89] Mehmed Pasha, the chief of staff, defined geology as a "grand science that brings many benefits" of which Turkish speakers had previously been deprived. But such a brilliant translation had finally been possible thanks to the sultan, the "protector of learning" whose kind attention to knowledge and the people involved with it was well known.[90] Fuad Pasha also congratulated Fethi for his contribution to the "gems and glories of learning" that had come out during the reign of the sultan—a time characterized by learning.[91] Finally, the future minister of education Subhi Bey thanked the translator for introducing into Turkish language a new science with such abundant uses, and he expressed his hope that more "useful works" of this kind would be published thanks to the sultan.[92]

The "stately" introduction of this volume on geology is illuminating in its symbolic meaning: this new, noble, beneficial science that speakers of Turkish had long been deprived of can now be accessed under the sponsorship of the sultan, the patron of science and knowledge, and his enlightened servants. Also note once again the association of the Turkish language with the new sciences.

While this great authority proudly backed the new sciences, however, it prevented the academy from working effectively. In his discussion on the end of the academy, Cevdet notes that academy memberships, just like posts within the bureaucracy and the high *ilmiyye*, were based mostly on personal relations rather than merit, and resentful ministers and bureaucrats who were not allowed to be members interfered with the efforts of the academy.[93] Cevdet Pasha's remarks can be seen as potentially subjective, yet it is critical that since membership criteria involved competence in at least one language other than Turkish and one branch of science, a reasonable amount of education appears to have been the true common feature of all the Ottoman members. As the set of individuals within the empire who would satisfy this criterion included more or less only the bureaucrats and the high-ranking ulema, it was unavoidable for the academy, a body with no institutional autonomy whatsoever, to manifest the appearance of yet another high council of the state. This would certainly raise questions about the reasons why any top bureaucrat was *not* a member. Similarly, while the regulations of the academy stipulated that members who failed to attend the meetings regularly would be expelled, such sanctions could hardly be used against bureaucrats of high rank.[94]

Problems of this kind remained inevitable as long as the Ottoman man of science and art was also, and primarily, an Ottoman statesman.

Another member of the academy, Derviş Pasha, published the first chemistry textbook in Turkish in 1848. Derviş Pasha (1817–1879) was a graduate of the *Mühendishane* and a student of Ishak Efendi. Following his graduation, he was sent to London in 1834 to further his training in preparation for a professorship at the Imperial Military Academy. After London, he went to Paris and followed courses at the École des mines. Before his return, he was authorized to purchase materials for the Imperial Military Academy. In addition to numerous general volumes and diction-aries on physics, chemistry, and medicine, he bought collections of scien-tific journals, laboratory instruments, and fossils.[95] In the decades follow-ing his return, he assumed many different posts including professorships at the Imperial School of Medicine and the Imperial Military Academy, diplomatic envoyships on numerous occasions, the ambassadorship at St. Petersburg, and the Ministry of Education. In this respect, Derviş Pasha was a typical member of the new generation who was able to take up many different but always prestigious roles thanks to the new type of education he received. The later court historian Ahmed Lütfi, in his discussion of Derviş Pasha's appointment to St. Petersburg, was probably speaking on behalf of many other officials who were disgruntled with this new order:

The reason why [Derviş] was chosen to such a sensitive and important post as the ambassadorship to Petersburg must have been the fact that he had for a while been educated in Europe. But can one be appointed to such a post simply because of a superficial knowledge of French? The office of ambassadorship is founded on a grasp of the art of ambassadorship, which, in turn, depends on training within that profes-sion itself.[96]

Derviş's book on chemistry, the *Usûl-i Kimya*, is introduced by a preface with a strong Islamic tone. Following the classical Islamic model, Derviş classifies philosophy (*ilm-i hikmet*) into the theoretical and the practical branches of knowledge, with metaphysics/theology, a theoretical branch (*ilm-i hikmet-i ilahi*), as the noblest of all. Yet mathematical and natural sciences not only help the learning of metaphysics, but they are also es-sential for "bringing forth the desired novelties and discovering unknown arts." Chemistry, being one of these sciences, helps "the acquisition of the new industries and the attainment of numerous benefits."[97] Further-more, all the weapons that are needed for the holy war ordered by Islam are made of substances discovered and utilized by this science, making it indispensable for officers to study it. His own work is intended to be used for this purpose in the Imperial Military Academy, and is only possible thanks to the sultan, who demands everyone, but particularly the officers, to study the "absolute knowledges and the partial sciences, thus attaining religious and worldly bliss."[98]

Derviş bases his defense of chemistry on an Islamic categorization, but he emphasizes the independent worth of mathematical and natural sciences for the production of "novelties," which is a novel approach itself. The distinction between the religious and the worldly is stated clearly, and, due to its inevitability for the production of new weapons for holy war, chemistry is presented almost as the true protector of Islam in the new era. In this respect, the new sciences and the new industries they bring about not only lead to happiness in this world but are required in order to obey the command of Islam and so to reach bliss in the afterlife as well. Whether Derviş's approach was but an attempt to appease skeptical readers is not a relevant question for our purposes, as the consequence either way is that his case for chemistry presented this science and those trained in it as indispensable for both the state and religion. It is also important to note Derviş's ability to express his view in a traditional Islamic tone, which shows that a member of the new generation trained in the new schools as well as in Europe was still conversant with the classic paradigms and terminology of Islam. But equally important is that their new skills were the essential bases for distinction for Derviş's generation. Tellingly, the copy of his *Usûl-i Kimya* that I examined was an autographed copy, signed by the author for Edhem Pasha, a former student of the École des mines, in French, rather than Turkish.[99]

Conclusion

The first half of the nineteenth century witnessed the formation of a new elite group in the Ottoman Empire, with a new kind of cultural capital that they were gradually able to convert into statist capital. Some members of the higher-ranking ulema allied with the new group and contributed to the formation of a new discourse on knowledge and ignorance. This discourse represented the new sciences of Europe as a type of knowledge that was equivalent in worth to religious sciences: while the latter guaranteed bliss in the afterlife, the former would bring prosperity and well-being in this world. This was a type of knowledge that the new bureaucrats represented and the top ulema sanctioned. It was useful knowledge that rendered subjects productive and enabled them to understand and appreciate the state that protected them. This characterization involved the portrayal of the new knowledge as simply facts to be learned, facts that showed the learner the true order of things.

But this period also entailed significant legal and economic changes that were disappointing to members of the Muslim community and, in particular, to low-ranking bureaucrats and ulema. This disappointment gradually led to the perception and representation of the new bureaucrats as snobs who adored and humbly obeyed European powers, rather than defending the dignity of the empire. In contrast to this representation, the official discourse appropriated established ideas about knowledge and virtue, and portrayed the possessors of the new knowledges as virtuous patriots; indeed, they were even defined as those who were "truly human." Many members of the new elite were well versed in Islamic literature as well, thanks to which they were able to construct a discourse using the Islamic idiom and exploiting the connotations of concepts such as *ilm*.

It was also in this period that the new types of knowledge started to be associated with the "language of the people," Turkish, and the old sciences with Arabic. While this was an outcome of the efforts to centralize education and bring it under state supervision, it also enabled the new elite to represent themselves as those who truly served the people, rather than an aloof elite group with an esoteric language. Yet an unintended consequence of this policy would be the transformation of the debate on science into simultaneously a debate on the identity of the community. Indeed, all these trends of the early nineteenth century would flourish and be more explicitly discussed in the following decades, when new outlets emerged for the articulation of alternative discourses.

THREE

Consolidation of the Discourse: Science, State, and Virtue in the 1860s

Introduction

The new "men of knowledge" of the Ottoman Empire were mostly young men who justified their claim to state authority partly by referring to a special, novel type of knowledge that they possessed. Even when the actual amount of new knowledge they had was admittedly scant, they emphasized that what truly distinguished them from the "ignorant" was essentially their awareness that this new type of knowledge should be respected and imported. A member of this rising class did not need to know much about, say, chemistry. It sufficed to have some basic familiarity and to acknowledge that chemistry was indisputably beneficial. If one wanted to save the empire, one had to be aware that the Europeans dominated the world thanks to this new knowledge they produced. Those who were too ignorant even to admit this fact could not be fit to govern. This awareness, in turn, was bestowed primarily on those who had spent time in Europe and/or could speak European languages—French, in particular.

Moreover, and most significantly, the new knowledge produced by the Europeans was also represented as one that made people appreciate and support their state. Learning a sufficient amount of science would make Ottoman subjects hardworking, productive, but also obedient individuals who always prioritized the interests of the state. Familiarity with the new types of knowledge would lead to patriotism, and the shortcut to this set of values was the term "morality."

In other words, what we observe in the texts of the early to mid-nineteenth century is the gradual emergence of associations between the new types of knowledge and both good rulership and good subjecthood. In this period of transformation, the new sciences of the Europeans were promoted with reference not only to their practical impact on the daily lives of individuals or the military and economic power of the state but also to their moral implications.

It is in the 1860s, however, that these connections were discussed most overtly, and the new elite started to import the moral and political connotations of knowledge and ignorance into arguments on science rather straightforwardly. It was in this decade, after all, that members of the new elite class started publishing the pioneering and influential Ottoman *Journal of Sciences*, which printed numerous essays that defined the contours of the debate on science. In this respect, the 1860s can be seen as a period of consolidation of the discourse that tightly connected scientific knowledge to morality—morality defined ultimately as producing a useful and good subject who would not posit a threat to the social and political order.

But in order to locate some telling clues about the overall meaning of scientific knowledge for the new elite, we should first look at two developments in the field of public education: First, the emergence of two institutions that were to some extent connected in terms of organizational structure and that represented the decision of the "sublime state" to engage the sciences and arts of the Europeans more powerfully than perhaps any other; and, second, an elaborate official statement of the purposes and significance of a new form of education in a period of rapid political transformation.

A. Prominent Educational Developments

1. École ottomane

Between 1839 and 1856, forty-three Ottoman students were sent to France, the intellectual patron of the Ottoman Empire in this period.[1] Most of them were graduates of the Imperial Military Academy, and in France they usually attended military schools after some preparatory training. There were fifty Ottoman students in France in 1856, the year in which the French and Ottoman governments agreed on establishing a commission to organize and supervise their training. The general test administered by the commission proved that Ottoman students, even those who had spent more than two years in France, were able to perform nothing more than basic arithmetic and had only very rudimentary knowledge of history and geography—which is most important as an indication of their training *before* coming to France. The result was the establishment of the École ottomane in Paris, where students would receive some basic training in a variety of fields from French teachers. In the detailed curriculum of the school, we see courses on history, geography, physics, chemistry, mathematics, and cosmography. Yet the school was hardly a success. The reports of the disappointed teachers mention several problems such as lack of discipline and the rather sloppy selection of the students by the Ottoman government—even though they tended to be around twenty years old and hence older than stipulated, they were also less prepared than expected.

In 1864 Esad Efendi, the director of the school, wrote a report stating that it was certainly beneficial to send students to Paris in order to eliminate the Ottoman dependence on French experts. The Ottoman subjects who would replace them would also become the teachers of the new Ottoman schools. But for this purpose, it would be much better to send younger students to France and send them directly to French schools so that they would learn the language well. If this were done, the very costly École ottomane could be shut down, Esad wrote, as it obviously failed to produce a sufficient number of "knowledgeable men."[2]

The response of the Council for Military Affairs (*Dâr-ı Şûra-yı Askeri*), however, unequivocally asserted that it would be unacceptable to send younger students, as "the primary requirement of education is to raise children within the congenital creed and national customs and morals of their parents and nation." But it was also true that if the teaching of French was improved at the Imperial Military Academy, its graduates would benefit from following courses in French military schools, particularly with respect to the practical applications of the abstract knowledge they adequately learned in Istanbul. This was the way to achieve the "high purpose" of raising "knowledgeable officers" who could convey information on the military sciences to the imperial military.[3] A few graduates of the Imperial School of Medicine could also be sent to Paris, and after comparable schools of public administration and diplomacy were opened in the Ottoman Empire, their graduates could be sent as well. Under the current conditions, it was agreed that the École ottomane should be closed down, which happened in late 1864.

Raising "knowledgeable officers and officials" who would respect the "congenital creed and national customs" was the key purpose not only of the École ottomane but the entire Ottoman educational system. Science-as-knowledge would be learned by the empire's military engineers, commanders as well as physicians, and both put to use and taught to newer generations. The empire needed these cadres, and these "knowledgeable and virtuous men" were its "men of science." In line with the pattern we observe throughout the period, the production of new knowledge by these educated Ottomans was not envisioned. The products of an innovative school like École ottomane were imagined, instead, as humble vessels who would simply transfer the new types of knowledge to the Ottoman Empire.

Another indication of the scope of the Ottoman expectations associated with the École ottomane is the two figures sent to Paris in 1857 as teachers of Turkish for the non-Muslim Ottoman students of the school: Tahsin and Selim Sabit Efendis. According to the instructions they received, another duty of these two men would be to become professors of mathematics and natural sciences for the *Darülfünûn* to be opened in Istanbul as prescribed in 1845 and 1846 by the Council of Education. It was not determined which one would be which, however. They would let the Council of Education know which branch they chose and learn these "needed sciences" in Paris. They would be supervised by the Ottoman Embassy there and were required to send reports "about the sciences they had learned" at the end of their first year.[4]

Of these two, Selim Sabit was twenty-eight at the time of his appointment. He was a *medrese* graduate who had also graduated from the *Darülmuallimin*, the teacher's seminary. He did follow courses on a variety of topics in Paris,[5] but he would make a career as a pedagogue, educational administrator, and reformer after his return in 1861, not as a professor of sciences. Tahsin, on the other hand, who was also a *medrese* graduate, was forty-six years old when he arrived in Paris. He attended courses in physics and chemistry at the College de France and, according to the French reports, led a rather hedonistic life during the twelve years he spent there.[6]

The implications of the characteristics of the two people sent to Paris to later become professors of mathematics and natural sciences are hard to miss. These two members of the ulema—and, as it was hoped, the future professors of the House of Sciences—had only casual familiarity with the sciences they were supposed to master in Paris and were already much older than their fellow students. Their training was not planned and supervised rigorously,[7] and it appears that acquiring *some* knowledge of *some* sciences was what the assigned accomplishment amounted to. Şerif Mardin's description of the Ottoman government's expectation from this experiment as being a step in the creation of "a Westernized *ulema* elite" seems pertinent.[8] But if we consider the official instructions that Selim Sabit and Tahsin received, we can also see the case as a reflection of Ottoman bureaucrats' conceptualization of science as but another part of the "wealth of knowledge" that was simply to be learned. This static and authoritative idea of science that is central to the official discourse is inseparable from the idea of science as a gift from the Ottoman elite to the people of the empire, and as a type of knowledge that taught one the correct order of things and people. The state needed "knowledgeable men," but these men were to be modest possessors, not producers, of knowledge. Obedience to authoritative knowledge and to state authority were two sides of the same coin.

2. Darülfünûn: *Constructing the Ottoman University*

The establishment of the Ottoman University constitutes perhaps the most suggestive narrative about science in the nineteenth-century Ottoman Empire, since the idea of establishing an institution of highest education—and it is primarily in this sense that *Darülfünûn* can be translated as "university"—was, just like the idea of science itself, constantly brought up in the second half of the century, but the institution itself never became a durable entity until 1900. It came to life and died several times in this period but was never truly institutionalized: instead, texts were where the university stably resided. In a sense, this lack of institutionalization meant that the boundaries of science themselves were never stabilized.[9]

The opening of a *Darülfünûn* constituted a central proposition of the 1846 report on public education, and construction started under the supervision of an Italian architect soon afterward. The location of the building indicated its importance: situated right between the great mosque of Sultanahmet and the Hagia Sophia and very close to the Imperial Palace, the massive building would become a symbol of the renewed might of the Ottoman Empire. But the provision of funds remained erratic throughout construction, and contingent on international and domestic affairs. Scottish traveler and author Charles MacFarlane—a Tory who had worked with Lord Brougham in the *Society for the Diffusion of Useful Knowledge*—notes that construction had stopped during the European upheavals of 1848.[10] Then between 1853 and 1856, during the Crimean War, the building was turned into a hospital. Refugees were allowed to stay in it in 1860.[11] Finally, in 1863 it was decided that lectures should start before the building was completed.

The institution boasted a 4,000-volume library, along with tools and instruments for experiments as well as geological samples.[12] It was not the highest point of a seamless chain, however, as there existed very few schools that prepared students for higher education; nor were there an adequate number of textbooks or competent professors. As a result, the first incarnation of the university was in the form of a series of public lectures, which were promoted rather enthusiastically by the *Mecmua-i Fünûn* (*Journal of Sciences*):

It is not unknown that a sincere comprehension of matters pertaining to natural sciences depends on the witnessing of the necessary experiments. Hence these experiments will be performed, and simple terms will be used as much as possible, so everybody will be able to benefit from listening. These sciences certainly have a good influence on the expansion of the opinions of the people and the progress of the arts. So it is doubtless that all classes of imperial subjects—*medrese* students, civil servants, and men of arts and crafts will show keen interest in learning and understanding them.[13]

The first lecture was given by Derviş Pasha, the director of imperial mines, to an audience of three hundred on January 14, 1863.[14] *Mecmua-i Fünûn* reports that Derviş started by praising the sultan's interest in public education and, after a discussion on the benefits of physics and chemistry, gave some basic information about them. What followed these were demonstrations that stunned the audience. The journal described the event vividly:

The performed experiments were curious phenomena, and the majority of the audience who were seeing such things for the first time in their lives were flabbergasted. Especially the experiments on electrical force where sparks of fire emerged from special tools, and when said force was transmitted into a man's body via a thin wire, his hand or whatever part of his body that the wire touched emitted blue sparks, and where, thanks to certain chemical compounds, an iron wire became incandescent and fiery like an inflammable substance, left them in astonishment.[15]

Details about the conferences were also presented in the Official Gazette. We learn from the Gazette that Derviş Pasha, who was also made undersecretary of the ministry of education in the meantime, received in April 1864 a certificate of honor from the attendees due to his most beneficial lectures. It was stated in the certificate that those who listened to the lectures had "acquired further knowledge and understanding of the place of divine power and greatness within the essence of certain odd and curious matters in the universe."[16]

Not all lecture attendees reacted in the expected fashion, however.

Münif Pasha, the editor of the *Mecmua-i Fünûn*, wrote that some members of the audience were watching the experiments as if they were but entertaining games. Those ignoramuses who looked at the useful sciences in this way, according to Münif, should not be prevented from attending the lectures, but they should not occupy the seats of those who acted properly. Münif proposed that meticulous investigation should be carried out in order to identify those whose "condition, level of competence and skill" suggested that they would attend regularly and benefit from the lectures, so a group of seats could be assigned to them.[17] But the proposal was not enacted, as the *Darülfünûn* did not last long enough to teach its audience how to act and how to perceive what was happening.

Derviş Pasha was not the only high-ranking bureaucrat who lectured at this first incarnation of the Ottoman University. Chief of the Central Court (*Meclis-i Tahkik*) and Chief Physician Salih Efendi, a graduate of the Imperial School of Medicine, taught biological sciences; head of the Accounting Council (*Divan-ı Muhasebat*) Ahmed Vefik Efendi, who was educated in France, taught history. Director of Military Schools Safvet Pasha gave lectures on physics during Derviş Pasha's absences.[18] Geography was taught by Mehmed Cevdet, teacher of geography at the School of Public Administration, and astronomy by court astrologer Osman Saip Efendi. While the latter was from the ulema, he knew European languages and belonged to the ranks of established high ulema that supported educational and administrative reform.[19]

This demonstrated that the sciences taught at the university were fully backed by the Ottoman state. According to an official statement, sciences were progressing in the empire under the auspices of the sultan, the "embellisher of knowledge," and the many sciences to be taught at the university would "realize many great benefits and common interests."[20] That some of the most illustrious members of the state mechanism were among the lecturers was hinted as a strong indication of this support.

But journal articles from the period suggest that this very fact must have caused discomfort among the audience: was it truly becoming of esteemed statesmen to be teachers—and teachers of a still not-so-respectable kind of knowledge at that? This apparent unease parallels Münif's complaint regarding the inappropriate behaviors of the audience during demonstrations and lectures: the subjects were apparently not cooperating adequately.

As sociologist of science Harry Collins once noted, the difference between a scientific experiment and a demonstration is that while the former is messy and unpredictable, the latter is smooth, with a clear and well-rehearsed conclusion. It is due to these characteristics that demonstrations are the tool of choice for making a convincing statement to an audience, and for as long as members of the audience attribute what they observe to nature and not to the person they are watching, the case at hand is a scientific demonstration and not illusion or magic.[21] But precisely due to this condition, for a demonstration to be truly successful, a trained audience is needed—an audience with trained senses and using the "correct" categories of perception and interpretation.[22] Additionally, as many studies have shown, many eighteenth- and nineteenth-century European "men of science" were particularly ambivalent regarding popular demonstrations about phenomena like electricity (such as those conducted at the *Darülfünûn*), since, due to the very characteristics of these phenomena, it was not easy to control audiences' interpretations.[23] How, then, would Ottoman audiences be taught how and what to see when they observed a demonstration? What would enable them to appreciate what was being discussed by the speaker?

The answers, once again, had to do with the characteristics of the speaker himself. Münif Pasha wrote in the *Mecmua-i Fünûn* that far from putting their high status in jeopardy, trying to spread sciences in this manner further heightened the glory of the Ottoman statesmen, as Muslim faith itself stated that knowledge was the highest rank of all.[24] Moreover, that the individuals running the state were knowledgeable in science, or at least appreciated and demanded it, was particularly pleasing; in the new era, ignorant and incompetent people could no longer run states as they did in the old times. State positions should now be bestowed on people who know "various sciences, especially those regarding public administration and international relations."[25] That the sciences taught by the bureaucrats in question were not among these apparently did not bother Münif Pasha, as it was ultimately the appreciation of science that mattered.[26] It is also striking how he was able to imply that former statesmen of the Ottoman Empire could be regarded as "ignorant and incompetent." Advocacy of science not only involved imaginations of the future but the rewriting of the past.

The young litterateur Şinasi, too, made a similar point, reinforced with an Islamic reference:

At the zenith of Islamic learning, certain individuals such as Ibn Sina [Avicenna] who had reached the rank of Grand Vizier, and even in the Sublime State [i.e., the Ottoman Empire] those who acquired the highest titles in the path of learning[27] used to teach sciences to the public; currently European cabinet ministers also give lectures in this way, if they are able to. Because, just as a branch that reaches maturity scatters its fruit to the soil that nourished it, men of learning, too, develop thanks to the public within which they are raised, and sharing with the public the benefits of their knowledge is, in reality, the fulfillment of their duty of gratitude.[28]

Hence, lectures and demonstrations were to be appreciated because of the people who delivered them. It is worth remembering at this point the well-known contribution of Steven Shapin and Simon Schaffer to the genealogy of the idea of the "objective scientist": the concept "modest witness." The credibility of the seventeenth-century experimental philosopher was built on this notion—that he was not a manipulator or intervenor but one whose presence only allowed facts to emerge by themselves.[29] But as Donna Haraway noted, the "invisibility" of the "modest witness" is possible precisely because "his" is already a position of power that makes "his" attributes the "neutral" ones.[30] That is why, after all, a true "modest witness" was a gentleman in Robert Boyle's experimental philosophy. In the Ottoman case, however, nothing was implied, there was nothing "invisible" about the merits of the representative of science and the disseminator of truths about nature: he was there, conducting the demonstration on electricity or lecturing on chemistry both as a "man of science" and as a "benevolent statesman" simultaneously and overtly. The relevance and authority of the knowledge conveyed were inseparable from the acknowledged and accentuated authority of the conveyor.

Indeed, state dignitaries not only gave the lectures but were occasionally among the audience as well. Yusuf Kamil Pasha is known to have visited the university both in 1863 as prime minister and in 1864 as head of the High Council for Judicial Ordinances.[31] As the minister of war, Fuad Pasha also attended an 1863 lecture by Derviş Pasha. His brief speech at the end of the lecture not only is another indication of the official support for science but also marks the terms in which this support was justified. Fuad first noted that the duty of the state was to deliver to the subjects all kinds of boons within its reach, and as the greatest blessing in life was knowledge, it was incumbent on the state to spread it among the people. He then defined the knowledge in question:

Even though the science taught here is called natural philosophy, it is in essence philosophy of the divine. Because it reveals to us divine knowledge at a level that our [limited minds] can grasp. Philosophy is a means for this purpose, and the difference between ancient philosophy and new philosophy is like the difference between a sailboat and a steamboat. The latter takes one to the destination in shorter time.

We are also grateful to the person who undertook the teaching of it. While he holds a most exalted rank among the highest ranks of the state, he truly demonstrates that knowledge is the highest rank of all. He thus honors the work of the eminent ulema of the past, who, after leaving their official duties, would spread knowledge in *medreses*.[32]

The arguments of Şinasi, Münif, and Fuad are very lucid in terms of the continuity they affirm. Never in their statements do they proclaim that the new science is to replace the old one, but this is precisely what is implied. Those who spoke about physics in the 1860s were by all means comparable to the religious scholars, the "eminent ulema," of the glorious days of Islam in general and the Ottoman Empire in particular. Indeed, in Fuad Pasha's words, the knowledge presented by the new scholars was actually even more effective, as they rendered the divine almost tangible to some extent.

Furthermore, examples from the Islamic tradition strengthen the connection between knowledge and the state. Ibn Sina was both a scholar and a statesman, the ulema of the past were both men of religion and men of state. Hence, it is no coincidence that their contemporary replacements are statesmen as well: the men who possess the new sciences will be, in a sense, the new ulema of the new state mechanism.

Last but not least, it is a logical conclusion that if the new "knowers" are thus comparable to the ulema, then their moral soundness can also be assumed. Like the ulema of the past, these new men of knowledge can be expected to also know right from wrong and virtue from vice. Ottoman men of science are men of state, men of knowledge, and men of virtue all rolled into one.

In this formulation, science, religion, and state can be seen as the justifications and reinforcers of one another. Needless to say, the discourse that brings these three concepts together in such a plexus renders potential comprehensive criticisms of science exceptionally difficult to advance. Similarly, when the authority of science is identified with that of the state, criticisms directed at the representatives of the state are rather likely to also allude to the kind of knowledge that they represent—as the Young Ottoman arguments to be discussed in chapter 4 will illustrate.

Despite all these statements about the supposedly obvious benefits of science to people from all walks of life, its link to divine knowledge, and the role of demonstrations in the learning of science and the comprehension of this link, this first experiment of the *Darülfünûn* came to an abrupt end. When construction was finally finished in late 1864—almost twenty years after it had started—the building was handed over to the ministry of finance. Lectures continued in a mansion, with falling rates of attendance, and soon afterward the building burned down with all the books and laboratory equipment within it. But the more basic reason for the failure of the experiment can be found in a report from 1868 that elaborated on the lessons learned from it. Stating that the lectures *did* have many attendees, the report complains that

. . . a portion of the attendees were public officials, and they remained auditors only, as they were unable to attend regularly. . . . Another group had no duties or occupations, but they were also ignorant of the background that is required for comprehending the lectures, so they came and went in vain. Yet another group listened to history as if it were but legends, and attended physics lectures just to be entertained by the experiments. Hence, lectures remained lectures only, without any outside benefits.[33]

The new sciences could not be taught to people without the proper training and, in conjunction with that, an actual reason to believe that what they observed was beneficial or worthy of special respect. Clearly, the authority of the new statesmen was not sufficient to make the new types of knowledge authoritative. Meanwhile, the group that advocates of science always relied on, that is, young civil servants, proved undependable precisely because they were civil servants. In sum, science—represented primarily as lectures by great statesmen and occasionally in the form of demonstrations that were perceived as curiosities—failed to generate a public who would appreciate and endorse either the new sciences or the new ideas about what science meant.

For the sake of comparison, it is also worth dwelling briefly on the Iranian namesake of the Ottoman institution at this point: the Dar-al Funun founded in Tehran in 1851. Like the Ottomans, the Qajar rulers referred to two needs in the mid-nineteenth century: the construction of a loyal cadre of officers and bureaucrats for the centralizing state apparatus, and the dissemination of the new knowledges produced in Europe. The Dar-al Funun, a polytechnic that offered courses in military subjects and the new sciences as well as languages and arts, emerged as an answer to both of these needs. Just like the Ottoman institution, the Iranian Dar-al Funun was initially strongly endorsed by the shah and members of the Qajar dynasty; it was built near the royal citadel to demonstrate this support. Furthermore, the first popular scientific journal of Iran, the *Ruznama-yi Ilmiyya*, started publication in 1863 (soon after the Ottoman *Mecmua-i Fünûn*), under the editorship of the Qajar prince Ali Quli Mirza, who also occupied the posts of the chancellor of the Dar-al Funun and the minister of education. While it survived its Ottoman namesake and was absorbed by the new Tehran University in 1935, however, political and economic concerns, as well as the fluctuations in the shah's perception of the institution's effectiveness (particularly in producing a reliable and loyal cadre) significantly impacted the Dar-al Funun's fortunes. In many ways, its fate was similar to that of the prestigious Ottoman schools of the late nineteenth century: the loyalty of its graduates to the throne remained suspect, and the Iranian Dar-al Funun was perceived to produce "under-miners" of the state.[34]

3. Science and Ottomanism: The 1869 Public Education Regulation

The 1860s were characterized by the constant reiteration of the official ideas on science established in the previous decades in the Ottoman Empire. A memorandum dated March 15, 1868, for instance, stated that "knowledge and learning are the greatest bases of civilization and prosperity." The August 14, 1868, memorandum about the opening of the *Mekteb-i Sultani*, a very prestigious high school designed in collaboration with representatives of the French government, also refers to the "perfection of the means for all the subjects of the sultan to achieve maturity by benefiting from the light of science and education."[35]

But the 1860s were also a period where the official discourse on science, education, and ignorance was colored most visibly by Ottomanism, the dominant policy of the Tanzimat Era. Ottomanism entailed efforts for constructing "Ottomanness" as an identity that would be above all religious and ethnic identities, and the rights given to non-Muslim communities were a key part of this policy. In 1869 the Ottoman Citizenship Law was enacted for this purpose, asserting the equal status of all subjects of the empire as citizens, regardless of religion.

A momentous product of the same year for our purposes was the Public Education Regulation. This was an act prepared under the guidance of Victor Duruy, the French minister of education. Not only did the final draft, written by Safvet Pasha with the assistance of Sadullah and Kemal Pashas, outline the educational system that the Ottoman government intended to construct, but its introduction in particular resembled a manifesto for Tanzimat policies:

It is needless to state and explain that sciences and learning are the principal source of prosperity. The realization of the progress that mankind has a propensity for . . . and the production of the inventions and useful institutions concerning arts and industries all depend on knowledge and learning.

Learning and knowledge give rise to the birth of arts and industries, and arts and industries to the invention of many vehicles and useful works that facilitate the [satisfaction of] the basic needs of human society. Therefore, it is clear that the reason why nations that belong in the civilized world get their share from the treasures of wealth of the world is their access to the most perfect means of human education.[36]

The contemporary development of industries was based on "certain methods and principles determined by the subtle sciences" that had constantly been spreading in "civilized" countries. These sciences had entirely changed the processes of production, rendering mere natural talent useless.

While all this was known in the Ottoman Empire, the draft asserted, the system of education had remained ineffective, and key in this failure was the state of elementary schools. These schools where "instruction consist[ed] only of the basics of religious sciences" were under the administration of people of dubious qualifications.[37] Furthermore, due to the lack of a sufficient number of higher institutions, there were only two routes for the educated: civil service or the military. As a result, the Ottoman system failed to raise "people of profound skill in sciences and arts."

In addition to an across-the-board reorganization of the Ottoman educational system, the introduction to the act recommended the rapid translation of books on the new sciences, as each nation could progress in the sciences only in its own language.[38] In line with the Ottomanist outlook, the introduction also stated that in new high schools where non-Muslim and Muslim students would study together, religious courses would be taught to different communities by teachers assigned by their own community, while all science courses would be under the administration of the central government.[39]

It is needless to state the implications of the phrase "It is needless to state" at the introduction of an official document. We observe in the regulations of 1869 a clear intent expressed in the strongest terms to centralize, organize, and closely supervise all educational activity within the empire. The speakers of the state officially announced their reservations about the competence and merit of at least some of the teachers of elementary schools who were, by and large, *medrese* graduates. Industry, inventions, and prosperity are linked unprecedentedly tightly to the new sciences, thus clarifying which sciences the official discourse deems necessary for the welfare of the empire.

Indeed, while earlier documents on educational reform and the significance of the new sciences always referred to religious sciences, the afterlife, or used Islamic justifications as well, the 1869 act is striking in its nonreligious tone and content. As Somel indicates, the influence of Duruy on the final draft cannot be neglected,[40] but it also appears that the idea of science as beyond religious and political affiliation was a perfect fit with the Ottomanist discourse of the 1860s. The document specified that religious classes would be taught separately to students from different religious communities, but they would learn science together, as Ottomans. Similarly, teachers of religious classes would be appointed by the communities themselves, but science teachers would be appointed centrally by the state. In the new political system, religion was to belong to the communities (albeit under state supervision), but science was state property. Moreover, science was the realm within which the equal status of all religious groups would be made manifest. In sum, importing science was a matter of constructing *the* community.

Yet this effort to present science as the type of knowledge that, just like the Ottoman identity, would bring all ethnic and religious groups together was particularly precarious due to the prevalence of perceived relative deprivation among the members of the Muslim community. Indeed, as the following chapters will illustrate, the growth of the Ottoman press would be characterized by the increasingly common challenges to the very idea that science—or *anything*—could be discussed without a reference to religion in general and the relative statuses of the religious communities in the Ottoman Empire in particular.

While competitors with concerns for popular appeal would thus gradually emerge and disseminate views that were occasionally at odds with the official definitions and descriptions, the leading scientific journal of the period, a landmark in the history of the Ottoman press, was endorsed by the top bureaucrats and served as an outlet for elaborating the official discourse.

B. Engaging the Official Discourse in the Early Ottoman Press: Münif Pasha and the *Mecmua-i Fünûn*

On April 11, 1861, Halil Şerif Pasha, the Ottoman ambassador to St. Petersburg, wrote a petition to the prime ministry asking for permission to establish a new association under the name Ottoman Society of Science. According to the petition, the need for such an association stemmed from the fact that it was "science and useful learning that [had] steered European nations to the highest summit of civilization and power." Despite their best-intentioned attempts, the Ottomans themselves had not yet reached the desired level. As a result, Halil Pasha declared, "certain servants of the exalted sultan who succeeded in acquiring a good education in Europe or within the Empire" were willing to contribute to the spread of "sciences and the needed knowledges" as a demonstration of their gratitude.[41] The society would focus on the writing and translation of books on all branches of learning, avoiding politics and religion, and conduct courses for the public at large. Funding for these activities would be provided by the members themselves, and the society demanded from the state only that the authors and translators be awarded.

Note the emphasis on gratitude, and the significance of the "science as gift" metaphor in the petition. This notion that science is a thing possessed by particular individuals who, in turn, want to offer it as a gift to the people evokes the arguments of Marcel Mauss: the presentation of a gift triggers indebtedness and is basic to the construction of a hierarchical relationship when it cannot be reciprocated.[42] Their self-presentation as gift-givers is integral to the elites' conceptualization of social order in the Ottoman Empire: they are the enlightened "knowers" in charge of a docile but trained and productive people.

The board of the society was composed of eight people, with five Muslim and two non-Muslim Ottomans and one European resident of the empire: Halil Pasha,[43] Münif Pasha,[44] Said Pasha,[45] Mehmed Kadri,[46] Sadullah,[47] Karabet,[48] Andreas David Mordtmann,[49]and Istefan.[50] Of the remaining twenty-five members, eight were non-Muslim. Young employees of the Translation Bureau dominated the society, alongside three teachers of the Imperial School of Military Engineering and three officers. The representation of the ulema by one person, and only among the provisional members, is striking. As İhsanoğlu notes, it is also interesting that very few of the teachers of science in the prestigious European-style schools of the empire are among the members of the society.[51] Hence, while its membership profile demonstrated the religious cosmopolitanism of the society and reflected the cosmopolitanism of the empire itself in this respect, it also resembled an exclusive club for rising Ottoman bureaucrats in their thirties.

The Ottoman Society of Science was indeed an *Ottoman* society, along the lines of the broader policy of Ottomanism implemented in the Tanzimat period. As the Imperial Decrees of 1839 and 1856 indicated, and the 1869 Citizenship Law unequivocally asserted, the people living under the dominion of the Ottoman sultan were equal citizens regardless of their religion, and their identity as Ottomans should transcend their religious affiliation. But this also indicated an ambiguity regarding the prestige of the ulema as the "knowing class," as their knowledge could not but be based in Islam. Knowledge and religion were inseparable for the ulema. The Ottomanist bureaucrats of the 1860s, however, constructed a new discourse where knowledge could be talked about as autonomous. Religion could motivate the search for knowledge, guide it, provide hints for it, but could not be identified with it. This *had* to be true, for otherwise it would not be possible for the Ottoman Empire to import the new knowledge that was the "engine behind European progress."

The overwhelming majority of the bureaucrats who established the society grew up in the heyday of Tanzimat reforms, during the construction and propagation of the inclusive ideology of "Ottoman, before Muslim or non-Muslim." Science, for them, was rather similar to what "being Ottoman" meant: it was above religious or ethnic identity, and, as they implied in their petition, they deemed it possible to talk about science without referring to religion or community. And it was this new generation who could explain what science meant, because it was they who possessed, as they saw it, superior training rarely found in the Ottoman Empire.

The petition of the society was accepted on June 3, 1861, as its aims were found harmonious with the state's "glorious efforts for the education of the people."[52] One year later, the most influential and substantial product of the society came into existence: *Mecmua-i Fünûn*, or the *Journal of Sciences*. This was not only the first journal devoted to the new sciences, but the first relatively long-lasting periodical in Turkish.[53] As the vice-president of the society, Münif Pasha was in charge of the journal and his name would become identified with it.

In the first issue, the journal published the by-laws of the Ottoman Society of Science. Members of the society, according to this document, had to know Turkish, Arabic, or Persian in addition to French, English, German, or Greek. Permanent members were required to submit essays to the society's journal and/or teach public lessons on a science they were competent in. The journal would be devoted to "sciences, learning, commerce and industry," and it would not discuss religious or political issues. Even though members were not required to be fluent in it, the official language of the society was Turkish, and works written in other languages by the members would be translated by the society. Finally, collaborations would be pursued with other associations whose mission was also to spread science.[54]

This final principle also made sense with respect to the Ottomanism of the Society of Science, as Münif Pasha was also an honorary member of the Greek Philological Syllogos of Constantinople, a learned society founded in 1861 by leading members of the Ottoman Greek community. Other "men of the Tanzimat" such as Âli, Fuad, and Edhem Pashas themselves were also honorary members of the Syllogos.[55] The journal of the Syllogos mentioned the activities of the Ottoman Society of Science in its first issue, and the Greek members of the Ottoman Society were also involved with the Syllogos. While further research is needed to uncover the connections and interactions between these contemporary societies, the complexities of the Ottomanist project that a cursory analysis reveals are worth noting. First, the two societies' references to the spread of scientific knowledge were quite similar, with one key difference: for Syllogos the audience was the Greek Ottoman community, not all Ottomans. Indeed, George A. Vassiadis notes that even though leading Ottoman statesmen did become honorary members of the Syllogos, the society was not formally recognized by the state, due probably to the potential implications of recognizing a society that "aimed at encouraging the cultural advance of a particular [religious community]."[56] Moreover, it is striking that the Syllogos admitted as honorary members Hellenophile European politicians like William Gladstone who were hardly sympathetic to the Ottoman Empire. Second, while the Ottoman Society of Science had non-Muslim members, the Syllogos was more exclusive—a policy that also informed the contents of the journal of the society that, in contrast to the more cosmopolitan attitude of the *Mecmua-i Fünûn*, was focused on Greek language and scholarship.[57] The supposed universality of science and learning and the policy of Ottomanism were not necessarily sufficient for constructing working intercommunal learned societies, nor a discourse that could transcend matters of community membership in the context of the mid-nineteenth-century world.

In order to better understand what the Ottoman Society of Science and its journal represent, it will be helpful to focus on the two essays by Münif Pasha published in the first issue of the *Mecmua-i Fünûn*—the introduction and, in particular, the rather lengthy essay entitled "Comparison of Knowledge and Ignorance." It is possible to find all the key elements of the dominant contemporary discourse on science in these pages.

As "useful knowledges and sciences" are the sources of felicity in this world and the next, according to Münif, the journal would be devoted to useful knowledge on all sciences and arts, and avoid topics pertaining to religion and current politics. Remarkably, Münif uses the phrase *saadet-i dâreyn*, a well-established phrase in Islamic literature that refers to happiness in both worlds, but for him the "useful learning" behind this bliss has been purified of Islamic connotations—this is a knowledge that can be known without religious associations.[58]

If speech is what makes man superior to animals, Münif proclaims, those men who possess knowledge are superior in a similar way to those who do not. The contemporary world is proof of this fact: peoples of Africa and America who live in "blindness of ignorance" are defeated and enslaved by "civilized nations."[59] A small country such as England came to dominate a population twenty times greater than its own "only because of their power in science and industry." Strikingly, with such a sleight of hand, Münif equates "knowledge" to the particular kind of science that England, along with other "civilized nations," possesses.

But then Münif takes another bold step and brings in the traditional idea that knowledge is what enables man to distinguish good from evil and act accordingly. He who is ignorant does not know his duties and responsibilities and fails to protect himself from vices and dangers. Hence, knowledge is also the path to virtue and, accordingly, bliss in the afterlife. What this implies, however, is that "civilized nations"—which, in Münif Pasha's essay, are the Great Powers of Europe—should also be the most virtuous, which makes the point particularly difficult to bring into line with established Islamic Ottoman thought.

Münif does tackle this point, also touching on the issue of science and religion. Some ignorant people, he states, believe that knowledge corrupts faith. Yet it is these "mindless friends of religion" themselves who thus harm the very essence of religion, as their argument amounts to saying that religiosity requires ignorance. Those who understand the many mysteries of the universe are the true believers, whereas the faith of the ignorant is nothing but imitation, as it is not built on a sound foundation.[60] Münif's examples here are Socrates and Hippocrates, figures already respected in the Islamic tradition. But the unstated yet obvious conclusion of the argument is that contemporary European men of science, if not all those who possess scientific knowledge, should be treated as true believers, even if they are Christians.

The invisible audience that Münif Pasha's text addresses is clearly those who doubt the consequences and representatives of the knowledge coming from Europe. Hence, having shown the link between faith and knowledge, he further clarifies the link between virtue and knowledge. If it is possible to observe disagreeable scenes in "the civilized world," Münif contends, it is not because of their knowledge but because their civilization has not reached perfection yet. Furthermore, while there is poverty and suffering in London, for instance, the English also provide the suffering minority charitable services that can hardly be found in uncivilized countries. Similarly, if "the general state and behavior of certain individuals who claim to have knowledge are seen to be devoid of the virtue and righteousness *that should be the consequence of knowledge*, this must not give rise to doubts about the beneficial impact of knowledge."[61] Those individuals are the ones who know only some jargon, not the subject matter, purpose, and the manner of application of the knowledge in question.

Kindness to the poor is not the only important virtue, of course. People who know also know the need for government, along with all the religious and rational dictates regarding obedience toward the holders of power. So "[men of] knowledge do not cause detriment to statesmen and disobey their authority by any means." Knowledgeable rulers know this, and instead of resorting to violence like ignorant rulers who would not deserve obedience, they encourage the spread of knowledge within their dominion. Hence, just like religion, the state needs people who *know*.

It is worth repeating, though, that this is not just any knowledge. It is the new knowledge that the Europeans produce.

Around twenty or thirty thousand Europeans covered in a few months great distances and, with utter ease, declared victory over China, which is unique in the world in terms of territory and population, and . . . has a few million soldiers. They went all the way into the Chinese capital and set the terms of the peace treaty. The Emperor of China was undoubtedly displeased with his soldiers who were ignorant of new military methods. Had the Chinese not insisted on preserving their ancient methods and imperfect civilization, would they have fallen victim to the insult of a few thousand foreigners?[62]

Those who "know," for Münif, are those who study not only history but also people, animals, plants, minerals, the earth, and the sky. As their minds are filled with true knowledge, they cannot be deceived by charlatans and their superstitious claims. They know, so they are also able to produce more than the uncivilized in much less time, and, as a result, they may be said to live fuller, longer lives. Knowledge is not only for the acquisition of material goods, however: it is for becoming virtuous. God Almighty created man not simply for him to live, but to "extend the limits of the virtues and faculties within his natural disposition, and improve his worth and rank in a way that befits the glory of humanity."[63]

Münif Pasha's essay demonstrates the intricacies of mid-nineteenth-century official Ottoman discourse on science with all the silences, unstated implications, vague transitions and reasonings, and abrupt shifts of terminology that it embodies. On its surface, Münif's text is about knowledge broadly conceived, with ignorance as its antonym. He commonly uses the term *ilm*, which in Islamic terminology does mean knowledge, but with religious connotations, to the point of being almost a synonym for Islam.[64] Branches of intellectual knowledge that were not necessarily related to tradition or religion were called *ilm* as well in Muslim societies, but the classification itself that defined certain branches of knowledge as "transmitted" (*naklî*) or "intellectual" (*aklî*) was Islamic. Likewise, the knowledge that distinguishes the "knower" from the ignorant is not one that makes sense without a reference to religion, and the connections between knowledge, action, and virtue cannot be set without "passing through" Islam.[65]

It would be helpful to remember here that *medrese* students were called *talebe-i ulûm*, students of *ilm*, in the Ottoman Empire. Similarly, the word *ulema* is derived from the word *ilm* and literally means "those who know." The word for "ignorant" in Ottoman Turkish, *cahil*, is borrowed from Arabic (*jahil*), and is a word that signifies the opposite of "knowledge," with all its connotations—so much so that the pre-Islamic period is referred to in Islamic literature as *Jahilliya*, "a period of ignorance."

Münif Pasha, however, effortlessly transforms the sciences that the Ottomans were learning from the Europeans such as zoology, botany, ethnography, physics, and chemistry into knowledge per se, which puts those unaware of these sciences in a position of ignorance, with all its moral connotations. In other words, while Münif Pasha and the journal he was in charge of stated that their discussions on science would avoid religious topics, he made very effective use of a particular Islamic interpretation linking the acquisition of knowledge to virtue in order to make a case for the sciences of the Europeans. We see the same strategy at work throughout the text in the shape of references to bliss in both worlds or to the natural disposition instilled in man by God and the duties and responsibilities this imposes.

Münif's numerous references to the branches of science that study the layers of the earth, animals, plants, and humans also indicate an important difference between Münif's generation and the authors referred to earlier in terms of what they perceived as science. After all, Münif was writing at a time when geology, following the contributions of Georges Cuvier and Charles Lyell, was constantly offering new findings and suggestions about the age of the Earth—findings that also influenced zoology and botany and paved the way for Darwin's arguments on evolution. Not only the sciences but also ideas on science were changing rapidly in this period, with a growing emphasis on the "unstoppable progress of science," which was part and parcel of the "civilizing mission" discourse.[66] Like his contemporaries in Europe, Münif was particularly impressed with developments in geology, but, once again, his arguments on this science were simultaneously about virtue and vice. Noting the rapid advances in this young science, Münif asserted that those who were not familiar with the findings of this science were like people who were ignorant of and indifferent to their own home, for which they should "feel ashamed and disgraced."[67]

We observe in Münif's texts on science, then, direct references to virtues and vices as well as rhetorical mechanisms that identify the new sciences with knowledge as such and attribute to them the moral connotations of a particular Islamic understanding of learning. The programmatic statement on science in this journal that declared that it would stay away from religious issues is thus enmeshed with subtle and overt references to the Islamic tradition. Likewise, while science is supposedly a topic that the journal is to discuss without reference to politics, in his essay Münif Pasha brings up a theme discussed earlier: science teaches people to be obedient to their government, if it is "those who know" who are in charge. Such statesmen, thus, should not only not fear science but endorse and take advantage of it. The need for a supposedly autonomous sphere called science is thus asserted entirely with reference to religion and the state. The authority of science is based on its benefits for the state and its endorsement by religion.

While Münif was thus linking knowledge to virtue with references to both religion and state, and attempting to show his readers that knowledgeable people were a treasure rather than a potential threat, Âli Pasha, Münif's supporter and one of the most influential statesmen of the era, felt the need to insert, so to speak, more *raison d'état* into this discourse. Foreign secretary at the time, Âli Pasha sent a letter of congratulation to the society that was published in the second issue of the *Journal of Sciences* and subtly corrected Münif.

Civilization, which involves the prosperity and security of mankind, is a God-given requirement of human existence, Âli Pasha asserted, and progress is its very nature. The level of civilization and progress of each human community is dependent on the education and awareness of its members; knowledge is the nourishment of the soul and the foundation of civilization.[68] But learning should be accompanied by good morals, as the ultimate issue is striking a balance between rights and responsibilities: "Knowledge without good morals is certainly the cause of endless harm," as he who lacks decency is likely to use his knowledge for evil.[69]

Münif and Âli Pasha's texts clarify the question: Knowledge—in this particular context, we can consider this the scientific knowledge imported from Europe—*is* beneficial, but can it be guaranteed that those who possess this knowledge will not constitute a threat to the stability and well-being of the state? Clearly, the importation of science constitutes a question of social order for both authors. Münif's essay attempts to demonstrate that the holders of power need not be concerned, while Âli Pasha expresses precisely this concern with his letter. There is no rift between the two in that they both regard social order and the stability of the state as the chief issue, and the question of morality-cum-obedience is equally fundamental for both. That disruption should be avoided by all means is what is shared by the competing arguments. As *the* symbol of state power in the 1860s, Âli Pasha emphasizes that assuming a direct relationship between knowledgeability and virtuousness can be dangerous for the state. Münif, on the other hand, becomes the representative of the discourse that associates the spread of scientific knowledge with the preservation of social order.

It is quite striking in this context that Hyde Clarke, a member of the British Association for the Advancement of Science and at the time cotton commissioner in the Ottoman Empire, refers to Münif Pasha as "the Brougham of Turkey," after Lord Brougham.[70] Henry Brougham's *Practical Observations upon the Education of People* (1825) was a fervent publicity for the Mechanics' Institutes—the embodiments of the idea of the spread of scientific knowledge presented in an authoritative fashion as the solution to the problem of social order.[71] For Münif, as for Brougham, scientific knowledge, properly taught, was the route toward the construction of a productive yet docile populace.

Very soon after the publication of the *Journal of Sciences*, two more periodicals appeared declaring that they would cover issues on science as well: *Mecmua-i İber-i İntibah* (*Lessons and Awakening*) and *Mir'at* (*The Mirror*). Their editors were also young bureaucrats: Ali Haydar, the editor of the former, was twenty-seven and had studied in Paris. He was working as the translator of the Tanzimat Council at the time.[72] *Mir'at*, on the other hand, was published by Mustafa Refik, a nineteen-year-old clerk in a governmental office.

In the first issue of *Mir'at*, the first illustrated journal in Turkish, Refik, too, wrote a long essay on civilization, education, and knowledge, with an emphasis that resembled Âli Pasha's. Civilization involved the permanent happiness and prosperity of a people, Refik wrote, and the basis of civilization was "good training."[73] As civilization brought to people certain liberties in addition to knowledge, the only way to prevent them from abusing these liberties and knowledge was to insure that they had good morals, and this is what could be achieved through good training. The following section illustrates the difference from Münif's argument very clearly:

Once it was thought that the correction of morals could be achieved through knowledge and education, but this should not be the case. Truly, knowledge is the essence of the life of civilization, but it is not the sole cause of the purification of morals. Because knowledge is certainly the guide of the mind in comprehending, using and implementing everything, be they good or bad, and mankind tends towards evil in most of its actions and attitudes. Particularly, certain sciences and knowledges have always acted as intermediaries towards malice and evil, in accordance with the level of depravity of the morals of individuals. And it is proven by many incidents that a person, once his morals are corrupted, cannot help but use his knowledge and skill for harming his state and nation.[74]

Hence, for Refik, if people acquire knowledge and skills without correcting their morals, they will threaten the security and well-being of their country, and it is in cities where science and learning are most improved where godlessness and sacrilege are most common. He then shifts the discussion entirely to the topic of morality and learning. If the morals of a certain people are untainted, he argues, they will perform their duties toward their state well, provided that the laws of the country are applied equally to every individual. Training and the purification of morals involve the work of both the state and the individuals: it is the duty of the state to spread good morals and punish those who do evil deeds, while the duty of individuals is to strengthen their religious faith and protect their honor. This they can do by studying religious and literary sciences and associating themselves with respectable, learned people.[75]

Refik's journal further clarified its broader conception of science, as it published essays on literary composition and translations from Montesquieu along with texts on the progress of agriculture and the characteristics of steam power. Even essays that gave information on new European sciences, like the last, emphasized the divine bequest for humanity that allowed such inventions to be possible.[76]

Münif's *Journal of Sciences* published lukewarm comments on both of these new periodicals. While it was pleasing to see the emergence of new outlets for the spread of knowledge, Münif wrote, the *İber-i İntibah* contained little information that could not be found in the already existing newspapers, and even its name was not syntactically correct.[77] He had further dry comments to make after the emergence of Refik's *Mir'at*:

It is quite striking, and perhaps telling, that while the need for such scientific publications had been felt for so long, no attempt had been made; and then two newspapers of this sort have emerged only a few months after the appearance of our humble journal. Everything happens this way in the world: when a beneficial path is opened, there emerge people who follow it, and it is clear that this will encourage the competent and the generous to produce all kinds of works useful to our country.[78]

This brief comment was sufficient to offend Refik. He wrote in his response:

If we assume that [Münif] has actually seen our journal, his disparaging tone seems to indicate that some of the matters we discussed in our first issue perturbed his thoughts. If we suppose he has not, it would be particularly unexpected of him indeed not to be interested in the perusal of such a publication [as ours], as his very decency and his violent attacks on the *Mecmua-i İber-i İntibah* would suggest.[79]

Refik's rather angry remarks included an overt ridicule of Münif ("Apparently his attitude is due to the ferocity of his desire and passion to spread learning"), and several examples intended to demonstrate that he was actually quite ignorant of the things he was writing about.

But this response from a very young clerk to the Translator-in-Chief of the Sublime Porte[80] and the former head of the court of commerce had a great cost. Âli Pasha found the way Refik expressed his opinions entirely at odds with the "official style" and indicative of an interest in oppositional politics, and he instructed Refik either to apologize or resign from his post. The parallels between Âli Pasha's and Refik's essays on civilization and the need for moral soundness are striking, but apparently Refik's unabashedly belittling tone was sufficient for him to sound like a dangerous youth. When confronted with an "angry young man" like Refik, the differences between Âli and Münif evaporated. Of the two options Âli Pasha gave him, Refik chose the latter and resigned; he also ended the publication of the *Mir'at* and became one of the founders of the Young Ottoman movement in 1865. He died of cholera soon afterward.

It is possible to see how Refik's arguments on knowledge and ignorance are simultaneously arguments about specific types of people: those who use the (nonreligious) knowledge they possess against the interests of the people, and those who need to defend themselves. The latter are encouraged to seek the support they need in religion, but the state is also imagined as an entity that protects the customs and morals of the people. A plausible interpretation of Refik's arguments is seeing them as a reaction against the Tanzimat bureaucrats—the elites whose characteristics were discussed in chapter 2. The merits of scientific knowledge are challenged precisely through a critique of these elites whose credentials and attitudes are condemned vehemently. Furthermore, the emphasis on religion not only is intended to qualify the authority of science (and its representatives) but to underline the precariousness of the Ottomanist ideal. Refik's question is simple in this respect: Is it possible to talk about science without talking about Islam and the interests of the Muslim community of the empire? I will take up these issues in further detail in chapter 4, in an analysis of the Young Ottoman arguments on science.

Both competitors of the *Journal of Sciences* thus disappeared soon after their emergence. The *Journal of Sciences*, on the other hand, remained alive for more than four years—a figure which includes interruptions that lasted several months. Even this constitutes a major success in the Ottoman Empire of the 1860s, with its very low rates of literacy and devastating financial crises.

The list of those who subscribed to more than one issue of the journal, which was published in its fourth issue, is indicative of the reason behind this success: the prime ministry, the office of the foreign secretary, the ministry of education, members of the high councils of government, ambassadors, the School of Military Engineering, and branches of the military were the main sponsors of the journal. If we also remember the affiliations of the founders of the society and the contributors to the journal themselves, the *Journal of Sciences* appears more like a state enterprise despite its quasi-independence. Indeed, the authors who contributed the most essays to the journal, after the founders Münif, Kadri, and Halil Şerif, are İbrahim Edhem Pasha[81] and Mehmed Cemil Pasha.[82] Muslim and non-Muslim members of the Translation Bureau, such as Mehmed Şevki Bey and Fardis Efendi (Alexander Themistoklis Phardys), also contributed numerous articles.[83]

That the society and the journal attempted to embody Ottomanism at work also put them in close alignment with the state and its official ideology of the 1860s. The displayed image was of Muslims and non-Muslims as joint contributors to the journal, with the common objective of strengthening the Ottoman state through the endeavors of equal citizens. Phardys, along with Alexander Constantinidis, another Ottoman Greek member of the Translation Bureau and a protégé of Safvet Pasha, submitted essays on ancient Persia, the history of Istanbul and its environs, and the Hagia Sophia. Alexander Karatheodori, who would rise to be the foreign secretary in 1878, wrote on insurance and the history of book production.[84] The Armenians Ohannes Vahanian and Sakızlı Ohannes wrote on economics.

This cosmopolitanism was also reflected in the essays of Muslim contributors. Münif Pasha himself wrote a series of articles on the lives and thoughts of Greek philosophers as well as on America; Mehmed Cemil and Halil Şerif wrote on ancient Egypt; Mehmed Şevki on Japan; and Kadri on England. Hence, information on new sciences such as geology, meteorology, and chemistry, and topics like telegraphy, photography, electricity, and magnetism were presented within such a context. The new knowledge belonged to the whole world, and knowing and using it was also about being a man of the world, a part of a universal history. This, however, tended to involve seeing the world through the eyes of the Great Powers and the *mission civilisatrice*. The racist element of this discourse was appropriated by Ottomans as well, as exemplified by Münif's statement that while "Caucasians, which include Turks, Arabs, Persians, Greeks and the Europeans" are most able in science, "Negroes, due to their creation, are reportedly unable to understand the intricacies of mathematical and rational sciences."[85]

Characterizations of this kind clearly indicate an effort to imagine the Ottoman Empire as on par with the Great Powers, and Ottoman Muslims as no less likely to contribute to "humanity." This approach also shaped the tone of the few essays pertaining to Islam that the journal published: in these texts, the subject was the contributions of Muslims to "civilization" (e.g., "Islamic Libraries," no. 45; "Services of Arabs to Geography," nos. 36, 41), or the lives and lands of Muslims in remote parts of the world such as Cape of Good Hope (nos. 9–11, 13, 26, 33), China (no. 8), and the Comoro Islands (no. 27). While a shared Islamic background was taken for granted in many texts—especially those written by Muslim authors, of course—we thus see in the pages of the journal a treatment of Islamic subjects as part of "general culture." Particularly in definitions of science, the Islamic tradition is referred to and revered, yet it is treated as a provider of incentives and clues for the production of knowledge rather than as a provider and definer of knowledge itself.

Another essay by Münif Pasha on the branches of learning is an obvious example. In this essay, Münif starts with the basic Islamic classification of sciences (*ilm*s) into the intellectual (*akli*) and the transmitted (*nakli*). While *ilm* literally means "to know," its technical meaning is "knowledge that is gathered and organized around a rule and studied specifically." But the term had been abused before, and "superstitions" such as astrology and geomancy had been sometimes been labeled as science even though they were unacceptable to both common sense and religion.[86]

Once again, at this point Münif shifts the discussion on *ilm* with its religious connotations to *ilm* as sciences imported from Europe. Even when he starts a section by referring to the intellectual sciences, as defined in the Islamic tradition, he drops the term at once to continue discussing them as science per se. The brief history of science he presents is based on the standard European narrative on the evolution of science among the Assyrians, Egyptians, Greeks, Romans, Muslims, and then Europeans, which is a narrative excluding Islamic religious sciences, of course.

Sciences are essentially composed of "some information and their implications," and they are all related to one another: physics and chemistry are dependent on arithmetic, medicine on anatomy, history on geography. Furthermore, as nature is the object that most sciences study, they do not change with changing times and places. Arithmetic does not change, as five multiplied by five is always twenty-five; physics does not change, as the characteristics of light and heat remain the same. Note that as Münif equates science to "knowledge of objects," his comments on the immutability of the facts of nature are simultaneously statements that sciences themselves are unchangeable: they simply have to be *learned* as they are—a statement that justifies the Lord Brougham analogy.

Another important phrase that we come across repeatedly in Münif's texts is "useful knowledge/science."[87] As early as 1860, when he had started to publish the *Ruzname*, the supplement to the newspaper *Ceride-i Havadis*, he had defined one of its tasks as providing the reader with "entertaining stories as well as [texts on] useful sciences such as history, geography, physics and political economy."[88] Note that some of these are branches of learning that were already practiced and taught in the Ottoman Empire, such as history or geography. But they were now being discussed and promoted differently, as sciences with unique benefits. The case was not simply about acquiring new types of knowledge; it was about constructing new ways of thinking about knowledge itself. Münif made this clear when he argued that "in Oriental countries . . . the instruction of the science of history is not customary; those enslaved by customs see [classes on history] with surprised eyes. . . . Indeed, if the science of history is seen as entertainment, and as parables and stories, there is no need for the teaching of it."[89]

What made the new sciences and, more generally, all sciences as practiced by the Europeans special was that they were *useful*: this was clearly a central theme of the official discourse on science. Sciences were useful for the development of arts and crafts, for the progress of civilization, for social order and moral purification. We see the phrase frequently in official documents on education from this period as well. An 1863 report on the problems of elementary education starts by asserting that "various useful sciences . . . are the basis of civilization and prosperity."[90] "The spread of useful sciences" is also referred to in a notice on the transformation of the Council of Education in 1864 and a declaration of the Ottoman cabinet in 1870.[91]

On the one hand, this is yet another theme strikingly reminiscent of the English useful knowledge movement of the 1820s through the 1840s. When Henry Brougham and his Whig associates founded the Society for the Diffusion of Useful Knowledge in 1826, their aim was "to divert the attention of working-class readers from revolutionary radical prints to more 'rational' pursuits"[92] and, via scientific education, render them "peaceable, respectable and diligent."[93] The knowledge conveyed to the readers was useful, thus, because it would enable them to be more productive and self-sufficient, and at the same time contain their dissent. We see both in official Ottoman documents and Münif's texts a similar theme, albeit with a different intended audience.

It is true that links between scientific education and the improvement of arts and crafts were referred to in these texts, but the journal's true audience comprised both the new generation of students attending the new schools and the prominent as well as young civil servants of the empire. The training of these groups symbolized the hopes for the empire's future as well as potential threats: Europe's sciences had to be imported and taught, but the many European political ideas that could inspire rebellion should be kept outside the empire's borders. Luckily, however, the former could actually make possible the latter, as Münif's arguments suggested. Respecting and learning "science as true knowledge" would teach the educated youth the meaning and necessity of order, the love for the fatherland, and, most significantly, the indispensability and greatness of the state. Hence, science was useful as well as true knowledge.

On the other hand, however, it should be noted that the concept "useful knowledge" (commonly in the form of *ulûm-ı nâfia*) was not new in Ottoman discourse at all, as it signified desirable knowledge in traditional *medrese* education as well. The unknown author of the *Kitâb-ı Müstetâb* wrote in the 1620s, for instance, that "Knowledge, sir, should be useful / It should ward off whim and ego."[94] This line of criticism directed at branches of learning that did not serve such high goals ("philosophy," as they were generally called) was well-established among the Ottoman ulema.[95] Similarly, Taşköprülü Ahmed wrote in the sixteenth century, "Useful sciences (*ilm*) are Qur'an, hadith, and fiqh / The rest are meaningless pursuits, fancy them not / Science is that which contains the words of Allah and the prophet / The rest are diabolical suggestions, may this be sufficient advice."[96] Even though he was critical of the contemporary state of the ulema class he himself had belonged to, the prominent jurist and statesman of the Tanzimat Era Cevdet Pasha maintained this sense in the basic textbook he wrote for elementary schools as well. His *Malumat-ı Nafia* (Useful Information; 1886) is a simple narrative on the history and principles of Islam.

As Muhammad Q. Zaman states, that knowledge should be useful was crucial for all Muslim scholars, but it was knowledge that assisted salvation that was indeed useful.[97] Indeed, this is the common interpretation of a well-known statement of the prophet Muhammad himself: "O Allah, I seek your protection from *useless knowledge*." In this respect, too, then, we see the appropriation and reinterpretation of a concept from the Islamic tradition. The phrase that signifies that knowledge should be useful is borrowed to be applied to a different type of knowledge taught at a different type of institution. But although it is evident that the context of salvation is not connoted by the new usage of "useful," it would not be accurate to interpret the transformation as a wholesale replacement of moral concerns by material ones. That the new sciences are useful because of their tangible products and the prosperity they create in this world is indeed a fundamental element of the discourse about science; but the moral connotations are kept intact, as the new sciences are deemed useful also due to their contribution to social order, virtuousness, and obedience.

Conclusion

The 1860s is thus a period in which the contours of the official discourse on science are established: Sciences imported from Europe are owned by the political elites, and it is their duty to offer them to the people as a gift. Scientific knowledge is knowledge endorsed by the state, and the authority structure in the Ottoman Empire is now to be seen in the shape of a triangle with state, religion, and science forming the three corners. The man of science, while the possessor of a unique type of knowledge, is to be a humble servant of his state (a knowledgeable official) and a beacon of morality (respectful of the congenital creed). Indeed, it is the uniquely useful knowledge he possesses that also has the power to make him a man of virtue. The subjects are to respect and learn from these men, and the knowledge they will acquire will enable them to realize how the world really works. The awareness that this will instill in them will, in turn, render them obedient and productive, thus bringing about a "good" populace. Ethnic and religious differences should not matter in such a context where scientific knowledge, just like the identity "Ottoman," is an umbrella under which all subjects can gather.

The identification of scientific knowledge with the holders of state power is a double-edged weapon, as we shall see in the following chapters. How could criticisms that would be directed against the rising class of bureaucrats avoid references to the new knowledge that they wholeheartedly advocated? Furthermore, how could matters regarding religion be paid lip service, the religious idiom simply appropriated, or science presented as the umbrella when perceptions of mistreatment and negligence were common among the Muslim community? As they asked such questions and offered answers, the generation of the Young Ottomans used the press and literature to construct alternative discourses on science, morality, and community that would in turn become building blocks for the official discourse of the 1890s.

While contextual factors shaped the specific concerns at play and the rhetorical strategies used, it is also important to restate before concluding that representations of science and "men of science" with references to morality and social order were not unique to the Ottoman case. Charles Lyell noted, for instance, that the "habits of thought" and "principles of reasoning" science instilled in its practitioners would contribute "both to the moral and intellectual advancement of society."[98] Such arguments are of importance particularly as efforts for the legitimation of new scholarly fields if not professions. In the Ottoman case, however, it is possible to observe an additional concern having to do with the sociopolitical context: the ease with which representatives of the new sciences in the Ottoman Empire could be portrayed as imitators of Europeans, with little actual contribution to the welfare of the Muslim community, gave rise to a heavier emphasis on morality in pro-science arguments. In this respect a relevant parallel can be found in the Greek case, where nineteenth-century "men of science" constructed a similar discourse, heavily laden with assertions about the link between science and morality. As the historian of science Kostas Tampakis argues, the representation of the man of science as a uniquely patriotic, selfless, hardworking servant of the nation was particularly common in the post-1840s Greek context due to comparable concerns. Hercules Mitsopoulos, professor of geology at the University of Athens, stated in 1845 that "natural sciences . . . develop and strengthen our intellectual powers, educate and improve the ethos and character of our soul and certify and stabilize our religious beliefs."[99] Münif Pasha could not have agreed more.

FOUR

Expansion and Challenge: Young Ottomans, New Alternatives

Introduction

In the center of the Ottoman Empire of the 1850s and the 1860s, "science" was an idea that was represented first and foremost by bureaucrats and diplomats in their thirties. Typically, these were men who spoke at least one European language, wore fezzes and pants (not turbans and robes), frequented the quarters of Istanbul where non-Muslims and Europeans lived, identified themselves with the Otto-man state, and commonly subscribed to Ottomanism. An awareness and appreciation of the new arts and sciences of the Europeans was a mark of distinction for these new elites. Consequently, it was not only knowledge and ignorance but "the knowledgeable," "the ignorant," and the respective moral qualities of such individuals that were being redefined in official texts and in articles published in the quasi-official *Mecmua-i Fünûn* (*Journal of Sciences*). The Ottoman man of science did not claim simply to be learned; he was a morally sound, reliable, and patriotic servant of the Ottoman state.

Yet while these new official meanings of ignorance or knowledgeability were becoming established and novel as-sociations between knowledge, morality, and political power were being formed, the reaction was also in the making. As the favored knowledge and experience gradually became more common among newer generations, the number of contenders for the prestigious posts increased. New actors possessing the exalted qualifications of the era began to assert their wish to enter the state field, and the "game of distinction" turned more difficult to play for the bureaucratic elites. As illustrated by Şerif Mardin's unsurpassed work on the topic, the frustration of these new generations with the political monopoly of the earlier generation lies at the heart of the Young Ottoman movement born in the 1860s.[1] And it was the emergence of this movement and the accompanying proliferation of discourses that gave rise to a full-fledged debate on science in the Ottoman Empire. The alternative ways of talking about science that emerged in this period will constitute the focus of this and the next chapter. But before delving into specifics, I will dwell on the broader economic, political, and social context within which the Young Ottomans arose and on the way these conditions shaped the way in which they debated science.

These younger, low-ranking bureaucrats were mostly products of the Translation Bureau; hence, they were by and large proficient in French and familiar with European politics, literature, arts, and sciences. Indeed, they were not dissimilar to the typical Tanzimat elite described earlier in terms of appearance, either. However, they were able to translate their own frustration into a much broader populist intellectual and political movement and to transform themselves into the representatives of a variety of disillusioned groups. Particularly the Muslim community, many members of which experienced the Tanzimat as a process of increasing economic and political regress, found a sympathetic voice in the diatribes of the Young Ottomans—a voice that expressed, in particular, the widespread reservations with the affiliations and lifestyles of the "overly Westernized" statesmen of the Tanzimat Era.[2] The Young Ottomans also allied themselves with the lower ulema and called, with a strongly Islamic tone, for a more participatory government. As aspiring members of the new bureaucratic middle class and "devoted sons" of the Ottoman Empire, the Young Ottomans did not reject the basic principles and goals of Ottomanism, but they strove to construct a more bottom-up ideology using the Islamic idiom.[3]

But while the Young Ottomans did cherish the cosmopolitan nature of Ottoman society and by and large supported "Ottomanness" as the common identity of all subjects, they also highlighted in their discourse the central location of the Muslim community within Ottoman society. Moreover, as their arguments were commonly expressions of the perceived deprivation of Ottoman Muslims in a changing economic and political order, the unique merits of the Islamic tradition and of Muslims constituted a leitmotif of Young Ottoman discourse. Advocating equal Ottoman citizenship regardless of religious affiliation while at the same time presenting the interests of the Muslim community as the ultimate concern did lead to apparent inconsistencies, however, and Young Ottoman contributions to the debate on science were marred by the same tendency, as we will see in this and the next chapter.

The unique role that the Young Ottomans played in the nineteenth-century Ottoman Empire is due, perhaps more than anything else, to their effective use of the press. After all, these young litterateur-activists were the owners and columnists of the first privately owned newspapers in Ottoman Turkish. Thanks to their rhetorical skills and the growing demand for alternative voices, the Young Ottomans were able to turn newspapers and journals into effective outlets for spreading alternative discourses. We see in their writings a harsh criticism of not only the policies of the Tanzimat bureaucrats but also their attitudes toward the "morals and customs of the people." The apparent acquiescence of the bureaucrats of the Sublime Porte to the whims of European statesmen, the overt power of European ambassadors in shaping Ottoman domestic and foreign policy, and the growing economic dependence of the Ottoman Empire on Europe were points they commonly raised.[4] But these points were rendered even more deplorable before the eyes of the Muslim readers with frequent references to the "degenerate" lifestyles of the elites.

Now while these biting criticisms arose in the 1860s, it is also worth noting that the Young Ottomans had, on the whole, no objection of principle to the fundamentals of the official discourse on science and its benefits. The idea of "science as savior" had, by the 1860s, acquired hegemonic status for the participants of the state field, and despite their success as journalists and authors, the ultimate goal of the Young Ottomans was political power. What *was* different in their approach had to do with the way they defined the boundaries of this prestigious type of knowledge called science. More specifically, they wished to expand the category of science-as-prestigious-knowledge in a particular way so as to distinguish themselves and other groups in whose name they spoke as "knowers" as well—"knowers" who were closer to the average literate Muslim Ottoman than the top bureaucrats could ever be.

But this was not all. A second, and arguably more significant, contribution of this generation to the debate on science was their definition of the truly respectable man of science as someone who struck a balance between the Western and the Eastern: a man who was at home with the new arts and sciences of the Europeans but who was also "authentic," that is, respectful toward his coreligionists, well-informed about the Islamic tradition, patriotic, and living in the proper fashion. After all, it was salvos against the super-Westernized lifestyles of the Tanzimat elites that constituted the core of the Young Ottoman discourse.[5] Consequently, the way a man of science lived was also a matter of utmost concern.

It is highly significant in this context that this period is also the one in which the laughingstock of many literary texts and the symbol of "wrong Westernization" in the nineteenth-century Ottoman Empire emerged: şık, or the fop, that is, a Muslim who learned to look, talk, and consume like a European, without any respect for the traditions and religion of Muslim Ottomans or any real knowledge about the topics he discussed— one of which, inevitably, was the benefits of science.[6]

The emergence of the fop as a powerful symbol—and the astounding popularity it gained—is indicative of widespread discontentment with the cultural and political changes brought about by the Tanzimat. The significance of the concept is particularly considerable in the context of the Ottoman debate on science, as the figure of the fop would ultimately come to haunt the debate. Even in texts that were seemingly only about the definition, boundaries, or benefits of science, the shadow of the fop always lurked in the background. Simply put, science was not a topic that could be discussed without any presuppositions or implications regarding the proper characteristics of the man of science.

This is hardly surprising if we remember that European science did not come to the Ottoman Empire on its own, so to speak. The discourse praising science emerged at the same time as the invasion of the Ottoman market by European consumer goods, the signing of treaties that guaranteed equal rights to non-Muslim subjects and the rise of non-Muslim Ottoman and European merchants who took advantage of the opening of Ottoman markets at the expense of Muslim subjects. The fop represented young Muslim men who, within such a context, wished to acquaint themselves with Europeans and live like them.[7] These were men who tended also to hold a public post thanks to family connections and/or some education in the new schools of the empire. That statements about the benefits and significance of science were made by this particular group unavoidably shaped the way alternative discourses were constructed.

Many examples regarding this figure will be presented in subsequent sections, but for one consider two characters from the play *İşte Alafranga* (*This Is "Alla Franca"*),[8] published in 1875—Hasan Bey, a man with a "powdered face, monocle, a very short jacket, a satin or velvet vest, wing-collar shirt and fashionable tie, with a thin cane and gloves in his hand," and his friend Mustafa Bey, who proclaims, while deliberating on himself: "We can no longer sit at our coffee houses like oysters in their shells, can we? I advanced on the path of civilization. Praise be to God, I am almost a *mon-*

sieur now. . . . Yes, yes! I am *scientifiquement géographique, chimiquement* radiant, I now look like a man!"[9]

Hence, European dress alone does not make one a fop: scientific gibberish uttered in French is part and parcel of this character. As a result, the fop is also a most handy device for disciplining those wishing to praise or practice science: these individuals also had to prove that they were not like the many "chemically radiant" Mustafa Beys one could find in the newspapers or plays of the 1860s and 1870s.

While the generation of the Young Ottomans would both bring Islam itself to the fore and introduce the uniquely effective character of the fop into the debate on science, the intellectual inspiration of many members of this generation was a litterateur who paid little respect to tradition and was precisely a product of the Tanzimat. As this seeming paradox in itself provides hints about the Young Ottoman response to the Tanzimat, I will focus on the contributions of this man before discussing the ideas that would characterize the alternative discourse on science more specifically.

A. The Seeds of a Critical Discourse

Ibrahim Şinasi (1826–71) was a young protégé of Mustafa Reşid Pasha, the very architect of the Tanzimat Era. Educated in Paris, where he acquainted himself with orientalists like Ernest Renan, Silvestre de Sacy, and Alphonse Lamartine, he wholeheartedly espoused the discourse on new knowledge and the virtues of its holders. Praising the direction of Tanzimat reforms in a poem he wrote in 1849, Şinasi asserted that the beauty of Europe was now being imported into the Ottoman Empire, turning it into "the envy of Frankish lands."

Mustafa Reşid, to Şinasi, was the chief enlightener of the Ottoman realms. Thanks to his patron, to whom Şinasi also referred as the "prophet of civilization," he joined the ranks of the enlighteners and became a member of the Council of Education in 1855. But he never rose further, as his patron was replaced by Ali and Fuad Pashas, and he failed to build the state career for which his education had prepared him. In the poems he wrote in the 1850s, he continued to praise Mustafa Reşid with verses referring to the "new knowledge":

Your justness and generosity could not be measured by the likes of Newton
Your reason and intelligence could not be grasped by the likes of Plato . . .
We were slaves to oppression, you freed us
It was our ignorance, the chain binding us.[10]

In a similar way, however, he also expressed his career exasperation: "O, president of the republic of virtuous people / Do I deserve to remain enslaved by men of ignorance?"

This advocate of new knowledge even wrote a couplet that hinted at the question of religious ideas conflicting with science: "Don't sell me worn-out Jewish beliefs, sir / How can I 'buy old stuff' with this new enlightenment!"[11] We do not see an elaboration or reiteration of this particular idea in Şinasi's other writings, and the Young Ottomans who appreciated Şinasi did not make similar pronouncements, at least not in this tone. Şinasi's effort to redefine the boundaries of religion and science—which likens him to the top bureaucrats of the Tanzimat—is evident: what does not pass the test of this new knowledge can no longer be labeled Islam; it is a "worn-out Jewish belief." This is an early illustration of a critical transformation in Islamic exegesis in the nineteenth century. While stories borrowed from the Old and New Testaments collectively known as *Israiliyyat* were a legitimate ingredient of the Islamic tradition, they were gradually thrust aside in the nineteenth and twentieth centuries by reformist Muslim authors, particularly due to their conflict with science.[12] Islam was indeed made central to the Young Ottoman discourse, but this was an Islam shaped by concerns regarding scientificity.

This emphasis on the merits of new knowledge and the person who possesses it can also be seen as a more general expression of the self-confidence of a new generation—a generation that is less likely to be "fooled" by the authoritative ideas of the past. We can see examples of this self-confidence based on a new kind of enlightenment also in Şinasi's journalistic output—works of particular importance, as in 1860, when Şinasi and his partner Agah Efendi became the publishers of the first Ottoman newspaper owned by Muslim entrepreneurs, the *Tercüman-ı Ahval* (Interpreter of Conditions). In the famous first editorial column of the newspaper, Şinasi wrote that it was certainly the right of the people, who were legally required to fulfill so many duties, to express their opinions on the state of their country. One only needed to peruse the political newspapers of the "civilized nations whose eyes were opened thanks to education" to be convinced of this fact.[13] The newspaper that he started on his own two years later, the *Tasvir-i Efkâr* (Description of Ideas), was introduced with a similar column. Şinasi made his aim clearer, though, when he wrote that newspapers revealed in civilized nations what the people regarded as "the way to achieve their interests," and his newspaper would be devoted to "news and education."[14]

Education was thus the tool that would help Ottomans achieve their goals, and Şinasi would enlighten his readers. But it was in his clarification of the essence of this enlightenment that this assertive, aspiring journalist came up with what we can call the "Young Ottoman synthesis." As he stated in a column he wrote two years later, the ultimate goal was "to combine the sage mind of Asia with the virgin thought of Europe."[15] New knowledge was thus to come from Europe, but it would be made sense of and put to use by the mature, responsible, wise Asian or Oriental. It is interesting to note that the generation of novel ideas itself is left to the Europeans in this proposal; the thoughts produced by Europeans are then to be imported and interpreted by the Asians. Also significant are the gendered metaphors Şinasi uses: in line with the central discourse of Ottoman Westernization, Europe is represented as a young female whose allure should be responded to, but with caution, by Asia, the old, wise man.

The fop, in this portrayal, can be seen as an in-between, effeminate character. Indeed, there are many condescending references to the delicate nature and chic appearance of the fop in Ottoman literature. As will be illustrated in the following sections, we see in late nineteenth-century Ottoman novels representations of the correctly Westernized, true man of science as a manly character as opposed to the delicate fop.

In any event, what we thus see in Şinasi, a "champion of Westernization"[16] and a "humanistic rationalist,"[17] is an association of fresh, innovative knowledge with Europe, and a mature line of conduct with Asia, and specifically, the Ottoman realm. This is a remarkable illustration of how the discourse on the best synthesis of the West and the East defined and, in a sense, froze these categories. This discourse stipulating that science is the West and wisdom the East not only fixed the "Orientals" in a position of constant importer, rather than producer, of new ideas, but also imposed on them the rather heavy responsibility of being the virtuous ones. The hierarchical social relationship that this virtuousness would help reproduce is the crucial aspect of this discourse that an exclusive emphasis on science fails to notice. Furthermore, the emphasis on synthesis almost necessitates the existence of a particular elite group that can delineate the ideal society and supervise the transfer of ideas into the Ottoman Empire or the East in general. It is in a sense the duty of this elite group to impose caution and establish a mature line of conduct among the people. Hence, the scientistic elitism of Ottoman bureaucrats was countered essentially by a different kind of elitism—one that we may refer to as moralistic elitism.

"The people" that the litterateur/activists of the 1860s imagined as their audience—and ally—is one that is better informed and vigilant about their rights vis-à-vis the high-ranking wielders of political power.

It is also one that is more industrious and productive, again thanks to the knowledge it possesses. But it is also a mature people composed of individuals with high moral qualities, as defined by what the authors would designate as "our values." In this respect the Young Ottoman contributions to the official discourse on science are, first, the emphasis on complementarity and synthesis (East and West, wisdom and science, tradition and novelty, values and knowledge), which, simultaneously, defines these categories as mutually exclusive and distinct, and second, the characterization of science as what enables productivity. The essentiality of the role of the enlightener is not questioned: the importation of Western science *should* take place, but under the control of gatekeepers who know both the virtues and vices of Europe. If the elite statesmen of the era, due to their political ambition, ignore those vices, then it is the role of the litterateurs to spot them and, in a sense, *better* enlighten "the people."

B. The New Themes of a New Debate

1. What Is Civilization?

Characterizations of civilization as a unitary, transcendent entity, best defined by a similarly timeless essence named "science," were common in the *Mecmua-i Fünûn* in particular and the Tanzimat discourse in general. Starting in the 1860s, these portrayals were challenged from a variety of angles. Some, like Şinasi, imagined a perfect civilization as the outcome of a synthesis between European knowledge and Oriental virtue. Others did not necessarily denounce the idea of a common civilization of humanity, but argued that the fundamental (i.e., scientific) contributions of Muslims to this civilization have been ignored for too long. As we shall see below, it was approaches of this particular sort that were commonly espoused by the Young Ottomans and would come to dominate the debate in the 1860s and 1870s. But we should also note the existence, albeit relatively ineffective, of another approach that resolutely saw civilizations, rather than Civilization, and highlighted the unique merits of the Islamic civilization—a position best illustrated by the arguments of Hayreddin Bey (Karol Karski), a Polish convert who published essays in various newspapers of the period.

In an essay entitled "Civilization and Turkey" and published in the newspaper *Terakki*, Hayreddin asserted his astonishment at the ease with which Turks surrendered to accusations of being uncivilized. The essence of civilization was community life and ties of mutual assistance and re-spect, which the Turks possessed perhaps more than the so-called civilized nations. He proclaimed:

Ottomans! No nation is superior to you in these respects, and I assure you, if some-body attempts to claim otherwise, you should respond that they are unaware of what civilization is. Civilization must not be confused with sciences and industry and ma-chines. Certainly, sciences and industry and machines are desirable powers. But they are material forces . . . whereas civilization is the totality of human virtues.[18]

Even though ancient Greece did not have access to scientific knowl-edge and its products, its people were more civilized than contemporary Europeans. Russians were stronger in terms of material forces, but they were less civilized than the Ottomans. Ottomans *should* learn the sci-ences, Hayreddin wrote, but in order to acquire the material power that their civilization deserved, rather than mistaking them for civilization itself. The author of a letter to the newspaper *Basiret* made a similar com-ment: "Our nation is civilized by nature. But is it only Ottomans, that is, Turks, who are innately civilized? This virtue is, in essence, but one of the endless virtues that result from being honored by the religion of Islam."[19]

This critical attitude toward Eurocentric conceptions of civilization—conceptions that were quite common among the Tanzimat elite, in-cluding Münif Pasha—was based not only on the broad lack of trust to-ward European powers but also on the attitudes of the Europeans who resided in the Ottoman Empire. When the Ottoman government banned smoking in Istanbul ferries during the holy month of Ramadan, for in-stance, a significant diplomatic crisis ensued.[20] Similarly, Ali Efendi—the publisher of the daily *Basiret*, one of the most successful enterprises in the history of the Ottoman press—complained that even when the ferries were packed, Europeans would let their pets sit on the seats, rather than "barbaric Turks." "We do not need a civilization that makes one consider his own kind less worthy than animals," he wrote.[21]

But probably the most upfront criticism of the understanding of civili-zation as material progress based on the new sciences was in another essay published in the *Basiret*. After arguing that civilization should be divided into "civilized progress"—that is, material products of civilization—and "civilized thought," the author stated:

If the progress of civilization has produced as its unique benefit the Armstrong can-non, it [has brought forth] only the word 'humanity' to counterbalance it. Therefore, civilized thought can be said to keep up with civilized progress only when the signifi-cance of humanity is as considerable for Europe as that of the Armstrong.

Imagine a simple community where people live in peace, the author continued. If they stayed this way, they would enjoy constant bliss and harmony.

But they are not satisfied with being protected by the Protector of humanity, and instead they place an Armstrong on the walls and bastions of their town. Each time these [cannons] open their mouths . . . , they utter words like "Disperse this commu-nity; because in these plains we shall fight the enemies of humanity who are envious of your bliss. We shall flatten these mountains and hills and with their debris construct new mountains in the plains you see now. We shall shake and demolish this earth from its roots. But know also that this zealous effort of ours is based on a sincere desire to preserve your happiness."

. . . But isn't it those very Armstrongs, the protectors of humanity, that invite other Armstrongs that are the annihilator of humanity and civilization? Civilized progress builds railways . . . , extends telegraph lines. It is only later that we find out that these trains were in fact for transporting our beloved children to the slaughterhouse one hour earlier, and the telegraphs were for bringing us bad news from them as soon as possible. What bliss![22]

It was necessary to import the sciences and arts of Europe, the author argued. But it was also true that Islamic teaching could not be confined, like other religions, to some commands on the safety and well-being of humanity; Islam also ordered that whatever prevents these should be ban-ished, unlike those that suggested turning the other cheek. Civilization, in short, should be evaluated according to Islamic, rather than European, criteria, and what does not pass the test should not be allowed to pass the borders.

But in a period where particularly the military might of the "civilized world" and the new types of knowledge that made this might possible were discussed with awe even by the critics of European policies, such wholesale criticisms of the consequences of material progress were hard to come by. Similarly, verses informed by Islamic mysticism like Muallim Naci's "Can those who learn the sciences also know the mysteries? / It is he who surrenders his self who knows the mysteries" were not particularly common in the poetry of the period.[23] The more viable alternative to the *Mecmua-i Fünûn*'s portrayals of civilization and science emerged rather in the form of a discourse that insisted on explicitly referring to Islam and morals while praising science or civilization.

2. Islam, Science, Civilization: Toward the Grand Synthesis

While its closeness with the powers that be and the conformism of its editors did not fare particularly well with the dissatisfied activists, the *Mecmua-i Fünûn* was nevertheless quite popular among the new generation of litterateurs. Ebuzziya Tevfik, a member of this generation, wrote in 1910 that "the youths who enjoyed reading [the journal] would carry it with them like an amulet" and virtually memorize its contents.[24] Classic studies on the era rightly emphasize the lasting influence of the journal on future publications.[25] But it is also a fact that the journal's cosmopolitanism and the tone of its contents that referred to civilization with a clearly Eurocentric focus were not entirely appealing. In a context where the policies and consumption patterns of the Tanzimat bureaucrats who were also the main patrons of the journal raised considerable dissatisfaction particularly among the Muslim community, civilization and science proved to be topics that could not be discussed without references to Islam. Neither subtle discussions on Islam's endorsement of science nor lip service to Muslim contributions to civilization were sufficient.

A striking example of this was the series of articles that the *Tasvir-i Efkâr* published between July 6 and August 24, 1866.[26] The author, Mehmed Mansur, a convert from Macedonia and teacher of French at the Translation Bureau, was infuriated by the oft-made argument that the glorious Library of Alexandria had been destroyed during the Muslim conquest of Egypt.[27] Strikingly, an article published in the July 1863 issue of the *Mecmua-i Fünûn* had also committed this error, according to Mansur. In this article on the history of books, Alexander Karatheodori indeed made a passing remark on the devastating impact of war on libraries, particularly in periods preceding the invention of the printing press, and referred to the case of the Library of Alexandria. While he avoided specifying the culprit in this controversial incident, he also noted that "after the invasion of Egypt by the Arabs, the connection that previously existed between Europe and Egypt was absolutely cut off, and as this also hampered papyrus trade, it became one additional reason behind the disappearance of ancient books."[28] This was no different than saying that had Islam not arisen, Greek masterpieces would not have been lost, Mansur asserted. After a detailed criticism of the view that it was Muslims who had destroyed the Library of Alexandria, Mansur stated that the relevant information could be found even in the works of the Greek historian Constantine Paparrigopoulos—"books in the author's own language." The less-than-subtle reference to Karatheodori's communal affiliation from a convert himself is hard to miss.

But it was not simply a particular author who was to blame. Indeed, "errors" of this sort had turned the *Mecmua-i Fünûn* into a means to "spread false rumors," Mansur proclaimed, and presented additional evidence in the revised version of his treatise. Apparently, "the authors of the journal were not satisfied with this, and as if only to confirm and advocate the false allegations of this author," they published another piece, this time in their forty-fifth issue.[29] Interestingly enough, this piece, entitled "Islamic Libraries," was one of the very few articles directly related to the Islamic tradition that the *Mecmua-i Fünûn* had ever published. Furthermore, it was overall a very complimentary account of the flourishing of sciences after the advent of Islam, and it referred to the destruction of books only by rebellious slaves and the invading Tartars.[30] Yet the allegation that the books looted by the slaves were later abandoned and remained buried under the sand for decades was unacceptable for Mansur: "Not a single person from within the Muslim community other than the authors of this article has ever claimed that the Muslims were ignorant enough to fail to appreciate the books written in their own language, and leave them under sand and dirt."[31]

Namık Kemal, the most prominent member of the Young Ottoman movement, published a flattering review of Mansur's treatise in the *Tasvir-i Efkâr*.[32] "Arab heroes" knew that the acquisition of knowledge was a command of Islam, he wrote, and their translations of the works of prior civilizations provided evidence for their "service to and esteem for science."[33] Yet while the false rumors about the destruction of the Library of Alexandria by the Arabs were no longer accepted even by many European scholars, they were constantly referred to by "those fanatics—the so-called enemies of fanaticism—who have an affection for Christianity or are set to betray Islam." Moreover, these same authors never mentioned the destruction of the Islamic treasures of knowledge in the Moorish Kingdom by the Christians. Mansur's excellent work would be the definitive refutation of the false rumors, Namık Kemal proclaimed.

Ali Suavi, another Young Ottoman and one of the most eccentric characters in nineteenth-century Ottoman history, went further than the other critics, and in another context referred to Münif Pasha, the editor of the *Mecmua-i Fünûn*, as a man who had, "by translating Protestants' raving attacks on Islam, revealed his apostasy and got loathed by the nation."[34]

Therefore, the reaction to the Eurocentric account of civilization and science that the *Mecmua-i Fünûn* propagated was an emphasis on "Islamic contributions" that, in effect, was built on the acceptance of the broader narrative. In this alternative approach, it was compulsory to state that Islam instructed and guided the acquisition of knowledge and that Muslims had for many centuries followed this order. Hence, in an article on the teaching of medicine in the Ottoman Empire, the *Tasvir-i Efkâr* referred to "the Arabic people who constructed a compound out of the transmitted and the intellectual sciences and gave a fresh life to humanity."[35] It added, however, that the Ottomans had not made as much progress as the Arabs in any branch of knowledge. In the first issue of the Young Ottoman newspaper *Hürriyet*, on the other hand, Namık Kemal asked, "Aren't the Turks the nation at whose *medrese*s the likes of Farabi, Ibn Sina, Gazali and Zemahşeri augmented knowledge? . . . Such a community that had once gained the title 'teacher of the world' now looks at the simplest product of knowledge and is fascinated as if it had observed a miracle."[36]

As we shall see below, references to the Moorish Kingdom of Spain or, in short, Andalusia, were commonly presented as the most convincing type of evidence for the claim that Muslims had made significant contributions to science. Münif Pasha did acknowledge the contributions of Muslims in his articles, but his emphasis was on novel developments, and, as discussed above, he did not address all branches of knowledge that Muslims had developed like the Young Ottomans now would.

Indeed, in addition to their uncompromising calls for inserting overt references to Islam in any discussion on science and civilization, the Young Ottomans also challenged the way "needed sciences" were delineated by the Tanzimat discourse. More specifically, they defined Islamic branches of learning such as *fiqh* (Islamic jurisprudence) as needed sciences as well. The Ottoman Empire was not simply to import new types of knowledge, it was to affirm and revive its own scholarly tradition, and this tradition was that of the *Muslim* community of the empire. Furthermore, the Young Ottomans also made it clear that arguments about science and civilization could not but have implications about the type of ruler and subject that the empire truly needed. In other words, for the critics of the Tanzimat, too, it was impossible to talk about knowledge without talking about virtue and vice.

These positions of the Young Ottomans can be observed together in their arguments over the "decline" of Muslim states in general and the Ottoman Empire in particular. As we have seen in previous chapters, the emergence of a new type of knowledge in Europe had been commonly referred to earlier as one significant reason. Vague phrases such as the one used in the Imperial Decree of 1839, "certain disturbances," were also commonly used by the bureaucrats, including Münif Pasha himself.[37] The key word for the critics, on the other hand, was "Andalusia."

While references to the Islamic state in Spain were not common in Ottoman texts before this period, they became ubiquitous after Ziya Pasha, a leading member of the Young Ottoman movement, translated sections from Louis Viardot's *Essai sur L'Histoire des Arabes et des Mores d'Espagne* (1833) with additional material from other sources and published it under the title *The History of Andalusia* in 1859.[38] The fourth volume of Ziya Pasha's work was dedicated to showing how Muslims introduced civilization to Spain and provided a long list of the "men of virtue" of Andalusia. The list contained scholars who contributed to a great variety of fields, including not only specifically Islamic branches of learning such as exegesis and *fiqh* but also fields like pharmacy and mathematics. It is worth underlining that this presentation was clearly based on the Islamic tradition that does not separate religion from specific fields of knowledge production. Also note that, based on this approach, Ziya Paşa was able to refer to these scholars as men of virtue.

After this publication in particular, the significance of the achievements of the Muslim scholars of the Moorish Kingdom—which, apparently, Ottoman authors learned mostly from European sources themselves—became a justification for emphasizing the Islamic element in the pursuit of knowledge. Andalusia proved that science flourished in a society where Islam guided knowledge production and government was based on Islamic law, with the science regarding this law (*fiqh*) protected by just rulers.

Thus, the discourse on the need for acquiring the new sciences that so far had been monopolized by the Europhile elites of the Tanzimat was now being infiltrated with Islamic references and, in a sense, popularized. But what was also remarkable about the way the Andalusia narrative operated in the arguments of the Young Ottomans was that it was in essence a parable: the tragic fall of such a glorious kingdom was portrayed as an example that demonstrated how the spread of degeneracy and vice among the rulers facilitated the work of the invaders waiting on the other side of the border.

The political cause of the Young Ottomans involved the termination of the despotic rule of the "enlightened" top bureaucrats and the expansion of the limits of political participation. It involved at the same time the expansion of the borders of science by reasserting the worth and applicability of Islamic sciences. But it also redefined the ideal Ottoman man of knowledge quite strictly. In an essay where he addressed the European critics of the Ottoman political regime, Namık Kemal stated that Ottomans wanted a parliament as well, and this was what Islam stipulated anyway, but he also delineated this ideal Ottoman in unequivocal terms:

You still declare our religion an obstacle to progress. Is progress possible under the tyranny of a few people? . . . Wasn't it Islam that preserved the glories of civilization after the decline of the Romans? Wasn't it Islam that advanced and revived rational knowledge? Some wise men among you cry, "The Arabs of Andalusia were the teachers of knowledge to Europe"; weren't they Muslim? If what you think is civilization is women going out in immodest dress and dancing in gatherings, that is at odds with our morals. We do not want that, we do not want that, we do not want that a thousand times.[39]

3. Redefining Ignorance

Namık Kemal made similar points in many of his essays, especially in those he wrote for *Hürriyet*, the newspaper Young Ottomans published during their self-imposed exile in Paris. European powers, admittedly, had a significant impact on the dealings of the Ottoman government, and, in order to gain their favors, the bureaucrats of the Tanzimat had invented the notion that religious fanaticism reigned in the Ottoman Empire, Kemal wrote. Yet "if the glory, prosperity and erudition of the Muslim world in Damascus, Baghdad, Egypt and Andalusia [were] taken into account," it would be impossible to argue that the religion of Islam impeded progress.[40] But the Andalusia example also proved the merits of the sharia, as this Muslim kingdom had become "the envy of the world" while it abided by this law.[41] Needless to say, the call for sustained commitment to the sharia also meant that the Islamic science of *fiqh* should be central to lawmaking in the Ottoman Empire.[42]

Thus, the boundaries of the category "needed sciences" as defined by Namık Kemal were certainly large enough to include the traditional Islamic sciences as well—the sciences that the *Mecmua-i Fünûn* had little to say about. This, as can be expected, also had important implications for the discourse on ignorance that had been developed in the earlier decades. "Our current cabinet ministers have read everything they have ever read in European languages," Namık Kemal stated, "so they are in effect ignorant Europeans who happen to know Turkish." People in power in the Ottoman Empire were, in Namık Kemal's words, "a bunch of scoundrels [who were] ignorant even of the catechism [of Islam]."[43] Ziya Pasha expressed the attitude of the Young Ottomans toward the top bureaucrats in his celebrated verses:

The zealous man is now accused of fanaticism;
Attributing wisdom to the irreligious, this is now the fashion.
Islam, they say, is a stumbling block to the progress of the state;
This story was not known before, now it is the fashion.
Forgetting our religious loyalty in all our affairs,
Allegiance to Frankish ideas is now the fashion.[44]

This insistence on the prestige and relevance of Islamic sciences thus involved a challenge to the new definition of ignorance that the official discourse of the early nineteenth century had established. For example, in the verse introduction to his anthology of Turkish, Arabic, and Persian poetry, the *Harabat*, Ziya Pasha wrote that ignorance and poetry could not exist together. The branches of knowledge he deemed essential, however, were figures of speech (*bedi*), comparison and metaphor (*beyan*), syntax (*nahiv*), eloquence (*fesahat*), and history, especially of poetry. In order to understand the world, a poet also needed to learn a European language: "That is where the sciences progressed / Do not keep yourself away from learning them." Learning foreign languages and new knowledges would not turn one into an infidel, he wrote, but, as can be expected, with an overt warning: "Acquire their arts and sciences / Leave behind their customs and vices / Forget not your essence with imitation / Do not despise your own nation."[45]

Strikingly, Ziya Pasha's recommendations appear less comprehensive than the way Fuzuli, a celebrated sixteenth-century Ottoman poet, had discussed his path to good poetry:

I regarded as unfaithfulness to keep my poetry devoid of the jewel of knowledge (*ilm*), and I detested poetry without knowledge, which is a mass without a soul. Hence I devoted part of my life to acquiring the rational and mathematical sciences. I placed in my verses pearls from a variety of masteries. . . . I then studied exegeses and the hadith and was convinced that the virtue that is poetry cannot be deplored.[46]

That Fuzuli was more resolute on the importance of "rational and mathematical sciences" than Ziya Pasha may be indicative of the transformation of the Ottoman conception of knowledge and science: in Fuzuli's world, the types of learning in question were still taught in Ottoman *medrese*s and deemed "native." In the nineteenth century, they had become European sciences that were rendered less foreign only with reference to Andalusia. Ali Suavi, on the other hand, blended in his own reaction to the discourse on ignorance a defense of both Islam and the Turkish community itself. Not only had Muslims made the greatest contributions to science; Turks themselves had excelled in all branches of science, according to Suavi. Indeed, great men of science mistakenly known as Arabs were actually Turkish. Ibn Sina and Farabi, two Turkish scholars, had made an enormous impact on contemporary Western science, and Turkish authors had surpassed even the Arabs in literature in Arabic. "It is true," Suavi wrote, "that with us arguments on nature were based on conjectural premises rather than experiments. . . . [Yet] the advanced physics of today has done nothing but confirm the results of those arguments and demonstrate them with experiments."[47] According to Suavi's interpretation, the Turks had produced proud representatives in all sciences: astronomy, economy, arithmetic, philosophy, natural sciences, as well as *fiqh*, and learning spread from the mosques and *medrese*s of the Ottoman Empire.

Suavi concluded, "It is nothing but buffoonery that some children in Istanbul and elsewhere assume that Turks are in ignorance and dare to give advice like aged men with some big words. It would be more in accordance with respect and reason that those types read and learn from their compatriots whom they are daring to advise."[48]

The idea of ignorance propagated by the Tanzimat bureaucrats was challenged also by the ulema, in an instance that also exemplifies the way those speaking in the name of the new sciences were commonly criticized. In 1872–73, the young journalist Ahmed Midhat published a series of peculiar essays in his journal *Dağarcık* (The Knapsack), on the emergence of humans on earth. Midhat's familiarity with the popular geological debates of his time is apparent; in one essay, he described not only the formation of the Earth but also the emergence of humans within an essentially materialistic framework, and he argued that the particles that make up each human being are pieces of the matter that has been in circulation since the formation of the Earth itself.[49] The sequel to this essay was even more remarkable in terms of its approach to the meaning of human existence, as it involved a conversation between the author and the voice of a brick in the wall of his room, which stated that it was once a human being:

I was born just like you. That is, my body had boiled for thousands of years among the minerals that had been burning and boiling in the center of the Earth, then turned into steam, intensified, and solidified, then remained within each layer of the Earth, broken apart among thousands of objects, and finally assembled within a drop of semen that dropped into my mother's uterus. All in all, I was present on Earth just like you, and lived for forty-five years.[50]

The voice then discusses how, after he died as a human, his body mixed with other matter in soil to later become the material used to make the brick, and tells Midhat that this is the way all beings, including humans, animals, and plants, come into existence.

The circulation of matter in nature had been a prevalent theme in European science since the eighteenth century, especially in relation to the work of Antoine-Laurent Lavoisier and Joseph Priestley. But it acquired additional popularity in the mid-nineteenth century, along with the developments in geology and the growth of organic chemistry, particularly with the studies of Justus von Liebig—studies that had also influenced the materialisms of Jacob Moleschott and Ludwig Büchner.[51] Midhat was clearly familiar with these discussions to some extent, and even though he did refer several times to the notion of soul in these texts, the almost exclusive emphasis on material processes in this discussion on existence is striking.

The voice from the wall that talks to Midhat in this essay is also clear about how the truth regarding their own myriad transformations can be best understood by the living: through science. Indeed, the reason it chose to tell this story to this young author was that Midhat was "a man interested in the curiosities and wonders of science," unlike most people, to whom he is implied to be superior.[52]

Science as a way of truly understanding the essence of all existence is a common theme of these essays. And as it was obviously clear to him that such scientific explanations could confuse the religious believer, Midhat tackled the question head-on, in another essay that came right after his dialogue with the wall. "Do not censure me for my thought," Midhat asserted in this essay. "My thought is ready to proclaim that any science that is not compatible with the Holy Qur'an is a lie. . . . But I have not found that the Qur'an is in conflict with and opposition to any science that I have applied to it."[53] Interpreting the ideas discussed in the previous essay with reference to verses from the Qur'an, Midhat noted that these ideas were no different than what the Qur'an taught. The only question that remained outside the scope of naturalistic philosophies was that of Judgment Day, which, if interpreted by a Muslim philosopher, would not but amaze the naturalists. In other words, while the authority of the Qur'an was uncontestable, Midhat suggested that the Qur'an and the findings and philosophies of science could help explain one another, even though the contents of the former could never be reduced to those of the latter.

Following these was yet another essay on a controversial matter, the emergence of man. In this text, Midhat offered a Lamarckian perspective—

note that Darwin's work remained unpopular in France for decades, and Ottoman authors were much more familiar with French debates—and argued that there was no reason to find the idea of a connection between apes and humans objectionable.[54] With many references to a French commentary on Lamarck, Midhat noted that orangutans' similarity to humans was undeniable, but also added in a confident way that rather than assuming that humans had somehow descended from orangutans, it would be more reasonable to assume that humans were simply a type of ape. For Midhat, thus, Lamarckian ideas on evolution were not satisfying; humans had emerged on earth simply as a type of bipedal ape, and gradually—thanks to the interaction and competition among them—acquired the skills contemporary humans have.

That Midhat's discussion makes no reference to a creator even in its discussion of the shortcomings of Lamarck's theory can be striking to the contemporary reader, yet the expectation that ideas on a topic like evolution can only be perceived in terms of a conflict with religion may be misleading. While Midhat's tone has an unusual kind of straightforwardness in its presentation of such consequential matters, the reactions to his writings, and his response to them, cannot be understood within a simplistic "religion versus science" framework.

Granted, the harshest reaction to Midhat came from a member of the religious elite. Harputlu İshak Hoca, a member of the ulema, anonymously published a scathing criticism in the *Basiret* on March 4, 1873. However, İshak framed his argument not simply as a comment on particular ideas about science or Islam, but also as a condemnation of individuals of a certain type: those "with bizarre opinions [who] have emerged in . . . the center of the Caliphate that worry and confuse the community . . . with their words, writings, attitudes and actions."[55] The views of the publisher of *Dağarcık* were in contradiction with the beliefs of all "people of the book,"[56] according to İshak, implying that Midhat was an atheist. But this was certainly not all; the problem was people with views that would weaken the state and the nation. Indeed, close to the conclusion of his essay, İshak's arguments took an overtly political turn with references to the story of the Paris Commune itself. Those involved with such dangerous views and actions had been eliminated from all countries of Europe, but while Islam protected the Ottoman state from such dangers, some "ignorant and feeble" members of the Ottoman community mistook those views for "civilization."[57] The fundamental question was, thus, not what Midhat or people like him wrote about scientific views on the origins of humanity; the existing political and social hierarchies within the Ottoman Empire were at stake.

In a note published in the *Basiret* the next day, Ahmed Midhat asked his critic to reveal his identity. The answer came in the form of an essay where İshak not only revealed his name but elaborated his views on different types of people within the empire. Some, according to İshak, referred to knowledge and learning, yet what they offered was but a bunch of false notions that could "misguide and offend the beliefs of common people." In such cases, the ulema were to intervene, as it was their duty to protect the religion, honor, and dignity of the people. Defining his role in these terms, İshak suggested that Midhat instead might simply be after "selling words and making money."[58]

This is a fleeting insinuation in a long essay, yet it is a clear indication of an important dimension of the conflict at hand: as Bourdieu noted, it takes a particular type of recognized authority (symbolic capital) for the actions of social actors to be perceived as disinterested.[59] With his reference to the likely differences between his objectives and those of his opponent, İshak, the seventy-year-old member of the ulema, opposes his young opponent's claim for symbolic capital. In İshak's formulation, the likes of Midhat sought political and economic power ("weakening the state and the nation," "selling words to make money") while İshak and the ulema on whose behalf he claimed to speak represented altruism and selfless service to the people. As we will see in the following chapters, this strategy was used commonly in the Ottoman debate on science, as arguments about science and religion, the new and the old, the beneficial and the dangerous, were ultimately arguments about authority—cultural, political, economic—and its legitimation.

Not only were İshak's opponents seekers of petty, if not perilous, goals; they were the truly ignorant ones. "When our pseudo-philosophers interact with the Europeans," İshak wrote,

> they take a needless trouble and discuss issues pertaining to religion and faith. They not only do not really know what they think they know, but they imagine and present themselves as learned men. As a result, when their addressee utters ridiculous words criticizing religion, these hopeless types are unable to respond. In order to relieve themselves of embarrassment, they then attribute their own ignorance to the entire nation . . . and cry "oh, our nation is in darkness; what are we to do?"[60]

The criticisms of Harputlu İshak, then, are directed against a group, not at Midhat alone. His diatribe is an attempt to reclaim the authority to label others as ignorant, revitalizing the discourse within which ignorance is quintessentially a religious concept, with epistemological but also moral implications. Midhat's arguments are unforgivable for İshak, but this is ultimately a matter of membership in the moral community. Hence, he concludes his remarks by making it clear where the likes of Midhat *do not* belong: "This country's thirty-five million people of the book, Muslim and non-Muslim, . . . the Protector of the state and religion and the possessor of the throne of the Caliphate, and the great men of the state share the same view and faith with us, whereas the publisher of the *Dağarcık* has nothing other than his journal and his cane."[61]

Accused of being not only ignorant but also a "cane-carrying" enemy of the Ottoman people and the state—which is particularly significant as canes were commonly regarded as accessories that the fops of Istanbul carried—and declared an outcast, Ahmed Midhat could not but directly address the issue of religion and the moral community. In his confident commentary, Ahmed Midhat elaborated on how his arguments had nothing to do with blasphemy and justified his analysis with references to the Qur'an and examples from the works of early Muslim scholars.[62] Islam, to Midhat, was the religion that encouraged the search for knowledge, the religion to which the admirers of the progress of science should be grateful. Unfortunately, Midhat wrote, the advocates of "old views" despised everything that came from Europe, even "European sciences."[63] What made matters worse, Midhat argued, were those who visited Europe without having received a sound Islamic education. Due to this ignorance, when these people witnessed the amazing products of the "new sciences and industries," they came to think that Islam was indeed an obstacle against progress, especially when they heard the hateful arguments of the advocates of "old views."

What was to be done, then?

[The problem] stems from the fact that the parties are ignorant of the sciences the other party possesses. Hence, the purpose should be introducing the two sides to one another, and showing to the holders of old views that European progress is a product not of infidel inventions (as it is believed to be) but, on the contrary, of old Muslims and that it is essential to appropriate it. Similarly, the holders of new views should be shown that progress and civilization are characteristics of Islam itself rather than the latter being an obstacle to them. This would make the terms "old views" and "new views" obsolete, and all [children of the Empire] would work together for bliss.[64]

The ideal Ottoman was thus defined by Midhat, in a way similar to the Young Ottoman formulation, as a person who did appreciate and learn the new sciences but who also knew that they were based on the findings of Muslim scholars, who, in turn, owed their achievements to the guidance of Islam. This guidance was to shape the conduct of contemporary

Ottomans as well. This solution did not save Midhat, however, as he was sent into exile along with the Young Ottomans soon afterward.

But Midhat's synthesis would survive. The fusion of the two discourses as characterized by his arguments yielded two sorts of science to be known and two sorts of ignorance to be eradicated in order for the Ottoman Empire to stay intact. Turning the tables on Tanzimat bureaucrats, the generation of the Young Ottomans redefined science, knowledge, and ignorance in a way that tied in with their own political agendas, and this approach gained many adherents in the following years. Science was to include both European and Islamic sciences (now increasingly defined as distinct), and one needed to have at least some knowledge of both to be able to call oneself knowledgeable. Furthermore, he who would deserve to speak in the name of science had to be a moral individual, unlike the Europhile bureaucrats of the Tanzimat and the fops of Istanbul, and this morality was based on Islam, not a simple consequence of learning some sciences.

Namık Kemal's take on the situation in the Ottoman Empire is also telling in this respect. To Kemal, the case at hand resembled a clash between a father and a son. The son, astounded by the beauty and prosperity of Europe but ignorant of the history and merits of his own country, came to the conclusion that everything European had to be imitated. The father, on the other hand, failed to understand that the way to transmit religious and moral values was to consciously teach them, not simply act on them. As a result, Ottomans had ended up imitating the depravities of the Europeans as well, rather than importing only the things that were truly needed, the "sciences and industries."[65] The representation of the enthusiast of European ways as immature and impressionable and the traditional Ottoman as a fatherly figure is a theme that became increasingly dominant in the following decades. Moreover, Namık Kemal's suggestions on how "the son" should be raised are strikingly reminiscent of the official discourse and policy during the reign of Abdülhamid II that chapters 6 and 7 will focus on.

A reference to the ideal man as represented by the generation of the Young Ottomans can also be found in Abdülhak Hamid's play *Sabr ü Sebat* (*Patience and Perseverance*), written in 1875. In a key scene, Mün'im Efendi meets Müyesser Bey, a young man his brother is considering as a potential son-in-law. Mün'im is glad to hear that Müyesser works at the Translation Bureau and has spent a few years in Paris, stating that his brother wanted a son-in-law who possessed new knowledge. After asking Müyesser about the schools of Paris and receiving a praising answer, Mün'im starts his lecture:

The thing about the Europeans that is to be imitated is their methods of instruction. Young men like yourself should see as a model those men of knowledge who reach perfection in those well-organized schools of Europe. Otherwise, what could be the consequence of acting like our fops today, and imitating the behaviors, accessories and clothes, prodigal customs of the Franks that they themselves refuse to appreciate, other than being unaware of one's own true nature and failing to preserve one's nation?[66]

C. Knowledge, Ignorance, and the Community

While their understanding of ignorance was thus broader, the generation of the Young Ottomans did not have any objection to the notion that ignorance was a chief cause of the hardships the empire had been going through. They argued that just as the welfare of an individual is dependent on his intelligence, a nation could not survive without knowledge: "The sole reason behind the progress of European states is knowledge and learning and the cause of our backwardness is ignorance and unawareness."[67]

But the Young Ottoman approach to the "problem of ignorance" was also framed in a way that was congruent with their broader concerns, that is, as yet another area in which *Muslim* Ottomans were at a disadvantage. In other words, it was once again the Young Ottomans who illustrated clearly how science, knowledge, and ignorance were matters that could not be discussed without reference to the community. Arguments regarding the (expected) spread of knowledge among the Ottomans could not but simultaneously be about what it meant to be a true Ottoman, and the question of knowledge could not be discussed without highlighting perceived intercommunal disparities that were deemed threatening to the Muslim core of the Ottoman population.

Tackling this question, the Young Ottomans asserted that "among [Armenians and Greeks] one could hardly find a ten-year-old who is unable to write in his own language and read newspapers. . . . Not even two percent of the Muslim community can write. [But] twenty percent of the other communities are literate."[68] Another essay published in the Young Ottoman newspaper *Hürriyet* noted that while Muslim children were able to read the Qur'an, they could not do much else.

Our child can only read; he writes, but it's more like scribbling. He can't express his thoughts on paper. [A non-Muslim child] knows at least some things about sciences like arithmetic, geometry, geography, drawing and music. Ours is utterly ignorant, and hasn't even heard the names of the sciences. So many ignoramuses all around are unaware even of the state of the world and the perilous position of our state and nation. And when it is heard that teaching methods will be enhanced, . . . some start rumors that the Qur'an will be banned and everybody will become an infidel.[69]

The constant interference of the Great Powers with the internal affairs of the empire and the impact of the unjust treaties were undeniable, Namık Kemal wrote, but the state of the non-Muslim communities within the empire proved that they could not be used as excuses for all failures: "The affairs of the Greeks and the Armenians are subject to the same treaties, but they do not cease to progress. . . . Thanks to their patriarchate, they are somewhat protected from oppression; and thanks to the organization of their schools, they advanced in education."[70] Clearly, if Muslims had the same kind of protection—protection that was denied to them by the corrupt top bureaucrats—they would be able to progress as well.

The *Basiret* also touched on the reasons behind the increasing wealth and power of the non-Muslim communities within the empire, with a less critical comment about the Tanzimat bureaucrats. While Muslim communities perceived the change as a sign of differential treatment by the government, the *Basiret* maintained the difference was due to education. Non-Muslim Ottoman subjects emulated the efforts of the civilized nations of Europe for "improving sciences and industries and expanding the means of commerce." They sacrificed their wealth for the education of their children, thanks to which they were able to better handle their affairs with the state. "Most Turkish tradesmen," on the other hand, "are unaware of the affairs of the world" and cannot express themselves even to an ordinary clerk. Hence, those who accused the state for attaching more importance to Christians should seek the blame in themselves.[71]

Ebuzziya Tevfik, on the other hand, stated that even the Jews, who were the most backward among the non-Muslim communities of the empire, now had better institutions than the Muslims. While their children now learned the "needed sciences," Muslim children still had to wrestle with "the thousand-year-old Grammar and Syntax, and even Logic, which was composed twelve hundred years ago in Greece."[72] But now that it was obvious that the government was unable to provide the necessary education, the Muslim community had to be set free like the non-Muslims, and private initiative should be in charge of the translation of the needed books and the opening of new schools.

Arguments such as these attest to the deepening insecurity among the Muslim subjects of the Ottoman Empire. Perceptions that non-Muslim Ottomans were consistently receiving special treatment from a treacherous government that ignored, if not overtly betrayed, Muslims were gradually spreading. The way the sciences of the Europeans and the Ottoman state's effort to adopt them were perceived thus cannot be understood outside this context of intercommunal relations and Muslims' perception of abandonment. If one wanted to talk about science, one had also to talk about the community and the meaning of being a good member of it.

Conclusion

The dominant Ottoman discourse about science in the first half of the nineteenth century was one in which science was portrayed as a new type of knowledge endorsed and represented by the high-ranking bureaucrats and diplomats of the empire. Their awareness of the meaning and benefits of this "wealth of knowledge" was among the qualities that contributed to the wisdom and virtue of the wielders of political power. In the 1860s, however, the claims of the representatives of the Ottoman state came under sustained attack from critics who were not unlike the members of the bureaucratic elite.

Their skeptical attitude toward the top bureaucrats also colored the ways in which members of the Young Ottoman generation interpreted the idea of science as a gift from the state to the subjects, which was very much alive in this period. *Mümeyyiz*, a newspaper for children, for instance, followed the technique established in the previous two decades when it started with an introduction about the importance of knowledge, but then it also emphasized that ignorance had become unacceptable in the contemporary era when there existed many new schools.

Up until ten, fifteen years ago the needed branches of knowledge had not yet flourished in our country, and the means for learning some sciences were not available. Thank God we now have such a sultan who . . . with so much sacrifice, opened many schools dedicated to the sciences and the arts. You should do your part to express your gratitude . . . and work to learn the branches of knowledge.[73]

Thus, the knowledge of the new, needed sciences was equated with knowledge per se,[74] and learning science was portrayed as a duty toward the sultan. Science was under his protection, and, in order to become a good subject, one had to learn it. We can observe such portrayals in books written for the new schools as well. In the foreword of a textbook translated for the Imperial School of Medicine, for example, it is stated that "the merciful gaze" of the sultan had revived the sciences, and the "coming across of his glance" had become the source of emergence of various knowledges and arts.[75]

Criticizing this idea in his analysis of an official document on education that used a cliché of bureaucratic writing and referred to official permission, Namık Kemal wrote:

What does it mean to say "The exalted permission bestowed unsparingly for the spread of science and learning and the perfection of public education"? . . . Do we need permission even to acquire knowledge? . . . Will our government, if it so desires, be able to deprive us of the light of understanding as well? . . . There are only twelve or thirteen thousand children in the public schools of the Ottoman sultanate which is one of the leading states of the world in terms of area and population. Yet official language unabashedly refers to the cause of education with such hoopla.[76]

What we see here is the approval of the basics of the official discourse, along with a call for expanding it. This is similar to the strategy the Young Ottomans used in their arguments about the meaning of "needed knowledge." It is worth remembering here that in their endeavor to become the voice of the discontented, the Young Ottomans made references to Islam and tradition central to their criticisms. As a result, while they were not hesitant to praise the new sciences of the Europeans, they made it nearly impossible for potential contributors to discuss this subject without, first, acknowledging that Islam was the basis of the acquisition of knowledge and Islamic disciplines were of at least equal worth, and, second, demonstrating that they themselves were true Muslims and not "shameless fops."

As we shall see in the next chapter, the Young Ottoman intervention would significantly enrich the debate in the decades leading up to the 1880s. This proliferation, however, made clear the ironies and the many complexities of not only the dominant but also the alternative approaches to science.

FIVE

Debating Science in the Late Tanzimat Era: Themes and Positions

Introduction

The Ottoman debate on science truly started in the 1860s with the rapid growth of the press. It was not only passionate critics like the Young Ottomans who made use of print media; young, literate Ottoman subjects in general started increasingly to imagine themselves as the potential writers of newspaper articles, as well as translators, if not authors, of books. A strict censorship mechanism was established in 1867, and many critics of the regime fled to Europe, but the proliferation of arguments regarding science continued. In fact, while censorship remained more the rule than the exception in the late Ottoman Empire, topics such as science and literature were deemed, by and large, safe matters to write about. As we have seen, statements on science were never anything but political in the nineteenth-century Ottoman Empire; yet under conditions where domestic politics in particular was difficult to discuss overtly, arguments regarding science acquired additional significance.

As a result, the debate about science became increasingly rich in the 1870s, and science was transformed into a topic that any literate Ottoman subject was, as it were, expected to be able to comment on. Indeed, when the objective of the Young Ottomans and many other reformists was achieved with the establishment of the first Ottoman parliament in

1876, one of the popular issues of parliamentary debate proved to be science as well.[1] An analysis of these particular discussions will conclude this chapter. First, however, I will outline the state of the debate in the late Tanzimat Era and focus on the themes that arose in this period in addition to the ones discussed in chapter 4.

A. Themes of the Late Tanzimat Era

1. "Spreading Science": The Young Litterateur's Burden?

The generation of the Young Ottomans was often critical of the elitist tone of the top bureaucrats' complaints regarding the ignorance of the people. But while their populist discourse entailed an emphasis on the prestige of Islamic traditions of knowledge, it was at best ambivalent. On the one hand, authors like Ali Suavi passionately condemned the arguments of the "children" in Istanbul who accused their own people of ignorance and simply imitated Europeans. On the other hand, these young activist-litterateurs did not conceal their concern with the advances of the non-Muslim community in the field of education, and they underlined the need for the Muslims to "awaken" as well. Similarly, they defined the emergence of a more aware populace also as an urgent need and the route toward the construction of a just rule.

In order to understand this latter attitude, an additional fact that we should not overlook is that from the 1860s onward, publishing became a potential source of income for many educated Ottomans. The generation of the Young Ottomans also included the first Muslim entrepreneurs in this field as well as the first individuals to turn journalism and literature into a career, usually as a result of a stalled career in the bureaucracy. Some newspapers reached sales figures above 10,000 in this period, and plays such as Namık Kemal's *Fatherland* attracted huge audiences. But these represent exceptional cases overall, and a career in letters was by no means a secure and lucrative one due to the low rates of literacy and strict censorship: hence the young authors' calls for spreading literacy. In a representative text, Ali Efendi, the publisher of the popular newspaper the *Basiret*, complained that because rates of literacy were so low, the sales figures of Ottoman newspapers were incomparably lower than those of European ones. This was lamentable, he wrote, as newspapers themselves were crucial for the spread of education.[2]

Such complaints notwithstanding, it remains the case that many educated young men attempted to launch journals and newspapers, and to share with their readers what they learned in school and/or from the commonly French works that they read on their own. As a result, these publications usually contained in their first issues a confession that the publisher-author was not an expert on the subjects he was writing about, along with a justification that he hoped his modest attempt would nevertheless be a contribution to the lofty objective of "spreading education."[3]

Ahmed Midhat, who later became a prolific author, touched on this subject when his journal *Dağarcık* received a question from a reader regarding "the formation of the earth according to science." That the question sent to Midhat, a journalist without an advanced education, was from a student of one of the military schools, which were known to be the hubs of European-style education in the Ottoman Empire, is a striking hint not only about the level of education in these schools but also of the role journals like *Dağarcık* were expected to play. In the absence of satisfactory institutions and resources for debate, instruction, and research, the press assumed the position of the teacher of the literate class. Aware of this situation, Midhat confessed to ignorance in such complex matters, and he complained that the current level of education in the country gave rise to an anomaly. He wrote:

Learning sciences and arts in our country amounts to accumulating some knowledge, and seeing in a book what one has heard of before and recognizing it, and, if one makes some progress, exposure to the basics of sciences, and later, getting a public office based on all this knowledge. . . . As a result, the sciences that we currently possess failed to enable us not only to raise one proper chemist or astronomer or admiral but to distinguish one group from among the men of letters . . . and authorizing them to report to their compatriots the direction and manner of progress of the sciences.[4]

According to Midhat, current authors were commonly around thirty years old—he himself was twenty-nine when he wrote this article—and they had started publishing usually after years of education within the muddled educational system and learning a foreign language, around the age of twenty-five. They all wanted to serve their nation but were still too few, and, thus, specialization was not an option. They worked diligently to improve the Ottoman language, publish novels and plays to correct morals, translate and write texts on history, science, and philosophy, and start newspapers and journals. When they had so much work to do, they could unfortunately not be expected to know all the sciences in detail.[5] Note that in this portrayal the young, self-assured litterateur is responsible for both spreading knowledge and correcting morals: the ultimate task, thus, remains the construction of a new type of person, in a way characterizing the entire Ottoman debate on science. Interestingly, Midhat was still complaining that the current level of education did not allow for specialization in his book *Alafranga*, published twenty-one years later.

Hence, the young litterateurs, students, and bureaucrats of the empire continued to talk about ignorance—so much so that the first issues of the usually short-lived newspapers and journals of the 1870s contained essays on education, knowledge, and ignorance almost as a rule. H. Nuri's *Revnak* (Glow), for instance, praised the new books on science published thanks to the encouragement of the ministry of education and argued that "our twenty-year-old youths who graduate after reading these books do not waste their lives after empty desires and whims as in the old times." These youths wrote new works themselves, assuring that the empire's future would be much brighter.[6] Similarly, Mehmed Cemil's *Sandık* (The Chest) published an essay on ignorance based mostly on Münif Pasha's *Mecmua-i Fünûn* article, "Comparison of Knowledge and Ignorance," and praising the knowledge that had enabled the English to dominate the world.[7]

But possibly the most pompous manifesto on knowledge and the young men who possessed it was published in another short-lived journal of the period, the *Afitab-ı Maarif*, "The Sun of Learning." In the first—and last—issue of this now-forgotten journal, the authors argued in a most verbose fashion that the sun of learning emanated its rays, which were then reflected by the sea of civilization that further spread the light. As a result,

Westerners are turning the West into the origin of the acquisition of knowledge by grasping the light of science on which the means of civilization depend. The Orientals are taking but ineffective steps to borrow this light which they had left behind the dark clouds of forgetfulness and negligence, and enlighten the Oriental lands, and, alas, turning the Orient into the West of four or five centuries ago, where the flame of knowledge was extinguished and the blaze of civilization had disappeared.[8]

While they themselves had made so many discoveries when the Europeans were entirely ignorant, Orientals now did not wish to learn the sciences even to prove Europeans wrong and simply belittled European inventions. Even the Ottomans, who were once the greatest warriors on earth, were in utter destitution due to "the catastrophe of lack of learning and arts." "How can we, in a world of progress where Western nations are competing to announce their scientific discoveries . . . , waste time reading and writing stories on love and affection?" the authors asked. There *were* wise Ottomans who wished to contribute to the spread of science in the empire, but they were not competent enough. As a result, "the Sun of Learning came into being . . . to radiate the light of accomplishment and virtue and illuminate the Oriental press."[9]

The journal published texts on archaeology, astronomy, zoology, and politics in this first issue. But this naively ambitious and ungraceful launch received one of the most brutal responses in the nineteenth-century Ottoman press when Süleyman Talat's journal *Kasa* (The Safe) published an issue ridiculing several Ottoman periodicals, with a majority of the pages devoted to the *Afitab-ı Maarif*.[10] This satirical issue of the *Kasa* contains an advertisement from a Society of Science welcoming the emergence of the *Afitab-ı Maarif*. This journal, according to the ad, conveyed information on such sciences as foppery and fashion and was published by four men who simply aimed to make money—note the similarity to İshak's criticism of Midhat discussed in chapter 4. Another essay noted, "Supposedly humanity was in darkness . . . and people were thirsty for the water of life, the sun of knowledge. And these men here have filled up their barrels . . . and now want to sell us water. Allegedly these four writers grabbed their lanterns and came out to illuminate our darkened thoughts."[11]

In addition to numerous mocking comments about the self-importance of the authors, the *Kasa* made fun of every single essay published in the *Afitab*. The one on astronomy gives an idea about the general tone:

Astronomy
There were three views in Greece: 1. Ptolemy said "In March 1874 a journal called *Afitab-ı Maarif* will emerge, but it won't sell, as it is stationary." 2. Plato said "*Afitab-ı Maarif* is stationary, so it will stay motionless in booksellers, but its authors will revolve around the stores." 3. [Tycho] said "Whether *Afitab-ı Maarif* is stationary or not, whether its authors revolve or not, the typists will demand all their fees."[12]

But it was in a more serious article in this issue where the publishers of the *Afitab* were condemned most harshly. "It is so strange," the author argued, "that some people publish translations of European absurdities that corrupt morals under the name 'science.'" Most specifically, it was the adulation of ancient Greek philosophers that was unacceptable: "Poor master of sciences! What is he to do? Had he read some Arabic as well, he would not have applauded Greek men of science so wholeheartedly." The counterargument concerned the by then firmly established narrative about the glory of the Arabs during the "Dark Ages" of Europe. Furthermore, the glorious Greece of the bygone eras was now suffering from the greatest ill of all: moral destitution—the reason behind the decline of even Andalusia. The conclusion, not surprisingly for us at this point, was not only about scientific knowledge, either:

No matter what anybody says. We prefer [the garb] of those old Ottomans who had appreciated the justice, wisdom, history, generosity, morality, and the greatness of Islam to the *alla franca* outfits of our new-fangled gentlemen with short jackets, tight slacks, . . . canes, chamois leather gloves and monocles, who lost their community and emulated Frankish manners. . . . While our youths try to learn clownliness in balls, dancing in theaters, the latest fashions of Paris, *alla franca* haircuts, bowing and bending while being introduced to a woman . . . , the qualities that adorn the moral virtues of the Ottomans are lost entirely.

It is true that learning is the water of life for any nation. . . . [But] we want to progress in the knowledges of civilization while maintaining the morals of our community, we do not need the vileness and degeneracy of Europe in the guise of civilization. We sincerely hope that we will see in our country a lot of progress with respect to the philosophy and experiments of the natural sciences, the science of law, the fundamentals of wealth, liberty, orderliness, freedom of the press, highways, railways, new weapons; but if our fops excuse us, we do not want to and will never see the . . . balls, dances, hubbubs and fashions of Europe. . . . We will progress within Islam, we will not be Franks in fezzes, with altered morals and manners.[13]

Ignorance *was* a problem worthy of discussion, and the Ottomans certainly needed the fruits of scientific knowledge. But nothing could be resolved if the arrogant fops prevailed. Who criticized ignorance was more important than what was said.

Indeed, Young Ottomans referring to ignorance did not receive similar criticisms. Nuri, an associate of Namık Kemal, for instance, complained in the Young Ottoman newspaper *Ibret* that there had emerged a view that condemning the ignorance of the people was an insult to the Ottomans. Nuri argued that the real crime would be to remain silent in such matters, and that campaigns for the spread of education would become all the more appealing if people were aware that even French newspapers complained about the ignorance of their own nation. "If we think a little even-handedly on this matter, we will realize that comparing our education to that of France is like comparing a child with a savant."[14] Similarly, Ebuzziya Tevfik argued that the children of Muslim Ottomans got their elementary education from teachers who were themselves "ignorance in material form." European children knew more than Ottoman children who were older than they: "When they hear the words 'history,' 'geography,' 'arithmetic,' 'geometry,' our children are astonished, and left wondering if those things are [the names of] humans or playthings."[15] Similarly, as we have seen, the Young Ottomans did at times appropriate the "ignorance discourse" of the Tanzimat bureaucrats; they used it particularly to argue that Muslim Ottomans were significantly undereducated compared to the non-Muslims.

But the emphasis on ignorance put those demanding political reform in a dilemma, as we can observe in their essays. If the Ottomans were too ignorant of the way the contemporary world actually worked and lacked the new types of knowledge about mankind and the world, then the implications could be that the Ottomans were indeed at a much lower stage on the civilizational ladder, as many European authors claimed, and that the knowledgeable few had the right to dominate, as top Tanzimat bureaucrats suggested. The latter implication made it particularly difficult to argue that the parliamentary monarchy that the Young Ottomans defended could actually work in the Ottoman Empire. As ignorant people were easier to lead astray, freedom of the press itself could be deemed dangerous, just as the Ottoman government resolutely did.

Examples of the resultant inconsistency are easy to come by. While the first issue of Ibrahim Şinasi's *Tasvir-i Efkâr* had referred to the popularity of newspapers in "civilized nations whose eyes were opened thanks to education," for instance, its fifth issue introduced a news story on superstition translated from an Austrian newspaper by stating that the story showed that "even though rational sciences have progressed remarkably in Europe, commoners are still so deficient in knowledge."[16] Similarly, advertisements about itinerant healers can also be found in its pages.[17] Namık Kemal, who wrote in the *Hürriyet* many essays where he indicated that Muslim Ottomans were less knowledgeable than non-Muslims, let alone Europeans, also asked in defense of an Ottoman parliament, "Do [critics] think that every peasant in Europe is able to distinguish good from evil, the idiot from the wise, the cruel from the just, the knowledgeable from the ignorant? No, they are just like our people. Only their affairs are better organized, and that is why they are wealthier."[18]

Yet in an essay published in the same period, the same Namık Kemal would argue that the people of Europe and the United States were much more than literate. "Even a common sailor or porter" in these places not only knew the basic principles of his religion but understood at least one foreign language and was familiar with the essentials of "geography, history, arithmetic, algebra, geometry, physics, chemistry, astronomy and natural history."[19] Similar comments about the astounding knowledgeability of the average European can be found in the letters he sent to Rifat Bey in August 1878. In one letter he stated that "the lowest peasant in those countries learns eight-tenths of all the classes taught at our Military Academy."[20] He even ridiculed Ottoman men of letters, along with himself:

Here, we are all literate. But we are as ignorant as peasants. Because I don't see among us anybody better at keeping his accounts than a peasant. I saw it with my own eyes in Sofia. A shepherd thrust a pole into the ground and measured the elevation. . . . We cannot measure slopes ourselves. We cannot measure the size of our house destroyed in fire in Istanbul. We don't know how to hold a match so our moustache won't get burnt. We are unable to save ourselves in case of an earthquake. We would hide under a tree during the rain and get struck by lightning.[21]

The outstandingly political nature of the struggles regarding the delineation of the meaning and consequences of ignorance and new knowledge as well as descriptions of the ideal man were thus made explicit by the very contradictions within the texts written by the generation of the Young Ottomans. The grand synthesis of the Young Ottomans defined this ideal person as one who would defend the rights and praise the traditions of *Muslim* Ottomans while espousing Ottomanism, spread new knowledge while emphasizing the matchless value of inherited knowledge, appreciate the new while avoiding foppery: the embodiment of knowledge and virtue, in short. These young litterateurs themselves were unable to live up to these standards, and this ideal person was ultimately to become but a cartoonishly didactic character that we see in numerous Ottoman novels.[22] Yet as we will also see in the next chapters, it was the disciplinary function of this imaginary character that mattered, not its realism. The much extolled but lifeless ideal man in these novels served to enforce norms such as patriarchy and patriotism and to invite any young Ottoman subject to prove that he was not a fop.

2. Useful Knowledge, Useless Groups

One aspect of the official discourse on science—as characterized by Münif Pasha's writings—that remained by and large unchallenged in the 1870s was its emphasis on usefulness. The prestige of the practical, material benefits associated with useful knowledge transformed this understanding of usefulness into a principal criterion for evaluating the worth of any activity. Indeed, this element became part and parcel of alternative discourses themselves.

A common target of criticism in this respect was literature. The early examples of this reaction, which would in the 1880s turn into a full-blown criticism of traditional Ottoman literature and its representatives, emerged in this period. In a letter published in the newspaper *Terakki*, for instance, an anonymous author complained that the Europeans constantly blamed the Ottomans for being lethargic as well as ignoring science and education. He asked, "Where are the books on the sciences that would be the source of prosperity for the people and development for the country that we can all read?" This was an even more conspicuous problem when it was known that the basis of European sciences were the books Muslim Arabs had translated. Accusing those who knew European languages of translating stories and tales rather than books on science, he stated, "They entertain us, instead of guiding us forward."[23]

The same argument can also be found in Talat's *Kasa*. In his condemnation of the translation of children's tales into Turkish rather than useful works, the author proclaimed: "Let us think not about the present but the future. Let us look at the West, the spring of achievement, and take warning. Isn't it a pity for us to write some ridiculous stories here while the Europeans contribute to . . . knowledge and the progress of the arts every single day?"[24]

We see a similar example in the chronicle of Ömer Faiz Efendi, the mayor of Istanbul who had accompanied Sultan Abdülaziz (reigned 1861–76) during his visit to European capitals in 1867. According to Ömer Faiz, the illuminated streets of Paris had impressed the Ottoman visitors, as they appeared as if the lampposts had stored sunlight in them. "We too have people interested in the light of the sun," Ömer Faiz stated. "They are our poets. . . . They compare their beloveds to the sun that turns night into day. They write volumes of poems and songs. [People in Europe] are interested in the sun, too, but here they are their chemists, scientists. That's the difference between us and them."[25]

While the 1860s and 1870s witnessed the efforts of the Young Ottomans to restore the prestige of traditional Islamic disciplines, the criterion of usefulness was also employed in assessments of the *medrese*s. Furthermore, this criterion led even the passionate supporters of alternative approaches to harshly criticize the curricula and students of the *medrese*s. According to an anonymous author of the newspaper *Terakki*, the true realization of human potential lay in learning and practicing "the most honorable and useful" of the sciences. In that case, "sciences that are interested solely in language, such as syntax and grammar," and those dealing with categories and representations, such as logic, could not be regarded as useful sciences. Learning Arabic grammar and syntax, the fundamental courses of *medrese* training, was useful if and only if it helped one to understand books on the truly useful sciences written in Arabic, according to this author. Languages—even Arabic, "the foundation of our language and religion"—should be learned only in order to make use of the works on sciences written in them. Furthermore, the way teaching was organized in the *medreses* enabled the reading of antiquated texts but not even contemporary newspapers in Arabic. As a result, those who learned Arabic this way were unable to read what was truly beneficial, and instead "deal[t] with some issues in *kelam* [Islamic theology] that are entirely irrelevant for the circumstances of our time."[26]

While it was true that *kelam* would be able to fend off the undesirable impact of certain new ideas, a new *kelam* was needed. The author complained, "Today's thoughts are different. Certainly the needs are different as well. The sciences learned do not serve our present needs; everything is in conflict with what is required." While he regarded studying all sciences in Arabic as not particularly useful, he argued that sciences such as *fiqh* and ethics as well as arithmetic had been discussed well in books written in Arabic, and in order to save *medrese* students from the miserable state they were in, it was such sciences that should be emphasized. Similarly, another author who wrote on the same subject claimed that many *medrese* graduates were unable even to perform basic arithmetic operations, thus betraying the legacy of the glorious Arabic scholars of the Golden Age of Islam.[27] In a poem he wrote in 1877, Abdülhak Hamid went so far as to refer to the *medrese* as a place where idiocy was taught: "People leave it more ignorant than when they first entered."[28]

For the Young Ottomans, the incompetence of *medrese* graduates was a pressing issue. While the need for them was remarkable, argued an essay in the Young Ottoman newspaper *Hürriyet*, many among the ulema

. . . could not understand anything from [a contemporary newspaper published in Arabic] unless they studied it for two hours with the help of a dictionary. They would admit to incompetence in *fiqh* if they were asked a question concerning it. Facing an argument on *akaid* [Islamic doctrine], they would grab a shield of bigotry and try to fend off their opponent by accusing them of blasphemy. . . .

If the discussion were on politics, they would be amazed at hearing that there exist countries like England, America, Japan or Morocco. If they needed to write a letter in Turkish, they would beg for help.[29]

According to the author, the real problem this caused was that *medrese* graduates could not be employed anywhere: "If it was said to them 'The government will offer you an official post. . . . Choose whichever field your knowledge is useful in,' one wonders what their answer would be." Their knowledge of Turkish was insufficient, they did not know anything about fiscal and administrative issues, and they could never communicate with foreigners. As a result, they could not do anything "useful for the state and the community" but teach in mosques, thus rendering years of education more or less pointless.[30]

Ali Suavi devoted one of his best-known essays to the same problem. In this essay where an imaginary *medrese* graduate discusses the training he received and its consequences, we see similar arguments on the incompetence of *medrese* graduates in all branches of learning: "I understand that these sciences are beneficial. But why haven't I benefited myself? . . . I am scared even to say 'I became a scholar, this is how I benefited.' Because all my acquaintances know that I am unable to write two sentences in a clear manner, even in Turkish."[31]

The idea that the empire needed a new type of religious scholar became a leitmotif of the essays written in the 1860s. Hayreddin Bey argued that the amalgamation of the true Islam with mistaken ideas had led to the derision toward novelty and, consequently, to Ottoman decline. To demonstrate what the religion actually instructed, the empire needed knowledgeable and virtuous *medrese* teachers and preachers: "Had it been prohibited for us to receive and borrow the estimable ways that did not exist during the era of the Prophet and emerged later, be they European inventions, then we should have deprived ourselves of all the beautiful things of our era and contemporary progress, such as cannons and rifles, steamboats and streetcars."[32]

Sadly, he wrote, some ignoramuses were unable to differentiate darkness from light. Similarly, according to an essay in the *Basiret*, while most preachers were commendable, there were unfortunately also those who told "stories containing delusions that reason and wisdom cannot accept."[33]

Thus, in the 1860s and 1870s the criticism of *medrese* graduates became a key element of the discourse on science. Science was admittedly useful knowledge and those who knew the sciences were useful people, whereas the knowledge that *medrese* graduates possessed did not turn them into useful subjects. It should be remembered, however, that the Young Ottoman line of argumentation highlighted not only the inadequacy of *medrese* education but also the state's own negligence of the *medrese*s in general and the traditional sciences taught in these institutions in particular. This approach also insisted on defining useful knowledge more broadly: the new sciences of Europe could to some extent be taught at the *medreses*, but the state also needed to make use of traditional Islamic sciences like *fiqh*. These sciences were not useless in themselves; they had become so only due to the betrayal of Tanzimat bureaucrats. Similarly, the reason behind the admitted and condemned uselessness of *medrese* scholars and students had to do with flawed state policy.

On January 30, 1873, an article published in the *Basiret* reflected on these issues and stated that there were suggestions that "sciences necessary for the present day" should also be taught among other courses in mosques. Such a demand was not reasonable, however, as the ulema were already under significant burden.[34] Less than two weeks later, the newspaper published a letter signed by "people who spread knowledge in mosques" congratulating the *Basiret* for its fairness. "The sciences [we are asked to teach] are most likely mathematics, geography, et cetera," the ulema asserted and then argued that they too realized the need for teaching them. But while it would not be intellectually difficult for them to assume this duty, it was also obvious that they already did not receive all the support they deserved, and they were assured that the sultan would ameliorate the situation the ulema were in. Unless their living and working conditions were improved, the letter implied, nothing more could be expected of the ulema.[35]

The *Basiret* brought up the same issue one month later, and stated that its publications on the need for improving the state of the *medrese*s and their graduates had attracted a lot of acclaim. Elaborating on the issues they deemed crucial, the author noted that the teaching of the needed sciences (*fünûn-ı lazıme*) in the *medrese*s was among the top priorities. The reorganization of the curricula would be insufficient if it only focused on the classes traditionally taught: "other sciences that are required for the needs of the era, such as mathematics, physics, medicine, along with grammar and orthography, and other necessary sciences" should also be part of the *medrese* curriculum. It is interesting, though, that the newspaper suggested that these classes on the needed sciences be taught in mosques or large *medrese* halls by appointees from the Imperial Military Academy, School of Medicine, and civil offices.[36] *Medrese*s would thus be saved from oblivion and obsolescence, but by the services of the graduates of the new schools. Similarly, the *medrese* graduate had to become a useful person, but in a way based on the model of the new man of science.

3. Science, Language, and Identity

Another ground of criticism directed at *medrese* graduates centered on language. Most *medrese* courses were centered on the Arabic language, as many of the books studied were Islamic classics written in Arabic regardless of the native tongue of their authors.[37] Even advocates of the significance of Islamic contributions to science like Ziya Pasha and Ali Suavi were critical of the failure of *medrese* graduates at Turkish prose, as seen above. In his critique of *Mecmua-i Fünûn*'s comments on the Library of Alexandria, Mehmed Mansur himself complained of the impact of Arabic on the spread of knowledge in the Ottoman Empire. Had the great Ottoman scholars of the past centuries used their own language, "they would have matched other Muslim scholars with respect to virtue and perfection and works, and they would have spread science and knowledge to the people in general, thus bringing the Islamic community, and perhaps the whole world, to an entirely different position."[38] The Arabic as well as European cases where the authors had for long written in their own language proved Mansur's point. "*Medrese* students are not a different type of person than the students of the Military Academy who learn sciences and write outstanding books in a short time," he wrote. The only—but very consequential—difference lay in the languages of instruction.

This aspect of the debate on *medrese* education added a new dimension to the overall debate on science, ignorance, and usefulness by hinting at an association between Turkish with the new sciences and Arabic with the old. The importation of the new scientific knowledge of Europe into the Ottoman Empire was ultimately a process of translation: many new concepts, categories, ideas were introduced into Ottoman discourse, and the existing ones were reinterpreted in this new context. The entire reform era could be labeled as the "period of translators" due to the fact that knowing French was the fundamental criterion for being a prestigious member of the new generation of bureaucrats and litterateurs. After all, many intellectuals and statesmen of the era were products of the Translation Bureau itself.

But, needless to say, translation from "the other" is a process that involves assumptions about "the self." And in this multilingual empire where higher (*medrese*) education for Muslims was based in Arabic and where elite literature had mostly been influenced by Persian literature, the translation of texts on the new sciences gave rise to heated debates about what "Ottoman language" itself was and should be. As the centralization and reorganization of public education coincided in the Ottoman Empire with the appropriation of the new types of knowledge, there emerged a strong connection between efforts for the standardization of language and the spread of scientific knowledge. In short, educational reform in the Ottoman Empire involved both a new language and a new type of knowledge.

A telling example of this parallel between science and language is Cevdet Pasha's speech regarding the establishment of the Ottoman Academy, discussed in chapter 2. This text from 1851 stated that it was the knowledge that it conveyed that bestowed honor on a language; it emphasized the need to transform the Ottoman language into a language of learning. If the aim was to spread knowledge, it should be expressed in a way that most people, not just the elite, could understand. Hence, the language of knowledge should at least be somewhat close to the language of the common people.

A second component of the association between language and science had to do with the alphabet. That the version of the Arabic alphabet used for writing in Ottoman Turkish was inadequate and, as a result, hampered the learning of literacy was another argument commonly brought forth by the advocates of the dissemination of science. Münif Pasha made a case for reforming orthography in his *Mecmua-i Fünûn* in 1863.[39] Hayreddin Bey published an essay in the *Terakki* where he asserted that devoting years simply to teaching literacy was unacceptable:

In that case, when will mathematics, philosophy, geography, chemistry and other sciences be taught? . . . Will there be enough time to teach the useful sciences that are demanded and needed by Turkey? Aren't letters the gate to the sciences? Why should the gate be so heavy that one needs to make a sustained effort to enter? It is complained that Turks aspire to nothing but getting a clerkship. But if children are taught nothing but calligraphy, it is obvious that nothing else is pursuable. Let the letters of the Qur'an remain eternal, just as the truths it contains. But the letters used for sciences, administration and trade should be made easy and simple.[40]

Ebuzziya Tevfik responded to Hayreddin's suggestion with a detailed analysis, accentuating the parallels between the discourses on science and language. Wasn't it the same Arabs whose alphabet was under fire who had furthered the major sciences, after all? "Were the Andalusians who reached the zenith of science and learning using a different alphabet?" Ebuzziya asked. Letters indeed resembled the key to the gates of science, and Arabs had grasped the complex matters of science using the same key as the one used by the Ottomans. Therefore, it was the methods of instruction, not the letters, that prevented the spread of useful sciences and arts, and such a condemnation of the Arabic alphabet would imply nothing but the denial of the Arabic contributions to science.[41] Ali Suavi went further and asserted that it was all Muslim scholars, regardless of their ethnicity, who had contributed to science, even in Andalusia. As these scholars had written in Arabic for centuries, Arabic was both the language of Islam and the language of science for the Muslims: "To those who ask us about our belles lettres and science, we shall present our works written in the language of Islam and science. Because it is we who wrote them. But, [it is said] they will think we are Arabs then. They will not be mistaken if they think that we are *Muslims*."[42]

Hence, the emerging portrayal was one that honored and claimed all contributions of Muslims to science regardless of ethnicity, but that also conceived of Arabs as, in essence, representatives of a different civilization than the Ottomans. A telling example is a letter published in the *Basiret* stating that Yemen resembled European countries in that the customs regarding dancing and veiling were much looser, and its writer invited Ottoman youth to this land. Ali Efendi's reaction was clear: civilization had nothing to do with these practices, just as it had nothing to do with champagne or tight pants. "It is true," he wrote, "that Muslims are authorized to acquire those aspects of civilization that have to do with education and industry. . . . But we certainly . . . are not required, neither by religion nor civilization, to acquire things that have to do with false belief, misconduct, and corrupt morals and customs, be they from Arabs or Persians."[43]

Ottoman debates about science were thus also debates about the identity of the community. They had as much to do with "Arabness" as "Europeanness," and language, religion, and ethnic identity gradually became inseparable components of the debate on the meaning of science. This dimension of the debate would be exploited further in the 1880s, as chapter 6 will show.

4. Science and Usefulness: What Is Beneficial Work?

While usefulness was certainly a criterion used to judge Ottoman subjects for all parties involved in the debate in the 1860s and 1870s, the new participants also emphasized another understanding of the concept, much more than the top bureaucrats ever had. They problematized the usefulness of bureaucrats themselves: they may have had a good education based in the new, useful sciences, but were they useful themselves?

Indeed, one of the defining characteristics of the 1860s and 1870s was the increasing emphasis on the links between the new knowledge and arts and commerce, rather than civil service. One fundamental thing that Ottoman manufacturers had to realize was that the hardship they were experiencing because of high-quality European imports was due to Europeans "carry[ing] out even the most ordinary art with knowledge," the daily *Basiret* wrote. Hence, the reason behind the decline of Ottoman industries, before anything else, was the lack of scientific knowledge.[44] Arts and crafts in the Ottoman Empire needed more knowledgeable people: just as in previous Islamic states, schools in the Ottoman Empire raised talented individuals, but arts and commerce were ignored. Not only had wealthy Ottomans abandoned the sphere of commerce to the Europeans, but those who had the required skills were devoid of the knowledge pertaining to their craft, leading industries to decline.[45]

By emphasizing the role of the entrepreneur and the craftsman, the *Basiret* problematized the prestige of civil servants. A member of the ulema joined the bashing with his own agenda. In a pamphlet written in the same period he argued that it was actually the new administrators who were incompetent. He stated,

These men are not clerks but ignoramuses. Only those who come from the ranks of the ulema deserve to be called clerks. An understanding of knowledge (*ilm*) is acquired only through years of study and exertion in the *medrese*. These are men whose drunken souls have seized on the present opportunity and been spellbound by the spoils afforded by the state.[46]

Soon an anonymous clerk wrote a letter to the editors of the *Basiret* where he argued that it had become popular to criticize bureaucrats for their inefficiency and dependence on state resources. Explaining how lower-level bureaucrats like himself had very low salaries and rather precarious careers, he asserted that despite the lack of public gratitude, he was proud of being a member of the bureaucracy, the class that did the greatest "service to civilization." Similar letters that supported the clerk's views were published in the following issues.[47]

The newspaper *Hadika*, which stated in its first issue that it would focus on the connections between sciences and arts, published a lengthy introductory essay that was also composed as a passionate call for the progress of sciences and arts in the Ottoman Empire. Reason bestowed on man by God had enabled the emergence of sciences and arts, and as the welfare and power of a state depended on these two factors, ignoring them would be equivalent to ignoring one's duty toward his country. The reason behind European progress was not that Europeans were more intelligent than Ottomans, but that they had long been striving to learn sciences, as opposed to the lethargic Ottomans. But now that the exalted sultan had opened so many schools to spread the light of knowledge,

why should we continue to meander in the valley of ignorance? This time of prosperity which is the origin of the light of knowledge and wisdom and the source of happiness and comfort has opened all the doors to science and knowledge both for us and for our descendants. Let us work day and night to counter those nations the natural advantages of whose lands are already much fewer than ours, and take steps to acquire our urgent needs for a life of contentment and comfort in this beloved country, so that the guidance of God Almighty will accompany us.[48]

The kind of work that the *Hadika* emphasized was crafts and manufactures. What is significant, though, is that while in the 1860s the idea that there was a relationship between the new sciences and the new industries came to be referred to more frequently, the exact nature of this relationship was rarely clarified. Furthermore, it appears to have been rather difficult for the Ottoman authors to imagine the man of science as anything but as a man who knows. An article in the *Basiret* argues, for instance, that it had become quite popular to publish essays indicating the importance of the development of industry in the Ottoman Empire. "It is true that the acquisition of wealth depends on industries," the author states, but "that, in turn, depends on learning." While this was a common way to introduce texts on the shortcomings of the empire's system of education, what is striking is the way the author elaborated on this point: "Even if, say, a clerk working at one of the bureaus of the state has remarkable aptitude to learn natural sciences, . . . it is almost entirely impossible for such a person to truly learn them."[49] Note that the essay that starts with an emphasis on the importance of sciences for industry swiftly turns into one on the teaching of scientific subjects to civil servants. It is not surprising then that the author praises the Imperial Military Academy for teaching the sciences properly without discussing how this could be related to the growth of Ottoman industries.

It is not that the weaknesses of this attitude went unnoticed. Indeed, the criticism of such ideas was a very common topic for newspaper essays. The *Basiret* published the following remark on May 2, 1873: "We have a rotten and harmful view. The lullaby we sing to our babies in their cradles goes 'My son will enter government service, he will be an *Efendi*.' We think to be an *efendi* one has to be a clerk. There are many among civil servants, however, who regret not having learned an art or got into commerce."[50] Ali Suavi's essay on education made the same point: The view that civil service was the only esteemed option for a literate person was the reason behind the state of the Muslim community of the Ottoman Empire, but this was due to a misconception about the meaning of education itself.

If education is considered, like we do, as learning superficially some terms and arguments, it is impossible for industry to come into existence. Similarly, if education is believed to involve certain knowledges to be learned not for the practice and manufacture of arts, but for receiving an undeserved income from the Treasury, then progress is impossible.[51]

Nevertheless, how exactly the new knowledges would be transformed into developed industries remained an unanswered question. Ultimately, it should be noted that even the critics of official definitions were products of the same educational system, who were themselves raised to be civil servants or members of the ulema. Despite their shortcomings, however, the new participants of the debate effectively challenged official definitions not only by asserting the relevance and usefulness of the Islamic sciences and by questioning the supposed links between the new sciences and moral virtues, but also by problematizing the very usefulness of the champions of the new sciences themselves. Theirs was an understanding informed by the hardship that lower-level clerks, ulema, small merchants, and craftsmen were experiencing in this period, and as Muslims were overrepresented in these groups, their discourse on science, usefulness, and virtuousness appeared at the same time a defense of Islam.

B. The Official Location of Useful Sciences: Reviving the University

While such alternatives were brought forth in the press, the official definition of "useful sciences" continued to appear in official documents, particularly regarding education. A typical example is a statement read during the sultan's visit to the Sublime Porte in August 1864. It was asserted in this document that the purpose of the failed initial attempt for the establishment of the *Darülfünûn* had been "the teaching and spreading of mathematical and other useful sciences."[52] Thanks to the sultan's well-known enthusiasm for the spread of "the light of knowledge," it was decided to commence the construction of a new building for the university.

The 1869 Public Education Regulation also stipulated that a university would be opened and, interestingly, referred to it as a "*medrese* of knowledge" (*medrese-i ilmiye*; article 79), thus appropriating one more word from the Islamic educational lexicon. The university would consist of three faculties: literature, law, and natural science and mathematics. The curricula were comprehensive: both Arabic and French literature, both Islamic law and French law would be taught, along with the natural sciences taught in French universities (articles 80–83). The medium of instruction would be Turkish, but in case of a shortage of qualified professors, it was allowed for professors to teach in French (article 84). Those who wished to enroll were required to take an exam in Turkish, history, geography, arithmetic, geometry, algebra, physics, and logic (article 90).

Hoca Tahsin, one of the two people who had been sent to Paris twelve years prior in order to be employed at the university on their return, became the director of the new institution. While many textbooks were ordered from Paris for the library, the current scientific periodicals on the list were later removed:[53] science as accumulated knowledge to be learned was preferred to science as contestable and changing knowledge.

The new university building was also much smaller than the previous one. Lands obtained from demolished city walls were sold in order to cover the expenses of construction. Harshly criticizing this decision, the historian and member of the ulema Lütfi Efendi wrote a decade later: "It was not worth destroying the thousands-of-years-old ancient tower and gates for such a trifling benefit. . . . Was there a shortage of *medreses*?"[54]

Out of the thousand-odd people who took the entrance examination, more than 450 were admitted. By itself this rate of success is an indication that the standard was quite low; that many of those who passed were *medrese* students is further proof that the examination must have tested quite basic knowledge of the sciences listed in addition to literacy. As it was established practice for *medrese* students to spend the holy months away from Istanbul, the opening of the university was delayed, and it was decided to revitalize the public lecture series of the previous university until the students' return. The lectures were on new topics, such as the development of industry, and the new sciences including biology, physics, and chemistry. I quote the schedule of the lectures directly from Lütfi's *History*, along with his own mocking comments in italics:

Lecturer	Topic
Cevdet Efendi	Magnetism *(Here is a new idea)*
Rıf'at Bey	Chain of animals *(What's in it, is in it)*
Tevfik Efendi	Machines *(Only for those concerned with them)*
Vahid Efendi	Cemeteries and health *(So useful!)*
Münif Efendi	The progress of industry and sources of wealth *(How about that!)*
Emin Efendi	The atmosphere *(Full of hot air)*
Aziz Efendi	Unification of the forces of nature *(I don't know who it's good for)*
Selim Efendi	The planets *(Requires arms and wings)*
Aziz Efendi	Climes and traits *(Not simply necessary, but essential!)*[55]

It is apparent that this member of the ulema did not find the topics particularly significant or, using the same criterion as the advocates of these sciences, useful. But another major reason behind Lütfi's criticism was that the lectures took place during the holy month of Ramadan. He was particularly annoyed with Hoca Tahsin's lecture on the universe and the concept of infinity. He wrote:

It is such a shame to corrupt the faith of the people with nonsense of this kind on holy nights. Such classes on the mysteries of natural philosophy are exclusively for the elite, and it is astonishing that those who deem themselves people of wisdom act against wisdom themselves, and fail to consider that spreading them among commoners is unwise.[56]

The university was officially opened on February 20, 1870, with speeches by Minister of Education Safvet Pasha, Head of Council of Education Münif Efendi, Ioannes Aristoklis,[57] and Jamaluddin Afghani, the leading Persian advocate of modernizing Islam and the solidarity of Muslims.[58]

Safvet Pasha's speech was a potpourri of the themes that had been well established by 1870, with borrowings from the alternative discourses. Because the Qur'an encouraged the learning of the sciences, Safvet Pasha argued, Arabs and particularly the Arabs of Spain had contributed greatly to the sciences initially developed by the ancient Egyptians and Greeks. The sciences he refers to in this context are medicine, arithmetic, algebra, astronomy, and the sciences of animals, plants, and minerals.

After the fall [of the Arabs] arts and sciences moved to Europe and developed particularly within the last century. Many curious discoveries and inventions that astonish the mind have been made as a result. . . . Remarkable inventions such as steam power, telegraphy, and electricity are among the great consequences of the progress of physics alone. . . . While [such breakthroughs] reveal the power and grandeur of God, those living in the darkness of ignorance, unable to comprehend the mysteries of creation that men of knowledge are aware of, only gaze at what is perceivable. . . .

Just as knowledge adorns the natural constitution of man with many a virtue, it is clear that the civilization and prosperity of a country are equally dependent on it.[59]

In line with the convention, Safvet Pasha blamed "the interference of certain impediments and difficulties" for the failure of the Ottoman Empire to keep up with the scientific developments in Europe. The most detrimental aspect of this period of "seclusion" was the absence of spaces for the communication of ideas, which was central for the progress of rational sciences in European countries. Safvet Pasha's example for the achievements of the new era in the Ottoman Empire was the emergence of such a space, but a quite peculiar one: the *rüşdiye*s, or the new middle schools. Indeed, what he presented as analogous to the spaces in Europe where scientific ideas were discussed was Ottoman schools for the bureaucrats of the future. "Before these schools were opened," he argued, "there were no places for learning sciences and skills for those who wished to be employed in civil service and the offices of the Exalted Sultanate, and those who searched for other means of subsistence." The university was the highest point of the new system of education: it was for those who, after graduating from the new high schools, "wish[ed] to further complement their scientific education and the diverse knowledge that they accumulated."[60] Fine arts and other crafts would also be part of the curriculum, however, and help those who rightly knew that state service was not the only means of subsistence.

Once again, however, the university did not live up to expectations and was shut down in less than two years. Among the most commonly cited reasons behind the end of the second attempt are reactions to the public lectures that continued in tandem with the classes. In one such lecture, Hoca Tahsin spoke about the concept of vacuum and performed a demonstration where he removed the air in a jar, causing the pigeon in it to die. The great public outrage that followed led to Tahsin's removal from office. Similarly, the audience found Afghani's lecture where he likened prophecy to art particularly offensive. Another problem with the lectures that we can identify is their style. In his classic study of the history of the university, Aynî examines Tahsin's lecture on water in order to demonstrate how simplistic it was.[61] Yet what is truly striking about the tone of this lecture is its utterly convoluted and embellished style—the style of traditional *medrese* books and bureaucratic texts, which would hardly have been comprehensible for the audiences the lectures were supposed to attract.

Ultimately, textbooks proved insufficient, students unprepared, qualified teachers lacking. The system spelled out in so much detail within the French-influenced 1869 regulations remained on paper, and the university did not evolve into anything other than yet another series of lectures. Aynî notes that the involvement of European university students in "rebellious" movements in 1848 always remained a cause of concern among Ottoman statesmen as well.[62] The idea of a university clearly sounded good in theory, but it is also clear that the leading Ottoman bureaucrats remained hesitant about the ultimate benefits and potential dangers of higher education. What exactly the teaching of the new sciences would help achieve remained vague, and, with an educational system that hardly prepared students to a high school level, the efforts to found a university appeared disorganized and confused. That the university never received sufficient funding is a clear indication of the priorities of the Ottoman bureaucrats as well.

Interestingly, another attempt was initiated in 1874, but without publicity and almost as a confidential project. This time the university was designed essentially as the continuation of the most prestigious high school of the empire, the *Mekteb-i Sultani,* opened in 1868 in collaboration with the French government. Schools of law and engineering designed along French models constituted the new institution that attracted non-Muslims as well, but after seven years of intermittent instruction, very few graduates and records, the final attempt of the nineteenth century came to an end in 1881.

C. Showcase of Discourses: Science in the Ottoman Parliament

After years of struggle, the first experiment with a parliamentary monarchy started in the Ottoman Empire in 1876. During this short-lived experiment, the issues of science, knowledge, and ignorance became topics of heated debate. In a sense, the parliamentary debates of 1877–78 represent the culmination of the decades of construction of discourse regarding the meaning and benefits of knowledge and the worth, duties, and responsibilities of its possessors.

In his opening speech, Sultan Abdülhamid II (reigned 1876–1909) himself indicated that the parliament was supposed to focus on these issues. As it was only through science and learning that "the progress of agriculture and industry" could be achieved and "civilization and prosperity" carried to the highest level, the sultan assured the parliamentary representatives that proposals regarding educational reform would be brought to their attention the following year.[63] The parliament's official reply to the sultan expressed the confidence of the legislative body that the opening of schools for spreading the useful sciences and arts would contribute to the growth of prosperity and welfare throughout the empire.[64] When it was suggested that debates on education should be postponed until the proposals were received, Speaker of the Parliament Ahmed Vefik Pasha, a Tanzimat bureaucrat educated in France, objected, stating that "no country can survive without science. In a country without science, no knowledge, no skill can be found."[65] But truly enough, the first year of the parliament was devoted to debates on administrative, military, and economic issues.

On January 14, 1878, Abdül Bey, the deputy from Yanya (Ioannina, in today's Greece), made a long speech about the state of education in the Ottoman Empire. Attempts for reform were not taking hold, according to

Abdül Bey, as the officials in charge were not well educated themselves. Asking the deputies to be fair, he asked, "What schools do we have for the people, other than some disorganized *medreses* and . . . military schools of some success? . . . How can we be civilized with so much ignorance? How can we progress? How can we preserve our religion?" His suggestions were fundamentally about the sincere implementation of all the principles of the 1869 Public Education Regulation, including the opening of a "complete," "European-style" university making available all sciences, such as law, medicine, mathematics, and agriculture—in brief, an institution "nothing like our present university."[66]

Abdül Bey's remarks and suggestions appear to have fallen on deaf ears, however, as the reaction consists only of brief comments. Ten days later, Mehmed Ali Bey, another deputy from Ioannina, made similar arguments:

While it is pitiful that due to our ignorance of contemporary sciences we are unaware of the true nature of things and the state of the world, what is most distressing and worth our attention is the poverty and misery that ignorance leads to. The Europeans beautified and cheapened their commodities thanks to science, and we are now forced to import almost all our indispensable needs from Europe.[67]

Mehmed Ali's arguments were in reference to an article published in the *Levant Herald*, a newspaper published in Istanbul in English and French, criticizing the ignorance of the Ottomans as well as the Ottoman laws on the buying and selling of land. Agreeing with the newspaper's comments, Mehmed Ali suggested that the trade of land should be facilitated, as selling their land had become the only viable option for many Ottoman subjects due to the poverty brought about by ignorance.

In these arguments we see the combination of a number of key factors that shaped the way in which knowledge and ignorance were perceived in the Ottoman Empire of the 1870s. Among the most noticeable is the "Europeanness" of science and the positive references to Europe in the arguments of those who praised the sciences of the Europeans wholeheartedly. Furthermore, in this specific instance, Mehmed Ali Bey's suggestion that land laws based on the sharia should be changed was directly linked to the ignorance of Ottoman subjects in European sciences. Hence, the reaction to his comments was, unsurprisingly, recourse to the alternative discourse on science and ignorance perfected in the 1860s and 1870s.

Asserting that laws based on the sharia could not be changed, Sadi Efendi, the deputy from Aleppo, also proclaimed, "We shall never accept the statement 'the people of the well-protected domains[68] are ignorant.' . . . Nobody can deny that Europe borrowed the philosophical sciences initially from the Arabs. I believe that the Imperial Domains have all the schools and knowledges, since the beginning of the Tanzimat."[69]

Other deputies such as Mustafa Efendi of Adana and Mecidiye Efendi of Konya made similar arguments. In addition, Hacı Ahmed Efendi of Aydın stated that while "it has become fashionable in our country to wish that education should progress," no reasonable suggestions were put forward. That *medreses* were the problem was preposterous, as most of the teachers of even the new schools still came from the *medreses*.

In response, Halil Ganem of Syria made the following remarks:

The gentlemen say we have the sharia, not ignorance. We have science, we have schools. Indeed, we have schools, and science, to some extent. . . . But do we have well-organized schools? . . . We need science, we need arts. So the Europeans borrowed them from us. I do not deny that, this is a historical fact. But our ancestors were rich, we are destitute. What does this indicate? And then they refuse the civilization of Europe. The civilization of Europe is undeniable. There are two civilizations in Europe. One is that of the youth, the fop. Puts on nice European clothes, does his hair and the like! . . . [But] there is also another civilization. That civilization is . . . law, order, industry, literature, wealth, and the science of wealth. . . . That is the civilization that we need.[70]

Ganem's support heartened the deputies from Ioannina. Mehmed Ali Bey argued that he did not comment on the issue relating to the sharia, and that he did not attribute absolute ignorance to the Ottomans. It was ignorance of contemporary sciences that was indisputable. Abdül Bey's speech was more passionate:

It was said "we do not admit ignorance, our civilization is perfect." Yes, now that we are a nation that originated from the Arab nation, we certainly are civilized. And just as we took knowledge and civilization from the Greeks, Europeans took it from us. But they took it in such a way that they left nothing for us! We studied history as well, and we know what our past and our present are like. . . . I am faithful to my word: we have no schools. . . . In my opinion, there are no schools, not only in Albania, but the entire Well-Protected Domains, even in that civilized Arab continent. There are *medreses*, but the sciences [taught there] are deficient. Are the sciences taught at the schools of Baghdad and Damascus six hundred, a thousand years ago still taught at the *medreses*? I don't think they are, otherwise we would have seen the graduates of the present *medreses* in a different way.

We are unable to bring forth anything on our own. If reform is demanded, then we should borrow and implement what Europeans first took from the Asians and then developed.

While Abdül Bey thus underlined that *medreses* no longer taught the needed sciences, Hacı Ahmed Efendi in his response indicated that *medreses* were for religious sciences. What he called "civilizational sciences" (*ulûm-ı medeniyye*), on the other hand, were needed for this world, and they had indeed been neglected due to the constraints of the times. Sciences such as medicine and mathematical sciences did not conflict with the sharia and so should be fostered, Ahmed Efendi argued. This could be done, but only if the real problem, that is, the constant interference of European ambassadors in the affairs of the Ottoman Empire, was prevented.

The debate continued in other sessions. In another memorandum on ignorant civil servants, Abdül Bey suggested that foreigners be employed in some public commissions. Sadi Efendi of Aleppo was furious:

This debate on skill and knowledge that has been going on for a while is a stale one. When two people meet on the street, they say the same things. However, our country is not in a state of ignorance as presumed. It cannot be stated that we have no skill and knowledge in our country. There are only certain things called industrial goods that are manufactured with machines in foreign countries. [A few years of peace is enough for us to have them], as there are more skilled people in our country. . . .

As for sciences and arts, in Arabia and other territories we have schools for all sciences, like philosophy (*hikmet*), logic, manners (*adab*). We have schools even in villages. . . . Yet this report portrays our ignorance at such a level that even in our parliament it seems necessary to have some men from foreign countries!

Ohannes Efendi of Istanbul, an Armenian deputy, took Abdül's side, stating that the Ottomans needed "new knowledges," not "old things" like logic. While stating that his hometown of Beirut was not in such ignorance, Abdurrahim Bedran of Syria also concurred with Abdül. In his clarification, he discussed the issue in detail:

Our education amounts to the sciences of grammar, syntax, logic, rhetoric. We have many men of knowledge, too. Yet these sciences are not beneficial for us, for this world. What is needed and beneficial is mathematical sciences. It is mathematical sciences that invented the telegraph. They found the power of steam with mathematical sciences. These cannons, rifles were produced with those sciences. I said we have schools in Beirut; yes, we do, but they are there thanks to foreigners. Protestants of America, and then the Jesuits came and opened them, that's why we have a few schools.[71]

Never underestimate the sciences of the Europeans, he concluded: "Nobody can ever deny the science and industry of Europe."

The debate did not last long, however, as Sultan Abdülhamid II shut down the parliament indefinitely on February 14, 1878, on the pretext of the Russo-Ottoman war. Nevertheless, that the issues were brought up in parliamentary debates in ways almost identical to the various representations in official documents and newspaper articles demonstrates how established the alternative discourses were by the end of the decade. The tone of the debate reveals that establishing the meaning of science and defining the characteristics of those who learned the new sciences were crucial, as the debate was essentially about who the Ottomans were and who they should be. Additionally, however, the parliamentary debate also indicates the introduction of another theme related to the broader debate on science: the missionary influence on Ottoman communities, a topic that would become much more prominent in the following decades.

Conclusion

The 1860s and 1870s witnessed two parallel developments. Leading bureaucrats of the era such as Münif and Safvet Pashas and the contributors to the *Journal of Sciences* reinforced the discourse on the unique benefits of the new sciences. They also characterized those ignorant of these sciences as parasites with no real understanding of how the world actually worked and no means for self-dependence. Science was defined as a sphere distinct from other spheres such as politics and religion, and while the idea of an inevitable antagonism between science and religion was never part of this discourse, traditional religious sciences like *fiqh* or *kelam* were never treated as deserving the same esteem and attention as the new sciences like zoology or geology. Useful sciences were those useful for the material world, and religion was referred to primarily as a provider of incentives for pursuing these useful sciences.

The alternative discourse that was constructed by figures like Namık Kemal, Ali Suavi, and Ziya Pasha, on the other hand, was one that fully acknowledged the significance of the need for the new sciences but that also demanded the same appreciation for the "old" sciences taught at the *medrese*s. For them, the category "science" should continue to contain the old sciences since they constituted not only the fields Muslims had developed but were also still beneficial. If the old sciences seemed inadequate, it was because they had been neglected by the culturally alienated new bureaucrats; they appeared not to be useful anymore simply because the ruling elites no longer made use of them. This attitude had also given rise to the emergence of a new type of individual who believed that memorizing some words and parroting European authors was sufficient for being considered a knowledgeable, wise, respectable man. So this alternative discourse was not critical of the new sciences; it simply demanded that those speaking in the name of the new sciences be "authentic" Ottoman subjects. Their lifestyles, consumption patterns, and aspirations should not differ from the established norms. Simply put, science should not be represented by snobs.

While the common perception of science was thus to a considerable extent based on those who spoke in its name—fops—we should also note the European representatives of science within the Ottoman Empire as well. A remarkable case concerns the opening of the Imperial Meteorological Observatory in 1868, under the administration of the French engineer Aristide Coumbary. Coumbary, who was in the empire for the maintenance of telegraph lines, also made astronomical observations that he reported in European journals both before and after assuming this post.[72] While this observatory is referred to as one of the important steps in the spread of European sciences in the Ottoman Empire,[73] its benefits and purpose appear not to have been clear, at least to members of the Muslim community of Istanbul—in a way reminiscent of the reaction toward the public lectures at the university. Newspapers did report the information sent by the observatory regarding solar and lunar eclipses and the weather,[74] but they also published pieces criticizing or mocking the observatory. Indeed, the observatory was the subject of many essays in the humor newspaper *Diyojen*, including at least one written by Namık Kemal. The theme of these criticisms was that the institution failed to provide the most basic service Muslims needed: the firing of the cannon to declare the end of the daily fast in the month of Ramadan.[75] As such, there was no reason for Ottoman Muslims to be impressed with this institution that they did not even perceive as truly belonging to them.

Clarifying this perception are two pieces Ali Efendi published in the *Basiret*. The first was a dialogue between himself and an acquaintance of his whose "Ottoman light has been peeled off" and who was adorned with "Frankish gilt." While this fop extolled the observatory, he blamed "us Turks" for not appreciating it enough, even though, according to Ali, the so-called observatory was but a chronometer, the holder of which was paid sums he did not deserve.[76] In another piece that is based on a probably fictional discussion at an Istanbul coffeehouse, he pointed out that the cannon fired from the observatory was intended to tell the time, but in terms of the European rather than the Ottoman system. It was particularly confusing in the month of Ramadan, as people thought it was the end of the daily fast. Ali's opinion is possibly the one he sarcastically has one of the patrons express: this was not a major problem, as conversion tables were published in the French newspapers of Istanbul, and "the state spends around ten thousand *kuruş* a month for the salary of Coumbary, [as well as] the expenses of this observatory, and the ads in the newspapers."[77] While the state spent so much money on a scientific institution at a time of immense financial hardship, no one but "the Franks" benefited from its services, and only the fops appreciated it simply because it was an institution imported from Europe and run by Europeans.

The alternative discourse that emerged in this period is thus the expression of the overall dissatisfaction and estrangement of a sector of the young and educated class joined by the lower ulema. For them, the Tanzimat regime that had delivered much less than it had promised, rendered some groups obsolete and caused a broad sentiment of demotion within the Muslim community—as also made clear by the parliamentary debates. While it was absolutely not a discourse on the dangers or harms of the new scientific knowledge of Europe, it embodied a comprehensive criticism of the type of individual with whom these new sciences were associated. Hence, anybody who wished to make a case for the new sciences also had to demonstrate that he was not like the fops whom the Tanzimat had produced and empowered. It was less the arguments about science than the character of their makers that was criticized in the environment of the 1860s and 1870s. Opinions about science now had implications for membership in the community.

SIX

Inventing the "Confused Youth": Science, Community, and Morality in the 1880s

Introduction

The reign of Abdülhamid II (1876–1909) has long been the subject of heated debate in Turkey: Was Abdülhamid a ruthless tyrant, or was it his political genius that saved the empire from collapse? Was he but a champion of reactionary conservatism, or was he the sultan who laid the very foundations of modern Turkey? Such questions are directly related to the Ottoman debate on science itself, as the ambivalence that they indicate is central to the way science was discussed in the 1880s. What we observe during Abdülhamid's reign, in fact, is a peculiar amalgamation of the ideas of the early Tanzimat elites and the Young Ottomans, and the consequent emergence of a new official discourse on science.

In his interpretation of the so-called Islamism of Abdülhamid, the historian Selim Deringil argues that as the empire's power diminished, the sultan placed increasing emphasis on symbols of power and, in particular, symbols with Islamic significance. Following the territorial losses of the 1877–78 Russo-Ottoman war, which also led to a drastic decline in the empire's non-Muslim population, the sultan started to "stress the Islamic religion as a new bid for unity against what he saw as an increasingly hostile Christian world."[1] While concepts and principles associated with Islam had always been central to Ottoman political culture, Abdülhamid's reign is a period in which the Islamic identity, or "Islamicity," of the empire, rather than being taken for granted, was asserted and made visible. In this respect, the process is similar to what Dale Eickelman referred to as the "objectification of Islam"—the reconstruction of Islam in modern Muslim societies as a well-bounded, coherent set of beliefs and practices, with specific instructions on specific questions that can be learned.[2] By attempting to make the empire more explicitly Islamic, the Hamidian regime thus transformed Islam into a system of ideas and values to be imagined as distinct from everyday practice, a label with which to mark statements and practices. And nowhere was this more obvious than in the field of education.

It is undeniable that Abdülhamid's reign witnessed an unprecedented rise in the number of Western-style schools all around the empire. The growing number of educated and ambitious young men also gave fresh impetus to the Ottoman press, and numerous short-lived periodicals were launched. As the following sections will illustrate, the popular newspapers were also filled with the essays, letters, and polemics of these young men—and, to a much lesser extent, women. What is also crucial to note, however, is that the change in the Ottoman educational system was not only quantitative but also qualitative. As the historian Benjamin Fortna argues, while Ottoman educational reforms of the early nineteenth century were essentially about "imitating the best attributes of Western European education," Hamidian reforms involved adapting these Western-style schools to the specific needs of the late nineteenth-century Ottoman Empire.[3] And this adaptation involved the implementation of a policy of emphasizing Islam and Islamicity.

As a primary step in his educational reform program, Abdülhamid II established several curricular reform commissions starting in 1885. According to the 1887 commission report, the new schools led students to take an extreme interest in Western ideas, which, in turn, gave rise to disobedience, immorality, and ignorance in Islamic matters. It recommended adding Arabic language courses to the curricula, increasing the number of classes on Islam, and monitoring schools closely in order to "fend off the danger [posed by students'] being occupied with Western works and writings that are harmful to Islamic morals and to the exalted sultanate."[4] A similar report from 1900 asserted that students needed to have all the essential knowledge pertaining to science, but they should also "obtain intellectual incisiveness and religious firmness [and] be faithful to the sublime sultanate and endowed with sound morals."[5]

A related and equally disturbing problem was Western encroachment via Christian missionary schools. The increasing popularity of these well-equipped schools especially among non-Muslim Ottomans alarmed the Palace. A special investigation on the state of missionary schools in the Ottoman Empire was carried out in 1893–94 by Minister of Education Zühdü Pasha, who stated in his report that the schools were a great threat because "the foreigners realized that they can achieve their political objectives by corrupting the minds of the students of these schools and leading them astray."[6]

Clearly, the growth in the number of young and aspiring men was a double-edged sword, and ensuring their loyalty became a central concern for the Palace. Indeed, we observe during Abdülhamid's reign the emergence of another "problem character" that would come to accompany the fop as a token of wrong Europeanization: the "confused materialist"—brilliant in the new sciences but oblivious to the religion and values of his own society.

As we have seen in the previous chapters, that students of the new schools—the future elite of the Ottoman Empire—be both knowledgeable and moral (i.e., patriotic and compliant) had been the chief concern throughout the nineteenth century. Consequently, the debate was essentially about whether men of science would by definition be moral individuals or whether morals should be inculcated separately. It was in Abdülhamid's reign that the latter became the official view and was put into effect rigorously. Likewise, while Islam had maintained its position as the ultimate reference in defining morality in previous decades, the school curricula of the 1880s and the 1890s made this connection more explicit than ever before. In a period shaped by the perceived missionary threat and the strategic use of Pan-Islamism as an ideology to prevent further disintegration, science was more overtly than before a matter that could not be discussed without a reference to community and morality.

Abdülhamid's efforts to win the hearts and minds of the students of the new schools in order to render them obedient did not simply involve teaching them Islamic morality, however. The reduction of Islam to a collection of moral instructions—a process commonly associated with secularization—was not deemed appropriate or sufficient, as we can observe from the policies of the period. That European science was in harmony with Islam, a theme also developed in the previous decades, became an official mantra in these decades. As we have observed in official documents already, the line of reasoning was as follows: If the authority of the state and the authority of religion were one, and if religion was the sole source of moral values, then the idea that science contradicted religion could lead young, educated men to question their faith and, consequently, disobey the state. Thus, Abdülhamid encouraged authors to "prove" that Islam and science were harmonious, and he was rather generous to those who proved this argument.

We can see many relevant examples in the writings of Ahmed Midhat Efendi, who had by then become, in the words of Carter Findley, the "collaborator and speaker" of the sultan.[7] As early as 1883, Midhat clarified the key point in an essay he wrote against European thinkers who argued that Islam and science were antagonists:

Had [a European author] argued not that Islam and science are in conflict but demonstrated that Islam *is* science, then he would have brought forth an argument that is truly worthy of investigation for serious men of science. . . . The belief "religion amounts to a few issues about morality" is now common in Christianity, and has led to the decline of this religion; yet if it can be shown that, in contrast, "religion means the totality of scientific judgments," then this will attract everybody's attention to this novel argument. And the relation between Islam and science is precisely that "religion means the totality of scientific judgments."[8]

Further attempts to strengthen this identity forged between Islam and science include the works of the Lebanese author Hussein al-Jisr, who was rewarded by Abdülhamid in 1891 for his efforts to demonstrate the harmony between Islam and science, and Ahmed Midhat's translation of and commentary on J. W. Draper's *History of the Conflict between Religion and Science*, published in four volumes between 1895 and 1900.[9] The latter was also particularly important as a treatise against Christian missionaries, as its emphasis was on depicting the true conflict as one between anti-science Christianity and pro-science Islam. This approach was increasingly popular in this period and would be further taken up by the mufti of Egypt and influential thinker Muhammed Abduh in his *Science and Civilization in Islam and Christianity*, published in 1905.[10]

While the attention of the Palace to science indicates a strong reaffirmation of the state sponsorship of science in the Ottoman world, it also further clarifies how the boundaries of science and the actions and personal characteristics of men of science were similarly policed by state authority. We learn from the memoirs of his daughter that Abdülhamid used to tell his children to believe in both religion and science,[11] but certainly this was a science that was in harmony with Islam (i.e., the officially sanctioned version of Islam) and represented by humble, patriotic men of science.

For the purposes of our study, the Hamidian era can thus be characterized by the increasing emphasis on two issues: the Islamic identity of the empire and the disciplining of the new generation. The emergence of the confused materialist as the villain stereotype is central to the latter, so I will analyze the birth of this character in the second half of this chapter and elaborate on its significance in the next. First, however, I will tackle the issue of Islamic identity, as in the 1880s the debate on science evolved in ways that would make it particularly inappropriate for this concern. As we saw in previous chapters, science had already proved to be a topic that could hardly be discussed without a reference to tradition and identity in the Ottoman Empire. Yet in the Hamidian era, the debate on science demonstrated unequivocally the precariousness of the idea of a common, binding Islamic identity for the new generation.

A. Debating Science, Defining Community

One does not need to look for zealous nationalist manifestos to see how arguments about science were inseparable from concerns regarding the definition of the community in the 1880s. "Harmless" brief exposés had as much potential in this respect as anything, as a short, seemingly insignificant essay on the science of pedagogy published in Ahmed Midhat's newspaper *Tercüman-ı Hakikat* will illustrate. Briefly discussing what this new science engaged in, the essay recommended that the Ottomans learn and make use of this science that was not well known "among us."[12] The reaction of *Tercüman*'s main rival, *Vakit*, was blunt—whether this statement was true depended entirely on who "we" were: "If, with 'the Ottoman world' [the author] means that [pedagogy] is not known in Turkish, I have nothing to say. But if he means that it doesn't exist [among Muslims], then I reject that." Muslim Arabs had developed a science on raising children based on Qur'anic principles centuries ago, according to the author, and it was "pitiful" that such comments that "insult[ed] the community" were made by those who read only European books.[13]

The response of *Tercüman-ı Hakikat* made it strikingly lucid what this discussion on a branch of science was also about, even with its title: "So What Do the Arabs Have?" For the author, it was indisputable that Arabs, along with Greeks and Romans, had made the biggest contributions to civilization. Yet it was the Christians in Syria who knew Arabic who presently made use of the books written by Arab scholars, not "us," simply because "we" did not speak Arabic. Moreover, there were very few "wise men among us who realize that we are a different nation (*kavim*) than the Arabs and thus we need different books." This misguided presumption that Arabic works constituted a scientific legacy that Ottomans (that is, apparently, Turkish speakers) could claim was leading many to turn down the attempts to import the new sciences. "What is truly pitiful is empty boasting. . . . When we say 'we don't have this science,' this is not an insult. It is expressing a truth to our compatriots." Those who were proud of Arabic works should write treatises and contribute by letting us learn the sciences Arabs have, *Tercüman* concluded.[14]

But *Tercüman*'s response did not end there. Five days later, it published sections from a reader's letter which, according to Ahmed Midhat's note, could not be published in its entirety due to the harshness of its tone. The reader's criticism was directed unequivocally against "the newspaper of the men of the *medrese*" that pointed toward a time six hundred years back, and did this to a people already passionately anticipating the future. *Vakit*'s reference to the Islamic scientific legacy was ridiculed in a most ruthless fashion: "We have two-wheeled ox-carts for covering distances, too. Then is it an insult to the nation if the *Tercüman-ı Hakikat* says 'We don't have trains. We should rush to make them'?" It was true that the Arabs had written invaluable books, but it was the Europeans who were presently making use of them. Yet this was done in order to determine the level of knowledge that Arabs had once reached, not because the present level of science was similar to what could be found in those books.

The author [of the essay in the *Vakit*] does not know that the people have a new awareness now. . . . In countries whose products of knowledge astound our vision, they prefer to books written six hundred years ago those that were written just one year ago. This preference stems from the hope that the more recent one contains something new. When freshness in learning is appreciated so much, the only impact of the words of the critics is to bewilder the people and make them ask: "Are we moving forwards or backwards?"[15]

What needed to be done was to translate Arabic works and turn them into "Ottoman works" in order to truly appreciate the past glory of the Arabs, along with the translation of the contemporary works of the Europeans.

For late nineteenth-century Ottoman authors, then, defining science was also a question of defining and coming to grips with a legacy and, ultimately, a process of self-definition. The debate regarding whether scientific works written in Arabic could be seen as the common heritage of all Ottoman Muslims made it possible to articulate the statements that Arabs and Ottomans were different, and that religion alone did not imply community—in contradistinction to the Hamidian policy. Furthermore, while the idea that Muslims in general and Arabs in particular had made so many contributions to science had been a crucial tool for the promoters of the new sciences, it also continued the association of Arabic with the old, with all its negative connotations in an age of progress.

The same year as the dispute between *Vakit* and *Tercüman*, the Albanian Ottoman author Şemseddin Sami published his book *Islamic Civilization* and started to construct his body of work, which came to epitomize the position of the critics of the discourse on Islamic contributions to science. Reiterating the common view, Sami argued that the pinnacle of the period between ancient Greece and the European Renaissance was the rise of the Islamic civilization. He was not hesitant to confess the source of his narrative, either: "We must not forget that it is European scholars themselves who show us the level Islamic civilization had reached and that contemporary civilization was born from it, and who even put before our eyes many works of our ancestors of which we are unaware."[16]

Needless to say, this confession problematizes the legitimacy of the Ottoman entitlement to the Islamic legacy: how could Islamic contributions to science be considered the tradition of which the Ottomans should be proud, when they learned about those very contributions from European authors? Indeed, a similar confession can be found in a letter of Ahmed Midhat: "When the situation arises, we brag, and say that we produced Ibn Sina [and] Ibn Rushd . . . Do we [actually] have any knowledge about what they said? Even though it is essential to know Arabic or Persian to know them, how many of us are able to truly understand one sentence in these languages?"[17]

Also crucial is the place ascribed to Islamic civilization in this linear progression. Şemseddin Sami spells it out: Islamic civilization was superior to Greek civilization, "just like contemporary European civilization is superior to Islamic civilization."[18] Finally, it was simply indubitable that Islamic civilization was not Arab civilization. Pre-Islamic Arabs were not civilized and occupied themselves only with poetry; the sciences were developed only after the birth of Islam. Furthermore, the majority of Muslim scholars were not even Arabs, and of the greatest scholars of the Islamic civilization, Ibn Sina was Persian and Farabi was Turkish.

Şemseddin Sami developed these arguments in a series of articles on civilization he published in the journals *Hafta* (The Week) and *Güneş* (The Sun).[19] While in *Hafta* he elaborated on the Eurocentric history of civilization, his lengthy essay published in *Güneş* in 1884 focused on the question of "transferring the new civilization to the Islamic peoples." He asserted in the latter that Islamic civilization had declined to such an extent that the current state of the Muslim world made one doubt that the Muslims had ever been civilized. Contemporary European civilization would not experience the same decline, however, as it had "protectors like the press, steam power, railways and telegraphy."[20] Islamic civilization was great indeed, but due to the absence of the printing press, even during the times of great scholars like Ibn Rushd and Ibn Sina, peasant masses were entirely unaware of their works and esteemed sorcerers and healers instead. Furthermore, the only patrons of men of science were the rulers, which made the progress of science subject to their whims. Freedom of thought, crucial for scientific progress, was abandoned by scholars who wished to please their patrons. It was true, as Muslims commonly believed, that moral corruption among rulers led to the decline of scholarship and civilization. But the relationship was only indirect, as it was obvious that debauchery was not making such an impact on the progress of civilization in contemporary Europe. The fundamental problem was leaving civilization only to the protection of rulers, as only under that condition would their corruption also lead to the decline of civilization. It was only when "half-savage" groups with no appreciation of science came to dominate the Islamic lands that civilization decayed. These new rulers separated religious sciences from rational and natural sciences and wished to protect only the former. But as the religious sciences of Islam cannot be truly understood without knowing all sciences, bigotry and ignorance had taken over the Islamic civilization for centuries.

Muslims had to admit the fact that their glorious civilization was now obsolete:

When there [are] the constantly growing products and illumination of contemporary civilization, referring to those ancient works or satisfying [ourselves] only with them is like facing the sun but trying to make do with the wick of an oil lamp. . . . It is true that religious zeal would impel a man to be content with the lamp which he knows to have been lit by his ancestors, yet it is essential that reason and wisdom should overcome any such feeling. Today however much the effort and expense is required to revive . . . the medicine of Ibn Sina, the wisdom of Ibn Rushd, the chemistry of Câhız, to extract their books from underneath the dust of libraries and translate them into the various Muslim languages, to publish them, and to found schools and colleges devoted to teaching them, we must make the same effort and go to the same expense to put into circulation among us the best scientific works of our own century. For just as we cannot cure malaria with Ibn Sina's medicine, so we can neither operate a railroad engine or steamship, nor use the telegraph, with the chemistry of Câhız and the wisdom of Ibn Rushd. Hence, we should leave the study of the works of Islamic scholars to the students of history and antiquities, and we should acquire sciences and arts from the contemporary civilization of Europe.[21]

It was wrong to think, like some among both the elite and the commoners in the empire did, that the current civilization of Europe had to do with Christianity. But it was equally wrong, even with the best intentions, to counter this misconception by emphasizing that the sciences of the Europeans were but those developed by the Muslims. While such efforts did indeed reduce the number of people who associated European civilization with blasphemy, they also led some to hate Europeans, thinking that their civilization was stolen from Muslims. They did learn a lot from us, Sami admitted, "but none of the things that they have today was stolen from our ancestors." Europeans had taken an oil lamp from Muslims but discarded it once they had illuminated their surroundings.

There is nothing here that we can be proud of. We should be ashamed of this. Because we dropped that lamp and caused it to die down, and now when we see the sun of civilization shining right before our eyes, we still don't want to take advantage of it. We choose to close our eyes and remain in darkness, with some of us saying "that's not the sun but an illusion" and others saying "that's just an imitation of our lamp that faded"![22]

This way of challenging the "Islamic origins of modern science" narrative was neither new nor unique to Sami. As early as 1873, a similar argument had been made by Ahmed Midhat, in whose view it was "[wrong] to conclude based on our precedence to the Europeans in science that we do not currently need [these sciences]. Because we have entirely resorted to comfort and ease in the last two centuries and failed to be worthy successors to our ancestors, and in comparison to Europe, we are awfully backward. Hence at present we are obliged to achieve the progress that we need by borrowing it from them."[23] Similarly, Ahmed Rasim wrote that while it was true that sciences had been introduced to Europe by the Arabs,

[One should not] look down upon the contemporary sciences. . . . While it is true that the introduction of science to Europe was thanks to the Arabs, Europeans studied and scrutinized the books they acquired after the conquest of Andalusia, and worked hard in order to benefit further from them, thus reaching their contemporary perfection. Here, too, the dissemination of learning can be achieved [only] through diligence, hence it is useless for us to pride ourselves [on] memories of the past.[24]

Hence, it is reasonable to argue that Şemseddin Sami's was a blunt expression of an idea to which others were also alluding. This directness, however, also makes clear the two dimensions of the statement that Islamic contributions to science belonged in the museum and the history book. First, while they were Islamic contributions, they were written in Arabic even though the majority of their authors were not Arabs—hence the legacy, even if it were still beneficial, had to be first translated into the Ottoman language, Turkish, to be utilizable. Second, the existence of a legacy did not even matter, as Muslim societies, including the Ottomans, were virtually at point zero vis-à-vis "contemporary civilization."

By suggesting that even religious sciences could not truly be developed in isolation from the rational and natural sciences, Sami deemed Islamic branches of knowledge worthy of respect but also proclaimed them and those who studied them ineffectual in their current state. Furthermore, by arguing that Islamic sciences themselves could only be understood by those who studied the new sciences, he made a case for both the need for and the "virtue and fidelity" of the new men of science. If the old sciences were no longer valid, the possessors of the old knowledge were no longer what the empire needed, and if understanding Islamic sciences themselves depended on grasping the new knowledge, then the masters of this knowledge would also be closer to comprehending Islam itself. The new men of science were not a threat; they were model Muslims.

While this was an uncompromising expression of the discourse of earlier Tanzimat bureaucrats, Sami's statements regarding the actual significance of claiming a legacy that comprised books in Arabic rather than Turkish were very closely related to a broader debate about the identity of the community at the time.

New works on rhetoric, and an innovative method for teaching Arabic developed by Hacı İbrahim Efendi, an ulema, were the focus of this debate, but science, religion, and identity were once again brought into the conversation as tightly related concepts.[25] Commenting on İbrahim's endeavor, Ahmed Midhat noted that something did need to be done in the area of language. But the needed task was composing a dictionary of the Ottoman language, with particular attention to scientific terms, rather than worrying about the teaching of Arabic grammar:

Even nations whose level of progress is higher than ours or that of the Arabs act in this manner. Even though words like "telephone," "microphone," "telegram" were not in their *dictionnaires* up until yesterday, now linguists include such words in their lexicons, as if they were their own words. . . . If our critics will lament us, calling us the imitators of Europeans, they are free to do so. When it comes to service, we don't see a difference between imitating Europeans or Arabs.[26]

Abdurrahman Süreyya published a similar essay arguing that without a language containing the names of new scientific inventions, it would be too difficult to make any scientific progress.[27] Hacı İbrahim, in a lengthy response, asserted that the Arabic language could never be treated so lightly, as Arabic constituted the core of the Ottoman language, and thus, those intending to construct a new lexicon and ignore Arabic would unavoidably fail. Ignoring Arabic grammar and similar fields, creating a new dictionary, and progressing in science were also unacceptable as a policy, when even the French continued to teach their children Latin.[28] Hence, Arabic was the essence of "our" civilization, just like Latin was that of the *other* civilization. For Midhat, such concerns basically amounted to a gross unawareness of the new world:

Yes, we say, "let us facilitate the instruction of the language, and then occupy ourselves with the other sciences and follow the path of progress." This is because we see all nations on this path. Our erudites cannot reach their goal and stop us by uttering buffooneries like "You idiots! What you call the path to progress is [inventing new words]. That cannot be allowed." Because we understand that progress has nothing to do with that.

We see progress where Krupp cannons, Gatling guns, torpedoes, hot air balloons, telephones, microphones, phonographs are invented. . . . There they occupy themselves with miraculous endeavors like hearing the sounds of ants using microphones, conversing between the two sides of the world using telephones, preserving sounds for years using phonographs. Their twenty-five-year-old engineers are trying to have trains traverse the English Channel. . . . Progress is precisely these. And these can be achieved by first having a language.[29]

When Ahmed Midhat underscored that it was the grammar and vocabulary of Ottoman Turkish that was needed to progress in science, not the teaching of Arabic grammar, İbrahim's reaction was clear-cut:

[Midhat] should not deny that Arabic is . . . the language of our religion, letters, and science. . . . Arabic, our language of science and letters has such a great wealth that it lent words to Spanish, Portuguese, Persian, Hindi, French, Italian. . . . It generously offered them sciences and knowledges. . . . [Hence] breaking and bending the rules of this language and changing its words with the excuse to compose a lexicon on sciences and arts are unacceptable.[30]

What made it inconceivable was also clear: the Arabic language was the language of the Qur'an, the foundation for understanding religion and the world. Therefore, the sultan, the protector of religion, would also be the protector of the language of religion.[31]

Ahmed Midhat found it appropriate to conclude the debate at this point. With a brief statement, he announced that it had become clear that an agreement could not be achieved; clearly, the tone of İbrahim had rendered this too sensitive a debate for Midhat. Indeed, just like his dispute with Harputlu İshak almost a decade prior, Midhat's debate with İbrahim, too, ended with his opponent's reference to religion and the sultan.[32]

Examples such as these are significant in their demonstration of how science and the meaning of Ottomanness were such closely associated matters in the late nineteenth-century Ottoman Empire, and how, via religion and morality, debates on these concepts ended with a reference to the authority of the sultan. As these texts attest, by the 1880s science had clearly turned into a subject that could not be discussed without an open reference to identity. The man of science was not to be a fop, but he was now defined and evaluated not only in relation to "Europeanness" but also to "Turkishness" and "Arabness."

The celebrated poet Muallim Naci published a poem in this period reiterating what I have referred to as the Andalusia narrative in order both to criticize the overly Europeanized Ottomans and to emphasize the Arab legacy: "We are proud to be the beneficiaries / of the good works of the Arabs . . . / It is from them that Europe attained learning / and now is busy selling it back to us."[33] The journalist Ebuzziya Tevfik, on the other hand, wrote in the same year that his only source of pride was the Turks, even though he himself came from an Arab lineage: "Because it was the Turks, not the Arabs, who made the Ottomans the shining light of glory in the East and in the West."[34]

This theme would become only more dominant in subsequent years. In 1897 Musa Akyiğitzade, a Tatar Turk, published a book entitled *A Glance at European Civilization*, where he proudly restated the idea that the contemporary civilization of Europe, while built on the contributions of the Muslims, was not the civilization of the Arabs. The great physician and philosopher İbn Sina, for instance, was "the Turkish son of a Turk."[35] Halid Eyüb espoused the same approach in a set of essays on Islam and science. Based on the works of orientalists, Eyüb discussed the works of Muslim scholars in chapters devoted to astronomy, geography, philosophy, chemistry, and the applied sciences. The book emphasized that not all Muslim scientists were Arabs, and it counted even the Mongols, along with the Turks, among rulers who supported scholarly activities.[36]

Even more significant was the publication of the first book dedicated exclusively to the "services of the Turks to the sciences and the arts."[37] Mehmed Tahir, the author, condemned European authors who insisted on portraying Turks only as warriors and attributing the works of Muslim scholars to Arabs, simply because they were written in Arabic. An analysis of the genealogy of these scholars revealed that a third of Muslim philosophers and scientists were Turks. Furthermore, in the field of religious sciences, Turks constituted almost half of the great Muslim scholars. Even this was not enough for Tahir: most Arabic and Persian men of science had lived in regions under Turkish control and been able to produce their works thanks to the patronage of Turkish rulers. Hence, they had, in a sense, been "Turkified."

But it was particularly the graduates of the new schools who reacted against the emphasis put on Arabic by *medrese* students and graduates, and it was this process that increasingly strengthened the associations between Turkish and the new, and Arabic and the old. A remarkable example can be found in the newspaper *Tarik*, which published in 1898 an article entitled "Arabs Have Many Sciences We Can Benefit from."[38] The author, a member of the ulema, noted the importance of Arabic for all Muslims, and he focused mostly on linguistics and literature but also mentioned the names of other sciences that Arabs had contributed to, such as astronomy, geography, and medicine. Soon afterward, the young journalist Hüseyin Cahit, a graduate of the School of Administration, published his response in the same newspaper. Written with an unmistakably sarcastic tone, the essay stated:

If we are supposed to be grateful to Arabs for those sciences, we can proclaim without fear our indifference. The feeble astronomy, mechanics and medicine of the Arabs are now but a plaything compared to the progress in today's binoculars, machines, geological discoveries, and anatomical investigations. If the books of the Arabs on these sciences have any worth, it is historical. If we want to become [men] of our age, we leave those books alone and embrace the books of today to fill our minds with the sciences of today. And we find those books in the West, not in the Arabs.[39]

This comprehensive attack on the idea of the legacy is in many ways similar to the repudiation of classical scholarship (*gezhi*) in China, particularly after the Sino-Japanese war of 1895, and the definition of Western science as an entirely separate category (*kexue*).[40] Similar arguments were rather common in India as well, where many authors described Western sciences as incomparably superior to traditional types of learning.[41] Note, however, that in the Ottoman case, the legitimate owner of the tradition in question was a contentious issue: Were Arab sciences and Islamic sciences synonymous concepts? In any event, could these concepts define the traditions of Turkish-speaking Ottoman Muslims?

Indeed, the reaction to Cahit's essay was, once again, expressed in the form of a discourse linking science to the identity of the community. Mustafa Sabri, a religious scholar, wrote: "Cahit says 'Let us not worry about the sciences of the Arabs, let us look at ourselves.' But who are we anyway? . . .

We are a perfect totality of components with an essence fashioned by Islam." As a result, regarding Arabic sciences like an outsider would was the mistake: Arabic sciences were *our* sciences, products of *our* civilization. "I do not deny the need for European sciences," Sabri continued. But these sciences were needed only in order to deal with the difficulties the empire was facing; they could not become an essence for the Ottomans like Arabic—that is, Islamic—sciences. It was the latter that would guarantee the future of the Ottomans.[42]

Hence, the identity of the man of science—his allegiances, what and whom he knew and respected—became an ever more crucial question in the 1880s and 1890s, as the debate about science was ultimately about the type of person that the empire required. In this respect, Mustafa Sabri's striking question "Who are we anyway?" had a normative tinge as well: what kind of people did an empire facing grave threats truly need?

Inseparable from this question was the issue of useful knowledge that had already been strongly established in the previous decades. In a sense, the primary "problem" with Arabic grammar was not necessarily that it was Arabic; it was that it was not useful, and the empire needed *useful* people. And it was precisely in the debate about this useful citizen that the "confused materialist" stereotype emerged.

B. Useful Knowledge, Useless Groups: Who Does the Empire Need?

1. Scientists or Poets?

a. Setting the Terms

In the journals and newspapers of the 1880s and 1890s one comes across many polemics regarding literature, some of which have been examined by historians of Turkish literature. What have commonly gone unnoticed, however, are the countless references to science in these debates. Indeed, it appears to have been hardly possible to discuss literature without referring to science. The reason is clear: these debates were ultimately about the good citizen whom the empire needed, and science and literature were commonly defined as two alternative routes toward its construction. Hence, the connection (or lack thereof) between science, literature, and virtue and vice was constantly brought up in these debates.

Literary arts and linguistic topics like Arabic grammar had always constituted a core aspect of *medrese* curricula, but such classes were offered—

to a lesser degree—in the new schools as well, since raising new bureaucrats with a good command of Ottoman bureaucratese was a key objective of nineteenth-century educational policy. In this respect, criticisms directed against those whose most valuable talent was in literary arts were as much against many civil servants as against men of literature. Arguments about the former group will be discussed further in the next section, but this fact is important to keep in mind when analyzing criticisms such as:

The nineteenth century is a genius, hence it wants as [its] companion nations that have progressed and acquired the sciences. . . . The absence of science in our country is putting those who fancy themselves litterateurs in a pitiful state. Ignorance is dripping from the pens of such litterateurs. . . . A man who wants to show off as a litterateur should at least know the uses, if not the contents, of sciences. Devoting one's life to learning Arabic, Persian, French and other languages, and turning oneself into a pile of grammatical and syntactical rules, and calling oneself a litterateur . . . Of what use is that![43]

While attacks against litterateurs in general were common in this period, the most heated debates tended to be about poets in particular, and it was poets themselves who contributed most to these criticisms. An influential work that ridiculed traditional Ottoman poetry with references to science was published by the prolific intellectual and litterateur Namık Kemal. In *Tahrib-i Harabat* and its sequel, *Takib-i Harabat*, published in 1885–86, he criticized an anthology of classical Ottoman poetry, *Harabat*, by his former comrade Ziya Pasha. Kemal clarified the basic principle behind his condemnation of the kind of poetry glorified by Ziya in verse form:

Aren't Nedim and Nef'i enough already?[44]
Is poetry of any use to us?
The brightest one is the biggest of lies,
Find a true one if you want to convince me.

. . .

I haven't seen a poem in Turkish,
Five verses of which were in harmony with science.
Its metaphors are delusions, fancies,
Compared to reason, they are insanities.[45]

In these polemical works, Namık Kemal advocated a new idea of poetry that would be based on facts, not filled with tired metaphors. Yet this was not for the sake of being scientific, as the verses above indicate: the ultimate purpose was to make poetry useful. Such arguments illustrating the changing criteria for assessing the merits of the cultural legacy are familiar to students of the history of debates about science, and similar examples can be found not only in the European but also the non-European context. In Iran, for instance, the merits of the giants of classical Persian poetry like Ferdawsi, Sa'di, and Hafez became the topic of a similar debate on useful knowledge in the early twentieth century.[46]

Leading Ottoman newspapers also started to publish texts assessing the relative merits of science and literature and, inevitably, their representatives after the 1870s. A letter sent to the *Tercüman-ı Hakikat*—and appreciated by the publisher—complained that more pages were devoted to literature than to science and argued: "Science is the nursemaid of literature. Literature, without science, is like a child without a governess."[47]

This patronizing attitude toward literature, along with the identification of science with reality and literature with childish fancy, was also common in the literary columns of the newspaper *Saadet*. Particularly important for the editors was to illustrate the "problems" with traditional Ottoman poetry and the traditionally inspired poems of contemporary romantic poets published in the journals of the period. A poet who wrote about the "blood in [his] veins [that] dried up" was ridiculed for his ignorance of "laws of nature," for instance.[48] Similarly, a poet whose poem referred to the "endless stillness of the night" was accused of insanity, as it was clear that nature never ceased to function according to its unchangeable laws.[49] Rather than succumbing to mysticism in the face of nature, the poet was recommended to "work for the eradication of ignorance, like the learned men of our century."[50]

These criticisms were commonly amalgamated with criticisms regarding the "degeneracy" of the adherents of traditional mystical Ottoman poetry, with its many references to wine and carnal love, as these poems were "causes of moral corruption and devoid of any benefit."[51] Depictions of love and pleasure were meaningless in the contemporary era, as "the entire nation [was] now the lover of the beauty called progress" and the lover needed to hear about its beloved—its benefits, beauties, and the pleasures of the path to its attainment.[52] Poetry, as Namık Kemal had also stated, needed to be useful and moral. Moreover, that traditional poetry was criticized both due to its unscientific imagery and its glorification of immoral activities presents another indication of the inseparability of science and morality in Ottoman discourse in the second half of the nineteenth century.

Saadet made this connection clear with a brief statement on beneficial poetry in response to several letters from the readers—a statement that would lead to the beginning of the most comprehensive debate regarding science and poetry in the 1880s and 1890s: "We are not against all poetry. But now that you have the temperament of a poet, why do you busy yourselves only with one aspect of poetry and ignore others, thus failing to strive to compose philosophical, scientific, moral pieces?"[53]

In its next issue, *Saadet* itself was put to the task by a reader who asked what exactly a scientific poem would look like. *Saadet*'s reply, though puzzling, may be regarded as indicative of the vagueness of the idea of scientificity in Ottoman discourse: science was always praised but rarely defined. "Indeed, just as you did not understand what a scientific poem would be like, we do not understand with what inattention we wrote that phrase."[54]

Saadet's reversal, in turn, was vehemently criticized by a young military officer named Beşir Fuad, who argued in his letter to the editor that scientific poetry was not only possible but most admirable. This kind of poetry was not about writing poems on chemistry; the idea was simply to avoid contradicting scientific facts in poems. Furthermore, Fuad argued, those who praised philosophical and moral poems should remember that philosophy and ethics were now sciences themselves.[55] *Saadet* conceded that Fuad was right. Yet the debate was not over, as soon afterward a new article entitled "Scientific Poetry" was published in the *Saadet*. This satirical piece was about a fictional poet who wrote "scientific poems" under the pseudonym *Fennî* (The Scientific). One of Fennî's poems on chemistry, the author wrote, went: "What are the components of the pure water you drink? / They're oxygen and hydrogen, oxygen and hydrogen!" The article concluded with another, much longer "mathematical poem."[56]

This mockery did not indicate a shift in *Saadet*'s approach, though, as in the same issue, a detailed criticism of another romantic poem was published. The author Faik Hilmi's target was a poem by Mehmed Celal containing a verse about how "springtime has left the universe." This fierce criticism discussed in fascinating detail how scientific books described the changing of the seasons, the movement of the planets, and the laws of attraction, all of which contradicted the poet's statements. "If the poet who is a lover of springtime can gather the sum needed to travel around the world, and tries to gain some familiarity with geography and cosmography, he [may be] able to spend his entire life in springtime," Hilmi concluded.[57]

Celal's reply to this pedantic criticism by "a lover of science" was brief, and stated that any reader could understand metaphors, except "men of science who aren't familiar with the pleasure of poetry." The conclusion was simple: "Science and poetry are entirely different things!"[58] Yet while Celal did easily fend off the criticism regarding the content of his poems, it is worth noting that Hilmi's criticism, like many others published in the *Saadet,* was as much about poems as about the persona of the poet—a man commonly defined using words like insane, drunk, or hysterical. The contents of the poems were unscientific, as poets themselves were irrational people with no benefit to society. Men of science, on the other hand, were sober, virtuous, and useful.

b. The Debate Heats Up: Beşir Fuad as Poster Child

These associations were always implicit to the debate initiated by Beşir Fuad's remarks. Fuad elaborated on his views regarding poetry and science also in his monograph on Victor Hugo, published in 1885–86. Hugo, the chief inspiration for many Ottoman authors, was portrayed by Fuad as a symbol of the key problem with Ottoman thought: a romantic negligence of science. Fuad wrote: "Bring the shiniest images of the greatest poets side by side with, say, topics from astronomy. You will see that those bright images are too dim when compared to reality. . . . If the works of the most famous poets were combined into one, the result would not be as amazing as even the branch of anatomy regarding the nervous system."[59]

Hugo's work did contain scientific truths, Fuad admitted, but he emphasized that while the most minuscule fact uncovered by science led to thousands of uses, pure imagination had never been beneficial for humanity. Similarly, fiction could be beautiful, yet the products of science were both useful and awe-inspiring.

Menemenlizade Mehmed Tahir, an old friend of Fuad's, reviewed *Victor Hugo* in his journal, the *Gayret.*[60] While appreciating the work, Tahir argued that science and poetry were both useful: science enabled people to spend their lives in peace and comfort, while poetry filled hearts with joy and bliss.[61] Fuad's response was unequivocal: even if both science and poetry should be seen as useful, the benefits of poetry could not even be compared with those of science. Indeed, Fuad wrote, the only useful parts of poems were those based in scientific fact, and thus, "as science is a source of light, it resembles the sun, while the light in poetry is but a reflection; hence poetry is like the moon." Furthermore, science was not simply about material progress, as material progress also gave rise to new ideas poets could never fathom. The question was simple: "Does intellectual progress occur in places where there are lots of poets, or places that produce many scientists?"[62]

Hence, the debate about the merits of poetry and science was a debate about the merits of *poets* and *scientists*, defined as two different kinds of people. While Fuad insistently defined poets as dreamers with little attachment to what truly mattered, Tahir rejected this portrayal of the poet as a parasite on society and strove to prove that poets were useful, too: they were creators of beauty. In turn, Fuad's definition of the scientist expanded to include this criterion as well: scientists not only found facts, but the facts were also beautiful, and in a higher sense than the beauty of a dream. Thus, even if the usefulness of objects or ideas should be measured in reference to their practical use or beauty, scientists would emerge as their true producers.

Tahir's final salvo against Fuad's emphasis on facts was in reference to morals, and brought forth the idea that romantic depictions of human misery were more influential than realist ones in helping inculcate moral values.[63] This effort to save claims of moral superiority from Fuad's definition of the scientist was also denied by Fuad, who argued that romantics only depicted unattainably ideal characters whereas realists described in minute detail the wrongs that should be avoided.[64] By examining vices in detail, realists served the elevation of a society's morals, just like scientists who, by uncovering the causes of illness, helped ameliorate public health. It was the ideas and actions of scientists that should be emulated if moral decline was the concern.

But this was not the end of this debate—indeed, it became increasingly antagonistic. Tahir's criticisms of Fuad culminated in a poem entitled "A Scientist and a Poet" that ridiculed the arrogance of an imaginary scientist and accused scientists of being superficial and emotionless.[65] Fuad's response was in the form of a sarcastic dialogue between himself and a "moonstruck" friend of his who "used to read . . . poetry while we studied our lesson in school."[66] It was poets who were arrogant, Fuad stated in this dialogue, as scientists, due to their loyalty to the truth, never denied their shortcomings. Furthermore, due to their comprehension of nature, scientists were much closer to a true understanding of the "Creative Might."[67] Rather than spreading myths, poets "should admit their true place in society" and start trying to illuminate minds.

As the debate got more and more bitter, Fuad furthered his efforts to expand the boundaries of the authority of scientists. Sociology and ethics were two sciences that identified the obstacles against and conditions for progress, therefore men of science had the right to judge which poems were useful and which were not. A scientist could criticize poets without being able to write a poem, whereas a poet could not criticize scientists without a sufficient knowledge of science.[68] While the debate then assumed the form of a knowledge contest with references to names like Archimedes, Newton, Lavoisier, and Watt, the allies of both sides also joined in. Fuad's very many critics ridiculed his ignorance regarding literature and argued that if they were followers of the truth as Fuad claimed, men of science could not be forgiven for any error they made, even when discussing literature.

Ahmed Midhat initially took the side of Fuad, albeit not vehemently, while at the same time presenting quite a rigid definition of science: "As science (*fen*) means reality, and poetry is about description, 'scientific poetry' is not like those works of insanity [that characterized old poetry], but is about uncovering the beauty of the reality that the poet writes about."[69] Scientists were to discover facts, and poets were to express these facts in a beautiful way, rather than write about imaginary entities and impossible events. Inspired by such comments, a student by the name of Tahir Kenan sent an essay on electricity for publication in the *Tercüman*, and noted sarcastically in his introduction that his aim was to give poets some basic information on ideas they carelessly referred to in their poems.[70]

Soon afterward Namık Kemal, one of the most respected poets of the era, weighed in and published a letter criticizing the attitudes of those who failed to appreciate Ottoman men of letters and "made a habit of leaning toward European ideas." Kemal, author of acerbic criticisms of traditional Ottoman poetry and an occasional political exile/prisoner himself, did not hesitate to condemn Fuad in an article he concluded with the following statement: "God willing, thanks to the auspices of our Master and Benefactor who deigned to steadily expand the means of learning in His land, we will see new works in our language that will [instead] demonstrate, if not contribute to, the soundness of the thoughts and the strength of the morals of our society."[71] The image of Fuad as a radical enemy of Ottoman social order was in the making.

What followed was a poem published in the *Saadet* that condemned those "charlatans" who "imitated Voltaire," to which Fuad responded with his first and only poem that commended Voltaire and his followers, just to show that anybody could write one, and he then challenged his critics to write a scientific article.[72] Yet another critic argued that Fuad's knowledge of science amounted to a familiarity with the French language, and that his appreciation of Voltaire, "an atheist," was despicable.[73] This was the end of the debate, as Fuad committed suicide soon afterward.

The details of Fuad's suicide, one of the iconic events of nineteenth-century Ottoman intellectual history, are outside the scope of this study. It is worth noting, however, that due to his interest in hereditary illnesses, Fuad was utterly worried about having severe mental disorders like his mother had had and, in his suicide note addressed to Ahmed Midhat, stated this as the reason behind his decision. He noted that he also hoped to make of his suicide a contribution to science and, indeed, was able to write a few sentences about his body's reactions after he slit his wrists.

Fuad's importance in the Ottoman debate on science cannot be over-estimated. But it appears more as a sign of the precariousness of the position of the ardent, albeit rather simplistic, advocates of the new sciences (to which they referred simply as "science") than that of their triumph. It is striking that in these debates few voices were raised in support of Fuad, and while his arguments about the superiority of scientists to poets were opposed, the ultimate point in these criticisms tended always to be the claim that Fuad was an atheist, an arrogant Europhile who was not only ignorant of but hated the traditions of his society. The weakness of Fuad's voice in the debate, and the constant pressure on Fuad to prove his adherence to the values, religion, and tradition of his society are striking. Most remarkably, Ahmed Midhat, who had been declared a "cane-carrying" enemy of religion himself following his brief remarks on evolution in 1874, would portray Fuad as a "lost soul" some ten years later. That an admirer of European science had to prove that he was not a Europhile fop was an established principle by the 1880s, and, after Fuad's death, the figure of the lost materialist would come to accompany that of the fop. It was not as much the boundaries of science or literature that mattered as the proper conduct of a man of science.

Columns published after Fuad's death are indicative of the way young proponents of the new sciences like him were perceived. Indeed, this suicide appears to have been an excuse for many Ottoman authors to reassert the importance of order, obedience, and religion as the route to morality. Many journals published essays and letters commenting on Fuad's suicide and argued that the reason why materialists like Fuad could commit such a horrible sin was their ignorance of their own religion. The weakness of their faith rendered them particularly vulnerable to harmful European philosophies that constituted a significant peril for the empire. A letter published in Midhat's *Tercüman* stated that it was essential to provide those learning the sciences with discipline, not simply knowledge: "If a man of science is regarded by the public as immoral, it does not matter in the least whether that person knows the sciences or not. . . . If a man of science attracts the detestation of the people, it will not be he who is condemned. . . . People will first and foremost say 'Shame on those sciences!'"[74] The author stated repeatedly that "discipline and good morals" should always come first, including in schools where the sciences are taught.[75]

But as a respected publisher and author who also knew Beşir Fuad personally—remember that Fuad's suicide note was addressed to him—it was the writings of Ahmed Midhat on this suicide that definitively set the terms with which the generation now represented by Fuad would be understood. Midhat considered Fuad a superb mind, but in a book published soon after Fuad's death, Midhat transformed his suicide into an example of how boundless appreciation of everything European coupled with estrangement from religion might lead to the downfall of young, well-educated Muslims. This narrative was in perfect harmony with the policies of Abdülhamid II, and it became so prevalent that it continues to color contemporary interpretations today.[76]

Midhat wrote that when he met Fuad, he admired this young man's erudition but advised him to "write about things that would be in harmony with the philosophy as well as interests of Muslims and the Ottoman Empire." The true service of an author was to avoid writing texts that would agitate youth; patriotism was best served by authors who simply provided sound information to young generations. Only such an author could be defined as a "hard worker of the world, lover of his country, grateful to his nation, servant of his state," whose efforts would be useful to the state and the nation.[77]

Fuad had great potential to become such a man, Midhat wrote. His knowledge of European languages and sciences were exceptional, his intelligence remarkable. Yet he was utterly ignorant in other matters, like Arabic, Persian, and the religious sciences. In fact, he had read the Qur'an only from an abbreviated French translation, and his knowledge of the works of Muslim scholars was also based on European sources.[78] The association he appeared to see between science and irreligiosity, inspired by materialism, was due precisely to this ignorance.[79]

As Midhat's analysis makes clear, the idea that men of science like Fuad could end up "losing their way" was not simply about the "errors" of some recent philosophical trend or about the wrongness of suicide. The "interests of the Ottoman state," the requirement that a young man be of "service to the nation and the state," and the dangers of "agitation" are concerns that all men of science are advised to take into account. The debate about science and poetry (or science and tradition) is more about the presumption that familiarity with and appreciation of the sciences, philosophies, and lifestyles of the Europeans could corrupt these young men and lead to the collapse of the Sublime State—the supreme concern. Moreover, as Midhat's assertions also indicate, in this formulation "useful" means "useful for the state and the nation," and this is the way in which possessors of "useful knowledge" are to define themselves.

During Abdülhamid's reign, this principle was made clear on countless occasions. In 1887, it was proclaimed from the palace that "the graduates of certain schools" demonstrated a weakness of faith, which made it essential to introduce religious courses into the curricula of these schools.

Furthermore, it was declared that anti-religious remarks would never be allowed.[80] In May 1892, the Palace requested the names of the students in Europe who did not "observe proper morals."[81] The importation of books on science and religion was also monitored carefully.[82]

Even in such a context, some among the new generations of students and graduates—especially from the military, medical, and administrative academies—continued to make a case for their intellectual and moral merit, albeit less provocatively than Fuad. Soon after Fuad's death, another young officer named Nabizade Nazım published articles espousing Fuad's views in a mild tone. In one article designed as a dialogue and entitled "On Poets," Nazım wrote that only poetry that did not contradict facts could be deemed useful: "Poetry as we know it . . . has rendered many men . . . unable to serve the public, to serve knowledge. Had Nedim, for instance, used his mind in the name of science and learning instead of devoting his life to poetry, he would have been a man of science. . . . What kind of a service is it to produce a *divan*?"[83]

It was the works of men of science that truly served the public, and Nazım did not hesitate to use striking comparisons to make his point: "I wouldn't trade a single page of [Laplace's] *Celestial Mechanics* for the great Divan of Nef'i. I can't consider a hundred Divans of Nedim equivalent to a single chapter of Duhamel's *Treatise on Infinitesimal Calculus*."[84]

Remarkably, Nazım argued that poetry ruined young people's lives, as imagination was harmful to the nervous system. Young people should devote themselves to science, as science was what the empire needed, and classical Ottoman poetry was nothing less than a public health hazard. Indeed, in a more formally written sequel, Nazım reiterated that it was science, not poetry, that could save the empire. Possibly as a reaction to critics who argued that the newfangled men of science did not know enough about their language and literature, he noted that it was in fact European philologists—that is, men of science—who truly knew and wrote about the intricacies of Ottoman Turkish, not Ottoman poets. The condescending attitudes of Ottoman poets were a sign of nothing but immorality itself.[85]

Nazım's emphasis on service and morality are impossible to ignore. But note also that while his views on science and poetry were similar to Fuad's, Nazım also published a piece entitled "Islam and Science" in which he mostly quoted Charles Mismer, the author of a book that made a case for Islam and was quite popular in the Ottoman Empire.[86] In other words, once again, we see a "radical" who, despite his criticisms of Ottoman litterateurs, proved that he was not an immoral Europhile atheist.

A few years later Mehmed Celal took up the issue in the journal *Maarif*. In two articles, entitled "Science in Literature" and "Literature in Science," he reminded his authors of the "Oxygen and Hydrogen" poem and, criticizing such "excesses," stated the argument that indicated the moderate point that now appeared as the dominant view: men of literature should not contradict scientific facts in their works, while men of science should not demean literature and could indeed learn from literary works the intricacies of their language in order to communicate their knowledge more effectively.[87]

Interestingly, in another article on the same subject Celal stated that the new era was that of scientists, not litterateurs.

Even though I prefer literature to all sciences, I say: the era of imagination has ended, the era of truth has started. . . . Fiction cannot maintain human existence. . . . I do not think that a poet can be as worthy of appreciation in the eyes of a person who is able to distinguish good from evil as Pasteur can. . . . Allow me to continue to talk about flowers, butterflies, leaves, the deep, dry coughs of a girl with tuberculosis. But how can I respond, my poet friends, if a man of science asks me what genus the flowers are a member of, what can be seen in the wings of butterflies under a microscope, how many veins there are in a poplar leaf, what causes lead to those coughs?[88]

Celal conceded that it was men of science who were indeed useful for the well-being of human societies, and, while literature was essential, it could not claim priority over science in the new era. This statement constitutes another association between science and usefulness, and it defines scientists as people whose service to humanity is more considerable than that of poets. But also crucial was Celal's warning, similar to that of Nazım, that literary works and the images they contained could confuse, and lead astray, young people and women of all ages. Reading serious— possibly scientific—works would give women a more decent sort of pleasure, and avoiding fiction and poetry while focusing on scientific works in their youth would enable men to be wise and acquire the knowledge that would help them make a living. In other words, learning the sciences *did* make one more likely to remain moral than reading poetry could, at least at a young age.

Warnings for youth regarding this subject made their way into textbooks as well. A collection of readings for high school students contains a piece entitled "Poetry and Science." This brief essay cautions its readers about would-be philosophers and would-be poets who despise each other's work and fail to understand that there need not be a contradiction between poetry and science, as phenomena can be described both scientifically and poetically. The emphasis is on moderation and the need for both.[89]

Şerafeddin Mağmumi, a graduate of the Imperial School of Medicine and later a leading member of the Young Turk movement, also revived Beşir Fuad's criticism of poets in the introduction of his work on physiology. His criticism, in the well-established manner, indicated that dreams (meaning poetry) were still preferred to "serious works" (meaning science) in most parts of the world, including the Ottoman Empire. But it presented this problem as a consequence of human nature, which causes people to "prioritize the need of pleasure over true needs." In other words, while poets' works satisfy a base, and thus more easily perceived, need, it is men of science who do the serious work—a characterization that ranks poets and scientists using criteria that have strong moral overtones as well.[90]

2. Scientists *or* Medrese *Graduates?*

While the portrayal of *medrese* students as ignorant of useful knowledge had started in the previous decades, the 1880s and 1890s witnessed even more direct attacks on them. In addition to their ignorance of the new sciences, *medrese* graduates were targeted because their only employment option was civil service. The more the association between the new sciences and the new industries was emphasized, the more parasitic civil servants appeared.

Şemseddin Sami, in one of his numerous essays on science, argued that as the collaboration of arts and sciences had led to great levels of productivity in industry, "today, a man who is familiar with arts and sciences appeals to nothing but arts and sciences." Contemporary science was not "simply about writing a couple of sentences or playing some tricks in texts. . . . Times that had once been devoted to syntax and grammar are now devoted to mathematical sciences."[91] This is of course more significant if one remembers that Arabic syntax and grammar were central to *medrese* curricula.

A more striking example can be found in three issues of the *Tercüman-ı Hakikat*: a long text entitled "Learning" written by Samipaşazade Abdülbaki—the son of Sami Pasha, a leading statesman of the era who had private tutors educate his children and whose mansion resembled an Enlightenment salon.

Abdülbaki argues in this essay that learning is about uncovering the laws of nature, and ignorance is failing to study and live according to them. Interestingly, he refers to these laws themselves as science and states, in the expected manner, that they had been uncovered first by Muslims. Yet, even though their religion itself advised against it, Muslims had later started to satisfy themselves with glorifying the past: "Had being a Muslim simply meant praying, fasting, and living like Adam, then there would not have remained on earth an Islamic government, the Ottoman state, or even a single Muslim." It was now time to reinvigorate learning. Abdülbaki's depiction of the purposes of learning should not be surprising at this point: serving the happiness of humanity, enhancing morals, and satisfying the needs of individuals and families.

"Now," asks Abdülbaki, "which of these purposes do the sciences taught in our schools and medreses serve?"[92] The courses taught in medreses— while up-to-date in the past, and thus the reason why Muslims produced so many great scholars—are no longer sufficient. After spending perhaps twenty years as a student, the medrese graduate is ignorant of the useful knowledges, unable to sustain himself, and can only rely on the state to have a job. Indeed, a medrese graduate "does not know as much about the new sciences as even an eight-year-old European schoolboy does." Realizing that this has been the case for three or four centuries now and that so many generations have thus been wasted "can drive one insane," Abdülbaki argues, and attributes nothing less than the "backwardness" of the entire Muslim world to the ignorance of medrese graduates.[93] But probably his most charged remarks are the following:

Even though for centuries the non-Muslims under our rule were much inferior to us in terms of learning, arts, trade and population, they too have realized the importance of education and opened their own schools, changed and reformed the sciences they taught . . . and succeeded in improving their learning. And for that reason, they took control over almost the entire Ottoman industry and commerce. . . . In the meantime, the so-called greatest ulema of Istanbul and the arrogant hodjas . . . or some beys and pashas who won't even greet you because they have such and such rank . . . are busy trying to get an advance on their July salary or pawning the chain of their watch.[94]

Abdülbaki was certainly aware that the new schools where the new sciences were taught also failed to produce anything other than civil servants, and his remarks occasionally do touch on the new schools and their bureaucrat graduates, yet his emphasis was on the medreses. One reason for that is that medrese graduates are unable even to translate "the exquisite books of Muslim scholars on the sciences and the improvement of morals" from Arabic into Ottoman Turkish, whereas the students and graduates of the new schools have at least been translating books from European languages on the useful sciences. The failure of the scholars of Islam was particularly appalling, Abdülbaki argued in a provocative way, when Christians had made so many great contributions to science, even though their religion did not even encourage it. As a result, "Muslims are among the most ignorant, wretched and backward communities on earth."[95]

Even though Abdülbaki concluded his series of essays by asserting his allegiance to Islam and the sultan, a letter signed by forty-eight ulema was received by the *Tercüman-ı Hakikat* in response. Interestingly, Ahmed Midhat, who had introduced Abdülbaki's essays with nothing but praise, introduced this letter by noting that Abdülbaki deserved a much harsher response.

The central argument of the ulema is that the sciences taught in *medrese*s—which are acknowledged by Abdülbaki himself to have raised great scholars in the past—are timeless sciences, as their subject matter is not one that can age. But most significantly, these sciences are useful as well, and useful in a unique way, as the religion these sciences are about is "the true basis of the Muslim community." Similarly, the ulema argue, sciences about the Arabic language taught in *medrese*s are about the language of these religious sciences. "So," the ulema ask, "which one is the mistake? To teach religious sciences in Muslim lands, and in particular, the center of the Caliphate, or to call this very education a mistake?"[96] Notably, too, the ulema complement this emphasis on the unique functions of their knowledge and expertise (cultural capital) with a reference to the economic sphere: Abdülbaki's insinuations about the undeservedness of the wages the ulema earn is nothing but an indication of "greed."[97]

This reply is another indication of how entangled ideas on science-as-useful-knowledge, religion, and community are. The common criticism that Arabic courses are devoid of any use is countered by the argument that a reaction against the teaching of Arabic is but a reaction against the language of Islam. Similarly, the response to the idea that religious sciences taught in *medrese*s cannot save the empire from collapse is that religious sciences are essential as it is religion that keeps the community intact. Arguments about useful knowledge and the needed sciences are not distinct from arguments about the identity of the community. Moreover, they are also about who is fit to lead this community, as the ulema proclaim that "those [like Abdülbaki] whose knowledge and understanding are inadequate and whose arguments are false should now understand that they cannot be the leaders and guides of a great Islamic community."[98]

Conclusion

The Ottoman debate on science intensified significantly in the Hamidian era. Themes established earlier, such as the usefulness of scientific as opposed to old knowledge, the definition of science as a matter of defining the community, and the emphasis on the virtues of a man of science were brought together in numerous new ways. Most significantly, the Tanzimat conception of science as *the* needed knowledge merged with the Young Ottoman emphasis on the requirement for the man of science to demonstrate his allegiance to the community to form the dominant discourse of the Hamidian era. Science was essential and should be regarded highly, but the virtuousness of a man of science could not be taken for granted. The fop ridiculed ad nauseam by the Young Ottomans was a character who did not even understand the sciences he so praised. The "confused materialist," however, was perhaps more of a menace: he *was* regarded as well informed about the new types of knowledge, yet his consequent sense of entitlement was a cause for concern. Indeed, the debate on science itself proved to be potentially dangerous, as the arguments about the worth of the Islamic legacy demonstrated. Hence, the disciplining of the man of science was vital.

Many accounts of the birth of the Young Turk movement toward the end of the nineteenth century rightly emphasize Şükrü Hanioğlu's findings: the spread of a superficial scientistic attitude along with insights based on German popular materialism, particularly among the students of the Imperial Military Academy and School of Medicine. With a naive faith in the progress of science, these young men were convinced of the obsolescence of "old views" (a phrase that, in some cases, probably did stand for religion itself) and, as Hanioğlu notes, they "made an impact on Ottoman intellectual life quite disproportionate to their numbers."[99] But as this chapter illustrates, this was indeed a marginal perspective at the time.[100] Thus, to complement these crucial findings, it is worth noting the significance, magnitude, and effects of the accusations of misguidedness that accompanied the depictions of these men. The "misguided (and dangerous) materialist" stereotype was used consistently to discipline these self-assured but not necessarily very sophisticated science enthusiasts.

Furthermore, while the claim that science made one a virtuous, modest, useful citizen was already central to the arguments of these "misguided" young men themselves—and in this respect they were but adopting the discourse that was established already in the 1860s—they highlighted this idea even more in their disputes with their critics. Hence, even though there was indeed an emphasis on the idea of "science as savior" in the arguments of these students, it is also worth noting their positive references to Islam—the genuineness of which is irrelevant—and, most significantly, their depictions of science as an enhancer of morality and a producer of harmless citizens.

SEVEN

Science and Morality at the End of the Nineteenth Century

Introduction

In the 1880s and the 1890s, science, community, religion, and political power were topics that were debated in the Ottoman Empire more overtly than ever before, and in close conjunction to one another. Characterizations of science as the type of knowledge that transformed its possessor into a not only intellectually but also morally superior individual, potentially *the* savior of the empire, were confronted by representations of the man of science as a misguided and lost soul who, far from being a savior, needed to be saved himself. The arguments were particularly heated because the debate on science was primarily on the qualities of the needed type of individual, not on the definition and boundaries of science.

Scientism and German vulgar materialism were rather popular among the students of the two uniquely prestigious schools of the Ottoman Empire, the Imperial Military Academy and the School of Medicine, in the last two decades of the nineteenth century—an observation that rightly informed studies on Ottoman intellectual history as well as those on the most influential political movement of the late empire, the Young Turks.[1] Organizing under the umbrella title "Committee of Union and Progress," these students would plant the seeds of much political transformation in the twentieth-century Ottoman Empire. Moreover, not only were many of the founders of the Turkish Republic itself former members of the political party that this secret committee later produced, but even those who weren't had also been shaped within the same intellectual and social milieu.[2]

These young men were certainly different from the early nineteenth-century Ottoman observers of science. They were products of new kinds of schools, familiar with different types of ideas, and aspired to positions within a new social and political structure. But what these men observed *as* science was itself also significantly different than what their predecessors had observed. This was the world of Darwinism, Maxwell's electro-magnetism, new calculations and debates about the age of the Earth, the periodic table, dynamite, the phonograph, and electrification. This, thus, also became a world in which it was possible for a young Ottoman author to argue that love was essentially an electrical phenomenon and base this argument on science, the source of "absolute authority."[3] Statesman Sadullah Paşa proclaimed his adoration of "the era of progress" and the "world of science" in his well-known ode to the nineteenth century—a poem filled with praise for a range of developments ranging from discoveries in geology and chemistry to magnetism and the steam engine.[4] Similarly, a textbook on the history of arts and industries devoted parts of its chapter on the nineteenth century—"the age of science"—to the changes in fields such as physiology and organic chemistry, with brief but positive references to the works of scientists such as Claude Bernard as well as Charles Darwin.[5] In addition to these remarks about the present state of science, excited statements about the future promises of science were common features of the Ottoman journals of the 1880s and the 1890s:

So many wonders have been observed thanks to science . . . but when one sees that the doors of scientific progress are opening all around . . . one cannot but ask "in what new ways will science serve humanity in the future?" There are great advances in all branches of science, but from which one will humanity benefit most in the future? While a decisive answer cannot be given, it is suggested that it will have to do with electricity which has produced the most amazing wonders. Thanks simply to the improvement in electric motors, today's motors have reached 400–500 horsepower while the most powerful of yesteryear were but 50–60 horsepower, which demonstrates that the day when machines of thousands of horsepower will be produced with electricity is not too far in the future, not even mentioning telegraphs or telephones. . . .

Who can claim that chemical operations will not be administered with electricity tomorrow, and that, as the growth of plants itself is a chemical process [involving electricity], our food and our bread will not be prepared with electricity? These are all scientific issues. The masters are working to resolve them. They will [all] be sorted out [in the near future].[6]

Likewise, an essay published in a journal that defined its mission as popularizing science in the Ottoman Empire stated that "all the means that feed, clothe, raise us have come to existence thanks to the progress of science," almost likening science to a mother for humanity. "We eat thanks to science," "we drink thanks to science," "we go wherever we want thanks to science," asserted the author, and argued that science enabled humans to meet all their needs, particularly thanks to machines and steam power, and to such an extent that without it human life would not be possible.[7]

While characterizations of science as a "fantastic" endeavor had been common in the early nineteenth-century Ottoman Empire as well, late nineteenth-century discourse had thus clearly been influenced by changes within scientific fields as well. Electricity, electromagnetism, physiology, organic chemistry, and particularly the growth of machine science were among topics Ottoman authors discussed, but these discussions tended to highlight more than anything else the "unstoppable progress" of science. And most significantly, in these types of formulations, science was represented more frequently than ever before as an agent, an authoritative entity that, so to speak, acted on its own volition. Phrases such as "from the point of view of science," "science states that . . . ," and "science makes possible . . ." became increasingly common while "what we owe to science" emerged as a particularly popular topic.

Yet as we have already seen, the glorification—and personification—of science by these mostly young, entitled men had additional dimensions. Science, in Ottoman discourse, was tightly connected to questions of virtue and vice; talking about science involved answering the question, "What does it mean to be a good Ottoman citizen?" All participants of the debate—that is, not just the overtly religiously conservative ones—conceived of science in relation to its potential contribution to the production of individuals with particular qualities. These qualities could be deemed dangerous or desirable, but the association of science with specific types of personal attributes remained central to the discourse.

For young "materialists," scientific knowledge enabled one to free oneself from superstition, assume a role not unlike that of a physician, and cure the social organism. In the Ottoman Middle East, this aspiration was observed among not only Turkish but Arab men of science. Indeed, thanks to the writings of Shibli Shumayyil, a graduate of the Syrian Protestant College and an ardent seeker of political and social reform, it was primarily in Arabic that the most passionate debates about materialism arose in the Ottoman Empire.[8] During the reign of Abdülhamid II (1876–1909), however, the official policy was shaped by a determination to reestablish the authority of the Palace by limiting the influence of high-ranking bureaucrats as well as to constitute an Islamic Ottoman identity to achieve social cohesion. In addition, the strong emphasis on maintaining the unity of what remained of the empire gave rise to relentless efforts to control the production and circulation of ideas. These entailed, among others, the use of a strict censorship mechanism, the manipulation of the Ottoman press commonly through financial incentives, efforts to influence the representation of the Ottoman Empire in the European press, and changes in school curricula. Furthermore, a set of rituals were invented and imposed in order to render visible the demanded obedience to the sultan.[9]

What we observe around the end of the nineteenth century, then, is that the question of keeping these young, educated, and ambitious men loyal to the throne became such a fundamental issue, and censorship defined the playing field to such an extent, that in especially the last two decades of the century virtually no text could avoid referring to these qualities.[10] Hence it is particularly important in analyzing these later years to ask what it meant for Muslim Ottomans themselves to write and to have a debate about science. When we approach the issue with this question in mind, we are able to identify the points that can be missed when the emphasis is exclusively on arguments for or against science. Indeed, as all authors who wrote on science were interested in making a case for specific virtues, and the potential of being portrayed as misguided and dangerous remained constant, even texts that praised science in a most simplistically passionate way were multilayered. Moreover, even though they received an allegedly secular education, students of the Western-style schools in the Ottoman Empire were systematically exposed to morality courses that were, in terms of both content and style, based on the Islamic Ottoman heritage.[11] It is reasonable to argue that the obsolete format of these courses and the textbooks' uncompromising emphasis on obedience to the sultan ultimately backfired, but they also enabled the students to use an Islamic tone in their writings.

It is worth underlining again at this point that the Ottoman case was not unique in terms of the ubiquity of references to morality in arguments about science and education in the late nineteenth century. In Western Europe, particularly in Britain, one of the responses to the growing power of professional scientists and their attempts to continue expanding their sphere of influence was the birth of the "bankruptcy of science" discourse: a discourse characterized by statements critical of either the amorality or immorality of scientists. States' efforts to establish more direct control over the direction of scientific research at a time of immense economic and military competition was not unrelated to the gradual popularity of this discourse.[12] In France, the transformation of government and citizenship in the Third Republic entailed the construction of *morale laïque* as a school subject.[13] Cultivating a loyal citizenry at a time of significant cultural and social change was an important concern in Russia and China as well, leading to the transformation of morality into a crucial topic to be discussed and taught.[14] A more thorough comparison among these and similar cases is certainly needed but outside of the scope of this study; yet it is possible to argue based on these observations that an emphasis on morality in arguments on science and scientists was a common consequence of partially comparable developments in different parts of the globe in the late nineteenth century.

The ultimate result of the specific factors that characterized the Ottoman case is that the texts of the 1880s and 1890s are quite ambivalent in their representations of science, and many texts with scientistic themes have additional noteworthy dimensions. In this chapter I aim to show this ambivalence and the difficulty of identifying two separate camps, one as pro-science and anti-religion/anti-tradition, and the other as anti-science and pro-religion/pro-tradition.

A. "Confused Young Men"?

It is undeniable that those who made the most passionate arguments about the uniqueness of the new sciences and the merits of young Ottoman men of science were the students and graduates of the Imperial Military Academy and School of Medicine. Yet the wholesale labeling of these young men as confused youths—ignorant of Islam and vulnerable to European philosophies as well as prone to immorality and, in particular, disobedience to the sultan—appears to be a disciplining representation that ignored the multiplicity of positions taken by these men of science. Moreover, while there did emerge a passionate scientistic discourse, it is also important to note that the official *and* hegemonic discourse of the period was one that consistently emphasized the danger of a corrupted youth and the virtues of a religious, modest, and obedient populace. Hence, many young men themselves—be they from a secular school or not—wrote on the deplorable arrogance of some men of science and their blind infatuation with European ideas.

The examples below are intended to indicate the complexity of the picture in question. What we have are students and graduates of the Imperial Military Academy who make a case for religiosity, young poets who criticize the degeneracy of Ottoman youth, and moralists who reiterate arguments about the need to combine scientific knowledge and "our" values. One thing remains constant, however: talking about science is talking about virtue and vice.

A letter sent to the *Tercüman-ı Hakikat* in 1881 by several students of the Imperial School of Medicine is a case in point. This letter stated at the outset that "the bliss and well-being of our nation depends on good morals," which are provided by religion, as "all that is good in the world is born out of religion, and all that is evil is a result of the negligence of religion." The students of medicine continued by making the by then commonplace argument that in the contemporary era, sciences and arts were essential for a nation's welfare, but for Muslims, learning the sciences was a religious obligation. The perfect teaching of the tenets of Islam was crucial for a nation of Muslims, the students stated, but they complained that, in some mosques, preachers discouraged Muslims from caring about worldly bliss and recommended worrying only about the afterlife. For the students, it was essential to remind people that the world could not entirely be abandoned, and the way to make bliss possible in this world was to learn the sciences, as Islam itself instructed.

Now while this argument appears to be a continuation of the "Islamic legitimation of science" discourse of the previous decades, what is worth underlining is how the students described what they saw as the ideal, most learned and judicious preacher in Istanbul:

[He] is a person who etches these points in Muslims' minds and strives to serve the bliss and well-being of our nation. He is a person who enlightens those young men who think studying the sciences is in contradiction with the sharia, and those who, while entirely ignorant of religion and its instructions, have learned some sciences, and pretend to be philosophers. He shows them that sciences are not in conflict with Islam, and that if these sciences progress and expand a little more, they will prove to be no more than the wisdom of Islam itself. . . . Hence, he saves them from the path of ignorance and corruption that they are on, and leads them towards the straight path of the sharia.[15]

Hence, it is students who study the dangerous sciences themselves who blame others of having ignorance in Islamic matters, and argue that preachers' efforts can save them from the dangerous path they are on.

Another example that challenges the irreligious materialism associated with the academies is that of Doctor Hüseyin Remzi (1839–1898)—a graduate of the Imperial School of Medicine who served in the Ottoman military as a physician and later worked as a lecturer in many of the new elite schools of the empire.[16] Most famously, in 1886 he became one of the three men sent to France to meet Louis Pasteur and learn about his cure for rabies. Upon his return, Remzi was first involved with the establishment of the Imperial School of Veterinary Medicine, where he later became the instructor of zoology; then in 1892 he opened the *Telkihhane*—an institution that manufactured vaccines. Now while this Ottoman man of science was famous for his works on zoology, microbiology, history of medicine, and public health, he was also the author of a pamphlet on morals as well as a book called *İlmihâl-i Tıbbî* (Medical Catechism). The latter, which was first published in 1887 and went through several printings, is a work that demonstrates how the instructions of Islam are in harmony with the findings of modern medicine. But what is probably more important is the way science, religion, and civilization are defined in its introduction.

Remzi's emphasis is on the notion of truth. The knowledge of truth enables people to understand and live with one another; hence truth is what makes humanity, order, and thus civilization possible. At the same time, truth is what "men of science strive to reveal day and night." And, finally, Islam is the essence as well as the criterion of truth. Thus, Hüseyin Remzi concludes, "Truth, humanity, civilization, Islam are one and the same thing."[17] The true foundation of truth is the Qur'an, which is the word of God, and as a result, it should be the guide of everyone whose purpose is to understand the truth. However, "as some blasphemers who fancy themselves Muslim and similar deviants happened to come across the Qur'an without any effort of their own, they act like a son who isn't grateful for the fortune he inherits, and they fail to appreciate Islam, which is the source of all truth."[18]

Anybody who fails to affirm that the truth sought by men of science resides in the Qur'an is thus a blasphemer. Also worth noting is that Remzi uses the word *fen* for knowledge that can change, knowledge that is the initial step to reach truth, which he calls *ilm*. Thus, sciences like biology, of which he is an expert in the Ottoman Empire, are essentially the path toward the truth, and, if we follow his reasoning, it is not before their findings are in perfect harmony with the Qur'an that they can be called *ilm*, or truth.[19] The complications of the definitions aside, Remzi's effort to portray science essentially as the handmaiden of Islam and his unhesitating labeling of those who do not approach Islam the way he does present us with a more complicated picture of the so-called materialist man of science commonly referred to in the columns of the Ottoman newspapers of the 1880s and 1890s.

As Remzi's work indicates, professors and students of the new elite schools themselves were among the staunch propagators of the hegemonic "Muslims as the true founders of modern science" discourse. The physician Mehmed Fahri (1860–1932), a graduate of and later a lecturer at the Imperial School of Medicine, wrote a similar book entitled *The Healthiness of Fasting and Other Foundations of the Religion of Islam*, where he made the comment that "research on the harms of germs for the human body" revealed the "religious and scientific truths in the glorious sharia of Islam."[20] Islam, with its emphasis on cleanliness, was the route to improving public health. In addition to his books on medicine, Fahri is also the author of a collection of religious poems.[21]

These physicians' efforts to construct Islam as a scientific religion paralleled the works of authors from Arabic-speaking regions of the empire as well as Muslim thinkers from other parts of the world. In the book for which he was rewarded by Abdülhamid II, the Tripolitan scholar Husayn al-Jisr (1845–1909) took pains to demonstrate how scientific findings and theories (including Darwin's arguments regarding evolution) were in harmony with Islamic faith. Similarly, Muhammad Abduh (1849–1905), a leading representative of modernist Islam, advocated an exegesis of the Qur'an under the light of new scientific findings and argued, for instance, that the Islamic concept *jinn* should be understood not as depicting supernatural beings but invisible creatures, such as microbes.[22] This suggestion for a scientific interpretation of the Qur'an was also exemplified by the works of the Egyptian physician al-Iskandarani as well as by the unfinished exegesis prepared by Sayyid Ahmad Khan, the influential Muslim Indian thinker.[23] The works of the Ottoman authors that offered ways of understanding and practicing Islam with references to science were thus contributions to a growing body of work in the Islamic world of the 1880s and beyond.

Yet there were also Ottoman men of science like Hüseyin Hulki (1862–1894) who were critical of attempts to connect Islam and science, and not because of an anti-Islamic attitude. A physician who had accompanied Hüseyin Remzi and met with Louis Pasteur, Hulki had also been sent to Berlin to meet with the other prominent microbiologist of the period, Robert Koch. Hulki then became a lecturer in dermatology and syphilitic diseases at the Imperial School of Medicine, and he published not only on medical issues but on Islam and science. His work on the scientific implications of Islamic rituals had a different take than those of Remzi and Fahri, however. For Hulki, it was simply wrong to argue that the benefits of fasting had been proven by modern science, as such arguments belittled Islam: "For us, the holy instructions of religion do not need the approval and recommendation of science. . . . Seeking [to prove] the splendor and holiness of Islamic commandments in reference to their harmony with contemporary sciences, or their health and worldly benefits, cannot be a token of devoutness."[24] For this leading Ottoman microbiologist, the authority of science could never even be compared, or considered complementary to, the authority of Islam.[25]

In addition to making such contributions to debates on the interpretation of Islam, young Muslim Ottoman physicians also expressed their complaints about the moral state of the empire, as a poem by Mehmed Cemil (1860–?), another graduate of the Imperial School of Medicine, illustrates. In this poem with allusions to a famous poem by Ziya Pasha, a prominent member of the Young Ottomans, Cemil wrote:

I passed through Ottoman lands; don't assume I saw mansions
I saw decaying, ruined houses of worship.
No sign of progress; some lands with no knowledge and learning . . .
Tradition and morals entirely corrupted, all out of joint.
Alas, no schools or *medreses* in sight; but drinking houses, I saw.[26]

Another example is a general who wrote books both on the military sciences and on morality, Mustafa Şevket Pasha. In *Burhan-ı Hakikat*, Mustafa Şevket argues that all departments of a state are interdependent, but collectively their proper functioning depends on one key factor: learning. Yet learning itself can be appropriate only if it is based on "good morals, which, in turn, depend on religion." Hence, if a state aspires to achieve progress, it should focus on these three components simultaneously: "Good morals cannot lead to irreligiosity. Some kinds of knowledge *can*, but that is not useful knowledge. Without knowledge, one cannot benefit from religion; nor can he find the path to good morals. Without good morals, one cannot serve learning or religion."[27] With this somewhat cryptic explanation, Mustafa Şevket demonstrates that good morals, religion, and learning are all interconnected, and he clarifies in subsequent pages that the word "learning" refers both to "religious sciences" and the "rational sciences." While the virtues of the religious sciences are clear in Mustafa Şevket's portrayal, the way in which he characterizes the virtues of the rational sciences is another clear indication of the inseparability of science and morality:

If the rational sciences are neglected and all efforts are spent on the religious sciences, it is obvious that this will contribute immensely to the national spirit and the protection of religion. Yet nations that concentrate on the rational sciences will, in the meantime, bring about new inventions, and using these, gradually appropriate the wealth of the nations that ignore these sciences, and make the latter dependent on themselves. . . . This material decline will unavoidably influence the spirit (*manevi-yat*) of the latter, and their morals will also start to degenerate. This, in turn, will lead to the abandonment of religious faith.[28]

Of course it is important also to note that Mustafa Şevket Pasha, like many authors this study has referred to, approaches the question almost exclusively from the point of view of the state, and his emphasis on science, religion, and morals is only because they help a state survive. In other words, what must be kept in mind is that the state needs religious people with proper knowledge and good morals.

In literature, too, we see a proliferation of poems critical or suspicious of the claims of science, or more precisely, of men of science. Ali Ruhi (1854?–1890) wrote in 1886, for instance, that

No one among those on earth know the beginning
Neither those who have left this finite world.
Is there anyone who has discovered the facts of creation?
Men of science, too, can only assume, I assume.[29]

Hüseyin Nazım Pasha (1854–1927), a bureaucrat who had studied law in France, wrote:

Can the mystery of the Almighty be discovered?
O God! What boldness, what revolt!
You, the ignorant of science, monger of heresy,
The science you brag about is but a result . . .
Studying the sciences each and every moment,
And saying "In the universe phenomena are numerous,
And God is great" . . . Alas!
Can His greatness ever be expressed in words?
Can God ever be comprehended? . . .
Can the Knowledge of the Creator ever be assessed,
Even if learning is expanded?[30]

Bıçakçızade Hakkı (1861–?), a *medrese* graduate, made similar remarks:

There is, indeed, a secret truth like the Phoenix,
I don't trust the witness and evidence of those who say "I know what it is!"
All myths and tales, with no accord
I don't trust Darwin, Hermann, or Leibniz.[31]

But an examination of the arguments of the young, educated critics of young men of science would not be complete without a discussion of one of the longest debates to occupy the columns of the *Tercüman-ı Hakikat* and the *Vakit* in the 1880s. On its surface, it is a debate about the merits of the old and the new, but clearly the more substantial debate is, once again, about virtuousness and the type of person the Ottoman Empire really needs in the late nineteenth century.

The debate started on August 6, 1882, when an author who identified himself only as "a student" published an essay entitled "The Superiority of the Virtues and Merits of the Moderns over the Ancients" in the *Tercüman*.[32] The main argument of this brief essay was that the logic developed by premodern scholars was no longer of much worth. In the next issue, the author revealed that he was Abdülhekim Hikmet, a student of the Imperial School of Medicine, and indicated that contemporary scholars had refuted the assertions of the ancients in many other fields as well, including astronomy and chemistry. But in the Ottoman Empire, Hikmet complained, the scholars of the past were still overly respected, which impeded progress.

As in most disputes of the period, it was the *Vakit* that opened its columns for response to the provocative essay published in the *Tercüman*. Law students Mahmud Esad and Ali Sedad replied forcefully that it was conceited of Hikmet, who was but a student like themselves, to assume he could so easily thrust aside such greats as Socrates. Hikmet failed to understand that the scholars of today owed their discoveries to the findings of those of the past; moreover, some contemporary "masters" had turned out to be nothing but charlatans.[33]

Hikmet's response to the reaction of these two law school students clarifies that his intention was not to belittle the ancients but to underline that their perspectives had been surpassed. As for the question of conceit, Hikmet's position was clear:

To improve the material and non-material conditions of the nation is the duty only of those [who] like us are in the lofty field of medicine, as they are well-informed about the material world, and the facts and issues that are considered to be the chief source of inspiration for contemporary thinkers. . . . Those who can assess [whether or not men like us have the right to do something] are only those who are physicians themselves, or people who are also knowledgeable about the material world. As even commoners know, a physician can be a philosopher, but a philosopher cannot be a physician.[34]

Esad and Sedad, in their reply to "a fellow student," noted that contemporary scholars owed most of their discoveries to the work of the ancients.

They also protested the argument that only physicians were responsible for the welfare of the nation and proclaimed: "We do not even consider worthy of a reply an argument that insinuates that we are ignorant of the natural sciences."[35] In his response, Hikmet sharpened his remarks by sarcastically referring to Sedad's father Cevdet Pasha's work on logic as proof that works on classical approaches were indeed unaware of the new physics.[36]

Esad and Sedad's next strategy was to highlight a major assertion that they had briefly touched on in previously published essays: the main source of inspiration for Hikmet's articles was actually a text by an American missionary—specifically, Cornelius van Dyck's essay with virtually the same name, "On the Superiority of the Moderns over the Ancients."[37] Thus, the authors asserted, their addressee was no longer Hikmet but van Dyck. The argument that the sciences did not exist before the contemporary era could only be uttered by a person like van Dyck, whose emphasis was on Europe, Esad and Sedad stated, and they underlined that in the East sciences had flourished long before they did in Europe. As for logic, "a science approved of by our religion," the authors argued that they knew it "much better than those whose interests it does not serve." In this reply, Esad and Sedad also directed attention to an issue that appears to have been another cause of concern for them: how could ancient scholars be considered inferior to the modern ones with respect to "merit and virtue"? Using these words in such a context could be excusable for van Dyck, a foreigner, but apparently Hikmet did not understand the significance of these terms, either, due to his ignorance of the language. Otherwise, he would have known that simply because contemporary men of science knew new things, the ancients could not be declared inferior in "merit and virtue."

Importantly, Hikmet's essay's title uses the phrase *Fazl ve Meziyet*, which can be translated as "Virtue and Merit." But the choice of these particular words is a further indication of the significance of the question regarding knowledge and virtue. While the connotation of *meziyet* is closer to "merit," the word *fazl* is defined in the classic dictionary of Ottoman Turkish, the *Kamus-ı Türki*, as "worth, merit, maturity, knowledge and erudition along with the right morals and faith." In other words, even if what is intended is superiority with respect to erudition, connotations regarding faith and morality are impossible to avoid when these words are used. This understanding, it appears, is behind the reaction of Sedad and Esad.

When, in response, Hikmet reiterated that Sedad and Esad were ignorant not only of the new sciences but also of the fact that van Dyck was an expert in the Arabic language, the respondents' reaction further assumed the shape of a diatribe against an ignorant, confused youth. Van Dyck's Arabic was not perfect, and, in any case, it was "our" duty to be more sensitive when using terms borrowed from Arabic, "our language of religion and literary arts." Furthermore, the progress of the sciences of the Europeans was due to the legacy of Muslim scholars, "our prides." Assuming that contemporary men of science were superior to the ancients would lead to such preposterous conclusions as Zambako should be superior "to our Ibn Sina."[38]

The contemporary doctor to whom the authors refer is Dr. Zambako Pasha (1832–1913), an Ottoman Greek educated in France who was rather popular in Istanbul in the 1880s and 1890s, and who served as the private physician of Abdülhamid II as well. Essentially, what the authors find preposterous is comparing a great scholar to but a physician. But even if their primary intention may not have been to stress that the physician in question is a non-Muslim Ottoman and the great scholar is Muslim (and is referred to as "our" Ibn Sina), this aspect of the comparison adds another significant dimension to their argument, particularly in an essay that is full of references to "us."

As the questions of identity and loyalty were thus introduced into the debate, Hikmet's tone also went through some transformation, as the following passage from his response demonstrates: "Our Sultan, the Refuge of the World, is the successor of so many sultans from the House of Osman, and thus is superior to them. This is because the progress of science and knowledge during his imperial reign which was devoted to learning is unprecedented." Similarly, the laws made during the reign of Abdülhamid were "the true basis of the material and spiritual bliss of all subjects of the Sublime State."[39] Such elaborate phrases were new to his writing.

Clearly, with this text Hikmet intended to manifest his loyalty to the sultan and the state: he was not a blind follower of an American missionary and a dangerous youth. Yet he did not repeal his main assertion: it was indeed true that Dr. Zambako knew more about medicine than the great Ibn Sina had; Ibn Sina's death had been due to an illness that Zambako, today, could have cured without much effort. Hikmet concluded with the remark that he would not be able to comment further as he was studying for his graduation exam—that is, he had more important things to do than argue with some ignorant men.

In yet another response, Esad and Sedad again made it clear that their opponent's remarks were "due to the influence of a missionary." His comments about Ibn Sina were things that could be said only because of "a pathological crisis." "One wonders," they wrote, if "Zambako, and *vice-docteur-en-médecine* Abdülhekim Hikmet himself" will not die in the end;

"maybe because they are doctors, they will die of diseases that cause the patients to have astounding shapes, like elephant disease or lion disease."[40] These allusions to Hikmet's French-style training and his alleged self-importance are quite clear references to the fop discourse that had by then become firmly established.

This essay led to a severe escalation of tension, and other students of the Imperial School of Medicine started expressing their support for Hikmet. One of these, Refet Hüsameddin, wrote that while Muslims had laid the groundwork for contemporary science, what existed today was a product of the Europeans and was indeed superior. Furthermore, "the individual who our opponents call a missionary . . . teaches Logic in Syria . . . and his work written in Arabic is an example of contemporary progress." The opponents of Hikmet had no knowledge of medicine, and they had no right to talk about issues of which they were utterly ignorant.[41]

Similarly, Mehmed Fahri was revolted by the "*vice-docteur-en-médicine*" remark, and argued that such attacks were directed against all students of the Imperial School of Medicine. Şakir Pasha, a teacher at the medical school, had been a student of Claude Bernard, and Hikmet was Şakir's assistant, which proved Hikmet's merit.[42] After these letters Hikmet published a brief note indicating that his opponents were nothing but charlatans, and, as he was not idle like them, he would no longer reply to their allegations.

Esad and Sedad's final response emphasized once again that van Dyck, a missionary, was a lecturer at a Bible College. They also stated that if nobody would be allowed to speak about issues in which they were not full-blown experts, Hikmet should also stop imagining himself to be an Auguste Comte or a Herbert Spencer and should never assume that just because he knew about physiology he could also talk about divinity, literature, or philosophy.[43] A man of science had to know his limits.

But the boundaries in question were clearly not simply those around science. As in several cases analyzed in previous chapters, the debate between these educated young men ended with the intensification of remarks questioning the faith, morality, and loyalty of one of the parties. In this specific case, Hikmet was portrayed as just a French-speaking physician who fancied himself superior to his peers, even though he was but an ignorant propagator of the ideas of a Christian missionary. As the target of these attacks, Hikmet ended up both adopting a convoluted tone and style and attempting to demonstrate his allegiance to the community and its ruler. The conflict in question, thus, involved not simply alternative ideas on science and tradition but much more consequential assumptions and implicit definitions about what constituted a true Ottoman Muslim.

This case indicates again the significance of the local context in shaping the ways in which science and its representatives are conceptualized.[44] Concerns regarding the role and future of the Ottoman Empire in an era of industrial capitalism and European imperialism were in the background if not the foreground of all arguments. Particularly after the 1870s, the influence of missionaries in numerous parts of the empire also augmented these concerns, as we saw in chapter 6. This was indeed a period in which references to science became ways of attacking or defending religions— and what religions represented in terms of communal or political affiliation. As Marwa Elshakry noted, missionaries in the Ottoman Middle East saw scientific education as a way to convert Muslims to Protestantism. James Dennis, a Presbyterian missionary in Syria, argued, for instance: "Contact with Moslem minds, so difficult through other means, is in a measure possible through education. Scripture truth may be inculcated in connection with science, and this when youthful minds are most susceptible to impressions."[45] In turn, and with the encouragement of the sultan, Ottoman Muslim authors strove to prove the opposite in their works on Islam and science and to combat what they saw as dangers to social cohesion.[46] Hence the meanings and implications of science, Islam, and citizenship and subjecthood could not but be widely overlapping topics of debate.

B. Science and Morality: Direct Engagements

1. The Press

The books, newspapers, and journals of the 1880s and 1890s contain many examples that demonstrate the ever-strengthening association between learning, science, and morality defined in Islamic terms. Most importantly, names that are commonly referred to in the contemporary historiography of nineteenth-century Ottoman thought as exemplifying the fascination of Westernized Ottoman intellectuals with science are also among those who wrote explicitly on the connections between science and moral values. For instance, Şemseddin Sami, an author who censured those who continued to exalt medieval Islamic sciences (see chapter 6), and who indeed wrote many pieces on the uniqueness of modern science, also wrote the following in his journal *Hafta* (The Week):

Learning (*marifet*) awakens one's intellect. . . . Don't we see that man . . . now overwhelms nature, transforms land into sea and sea into land, moves above waves and below mountains and turns the world upside down, is able to understand the properties of the substances in the core of the earth, the greatness and the movement of the stars in the sky, and the many secrets of nature? . . . But learning . . . also provides great services to humanity for the purification of souls. The survival or fall of a society depends on the improvement or deterioration of its morals and the improvement of its morals can be achieved only through learning and education.[47]

While the examples Sami provides are indicative of the kind of learning his emphasis is on, he makes the idea more explicit in the rest of the essay by referring frequently to European learning. Also worth underlining is Sami's division of the means for "the progress of learning" into two: first, new inventions and discoveries, and, second, spreading the knowledge on new inventions and discoveries. In its current state, Sami contends, the Ottoman Empire cannot produce inventors and discoverers, and the mission of educated Ottomans can only be the translation of "European books on various branches of science" and popularization of science via newspapers and books. In sum, while Sami's appreciation of European science and his association of learning with new discoveries and inventions are striking, it cannot be ignored that he also regards the spread of these products of the new sciences as a way to enhance morality. The definition of the new sciences as knowledge and unawareness of them as ignorance per se create a theme that is reiterated forcefully by Sami—who goes even further and states that "those who are oblivious to chemistry and physics, no matter what other sciences they do know, are to be considered ignorant"—and this statement is, as discussed in previous chapters, a way of forging a link between the new sciences and morality.[48]

One is indeed hard pressed to find areligious and amoral approaches in such texts on science. Similar examples abound in the series of cheap popular science books Sami wrote on a number of topics, where he not only presented information on new scientific findings but elaborated on the implications of being either aware or ignorant of them. In his book about recent astronomical findings on the age and nature of the universe, for instance, Sami noted that there still existed "crooks" claiming to be able to read the stars whom he condemned with a harsh Islamic tone: "When the size and movement of each [star] is dependent on an unchangeable law of nature established by the eternal power of God, it is such sacrilege, such blasphemy to turn this great creation of the Creator into entertainment for humans."[49] Moreover, Sami argued, while it was obvious that Islam only encouraged Muslims to study and uncover the truths of nature, the "superstitions of India, Iran, and Egypt" had crept into the Islamic tradition due to "certain ignoramuses entirely ignorant of science."[50] In this respect, then, the work of Sami and others like him was actually a religious mission as it entailed eliminating these un-Islamic latter-day additions. Referring to names like Newton, Herschel, and Georg Wilhelm von Struve as the giants of astronomy, Sami emphasized that for questions regarding the nature of the universe the authority to consult was contemporary science, but he also underlined that science had not yet been able to fully explain phenomena such as comets and would never be able to answer questions such as the creation of the universe. The latter, clearly, could only be known by God. But it is also worth noting that Sami highlighted the findings that suggested the universe had emerged gradually and made a case for reinterpreting the belief in a six-day creation in ways similar to the Day-Age theory that had become rather popular in Europe and the United States in the mid-nineteenth century.[51] Sami made similar remarks in his companion book on the Earth, where he focused on recent geological findings and introduced his readers to the science of paleontology.[52] In these representations, once again, the man of science is implicitly, if not explicitly, depicted as a better interpreter of the Qur'an and, hence, a better Muslim.

At this point it should not be surprising that in another issue of his journal Sami published an essay entitled "General Morality," which complained in unequivocal terms about the decline in the moral qualities of Ottoman people.[53] Most interestingly, this article is centered around the argument that learning and civilization may not necessarily stop the deterioration of morals—in fact, as the histories of ancient Rome and the medieval Arab empire suggest, progress in civilization may simultaneously lead to immorality. Indeed, it is immorality itself that caused the Ottomans to lag behind Europeans in terms of learning, and, as a result, the decline of the empire, according to the article. Sami's solution was simple: young men should be isolated from this immoral society, and they should be taught morality from textbooks in schools—a proposal that was not in conflict with the policies of Abdülhamid II.

What exactly is the connection between science and morality, then? Is it only that science is knowledge per se, and that ignorance itself is a vice? While this remained the common reasoning throughout the nineteenth century, references to another connection were gradually getting more and more popular: science not only uncovers facts but puts them to use, so in both of these aspects, it is closely related to hard work as opposed to indolence. This connection is clearly illustrated in an article by Mehmed Nadir in the journal *Mirat-ı Alem* (The Mirror of the Universe).[54] Discussing in detail how every object in one's room was manufactured by hardworking men in different parts of the world, and how one can read the works of Galileo, Newton, Lavoisier, and Buffon and "attain human perfection" thanks to the inventors of the printing press, Nadir argues that it is scientists, inventors, discoverers, manufacturers who should be glorified and inertia that should be damned.[55] In this account, sciences and arts are good because they save one from lethargy.

In a context where "Europhile" intellectuals like Sami and Nadir discussed science in such close connection to morality, albeit without explicit references to Islam, it should come as no surprise that similar arguments could be found in a variety of documents from this period. *Hadika-i Maarif* (The Garden of Learning), a short-lived journal published by Imadeddin Vasfi, argues in its first issue, for instance, that learning is essential in order to fend off moral degeneration. Yet not all branches of learning lead to the moral improvement of the people, and some sciences are useful only for particular purposes: "medicine is useful only for the ailed, geometry is useful for construction, and astronomy is useful for discovering stars and determining the time." It is only good morals that are useful for humanity as a whole, and, the editor notes, his journal will focus primarily on this issue, with additional texts on sciences that can also be illuminating in this respect.[56] Similarly, many other new journals of the period define their objectives in their first issue as "ameliorating morals" and "spreading the sciences."[57]

2. Advice for Ottoman Youth

Thus, texts about science consistently referred to morality in the 1880s and 1890s, even more commonly than in previous decades. As Benjamin Fortna's works demonstrated, furthermore, the reign of Abdülhamid II is characterized not only by the increase in the number of European-style secular schools throughout the empire but also, very importantly, by the increase in the weight of courses related to Islam and morality in school curricula. How did school books, and books specifically on morality, refer to science, then?

We should note at the outset that all books on morality refer to the acquisition of knowledge and science. Also important to note is that while the traditional, broad term *ilm* is commonly used in these texts, the equally common references to the recent rapid growth in *ilm* clarify that this term is now assumed to include the new sciences.

A work from 1881 entitled "Advice to Youth" states in the introduction that manners, decency, and good morals are essential for people who wish to be respected and appreciated. "In each era," the author asserts, "people who, thanks to their mental skills and knowledge, made discoveries in sciences and arts were people adorned with good manners, decency and sound morals, and avoided indecent attitudes and deeds."[58]

In the high school textbook "The History of the Arts," dedicated to a discussion on painting, etching, and architecture, the author Mahmud Esad devotes the final chapter entirely to science, as "the nineteenth century is a century of science." After passionately recounting the achievements of scientists, Esad concludes by arguing that "[science] now guides industry and commerce, even starts to organize politics. Furthermore, it has become the means to train the intellects and better the morals of all classes of society."[59]

After praising Abdülhamid's services to education in its introduction, the translator of the Arabic work *Mikyasü'l-Ahlak* (Criterion of Morality) from 1897 argues, for its part, that education is the basis of everything, and knowledge and science are the means toward all kinds of progress. It is the sciences and knowledges that they possess that make nations glorious. However, Mustafa Zihni, the translator, continues:

Education, knowledge and science have a fundamental basis . . . as well. This basis is morality. . . . If immorality spreads among the members of a nation, and God forbid, the number of immoral people surpasses the number of people with good morals, then that nation, no matter how many products of civilization, and even knowledge, it may possess, cannot avoid falling from the high summit of civilization with the speed of a train. It is also possible that its political existence itself will be extinguished.[60]

Zihni's suggestion was clear: works on morality should be translated into Ottoman Turkish along with European works on the new sciences. While Europeans' works on ethics were also important, Zihni contended, it would be more appropriate to translate Arabic works on morality, as morality "is based on religion." This, once again, indicates how interwoven arguments on science, morality, language, Arabs, Ottomans, and Europeans were in late nineteenth-century discourse.

Another book on morality, *Nuhbetü'l-Fezail* (The Highest of Virtues), from 1895 does not refer to "progress" like *Mikyas* does. Instead, it starts with an introduction stating that the worth of any branch of knowledge is determined by its subject matter, and that this is sufficient proof that the science of ethics is the most valuable of all. Ethics, for Sadreddin Şükrü, the author, is primarily about the duties of individuals toward God and one another and, most strikingly, "the primary duty instructed by the science of ethics is obedience"—obedience to God, to the teachings of the Prophet, and to the sultan.[61] The same argument is made in another textbook, *Fezail-i Ahlak* (Virtues of Morality), which defines ethics as the highest science and obedience as the highest virtue.[62]

While such comments were made in books for high schools and above, textbooks for elementary and secondary schools also discussed science, morality, and religion in similar tones. A poem on the importance of several branches of science (including religious sciences) in a book on morality states the following:

Have knowledge of the science of medicine,
So you can protect your health.
Learn it soon in the proper manner,
And appreciate the science of anatomy.
Yet when you learn sciences,
Do not distort the Word of God.[63]

In this book that emphasizes that the definition of *ilm* includes religious sciences as well, some benefits of learning are also listed as follows: "It is knowledge that saves us from the darkness of ignorance, delivers us to the world of civilization, prevents us from evil deeds and gives us sound morals, enables us to understand the might of God and strengthens our love and obedience toward him."[64]

A very popular book of readings for high schools also includes an essay on the virtues of knowledge and the relations between knowledge and virtue. "The Pleasures of Knowledge" uses the word *ilm* as well but makes explicit allusions to the importance of sciences such as geology, astronomy, and chemistry. The conclusion this text reaches, however, is also related to morality:

A man who finds the greatest pleasure of his life in learning science will, on the whole, not try to harm others. Let us think about the thing that he considers his sole activity in life. Who would such a man harm? . . . A man who devotes his life to learning science will immerse himself in a world of pure, spiritual pleasure and never tend toward worldly pleasures . . . that usually lead only to a guilty conscience. . . . The more one loves science, the more he loves the absolute innocence, highest virtue, and pure devotion to God that is essential for humanity.[65]

While the way science is linked to virtue and devotion to God could imply that the sciences in question are overtly religious ones, it is worth emphasizing that the essay starts with references to the new sciences. Moreover, another essay in this collection refers to names like Pythagoras, Kepler, and Newton as great discoverers of truth in a discussion on intellectual pleasures. Thus it would be safe to assume that the learning that keeps one from evil deeds does include the sciences associated with contemporary Europe.

Finally worth noting is an essay in another collection of readings, the *Mekteb-i Edeb* (1890). In an essay entitled "Learning the Sciences," the author invites the youth to learn the sciences in order to make a living and acquire happiness after the precious years of youth are over. Yet in harmony with many of the texts discussed in this book, the essay only emphasizes *reading* about the sciences, and concludes by saying that one should read not in order to look like a man of knowledge, but "to be a better, more virtuous, more mature person."[66] The ultimate purpose in reading about the sciences is to be more virtuous.

In the press, we come across similar arguments in newspapers and journals published specially for "the future generations." One of the earliest newspapers for young readers, *Etfal* (Children), started publication in 1886, defining itself as a scientific and literary journal for children. The introductory essay states that it is now time to acquire knowledge, and the state is doing everything it can to spread education. Yet, argues the editor, the sciences learned in school may not be enough, and newspapers and journals should supplement the training. Furthermore, "even though the paths toward sound morals and good manners are also taught in schools," the press should help in this respect as well by publishing stories with morals.[67] In sum, the duty of the press is defined as providing scientific knowledge and morality tales: two things that the new generations need the most.

Vasıta-i Terakki (Means of Progress) from the early 1880s talks about science and morality in an essay entitled "Greatness," and states that to be a "great man," what one needs is "first and foremost, good morals, and after that, acquiring science and knowledge."[68] These comments are addressed to "the last hopes of the nation"—the nation whose "high positions of service and officialdom" are awaiting the new, educated youth of the empire.[69] The path for the young Ottoman citizen is clear: acquiring sound morals and some knowledge of science and, finally, a position in civil service.

In *Çocuklara Mahsus Gazete* (Newspaper for Children), a long-lived newspaper filled with didactic texts on science, religion, and morality, an essay entitled "Acquiring Science and Knowledge" states the following:

It is science and knowledge that teach humans their humanity. . . . Can there be a better capital for us than acquiring science and knowledge? In order to be content in this world, one should perfectly attain all ranks of education. Learning good habits and avoiding bad ones is a consequence of this accomplishment, because all bad habits are due to ignorance. . . . Amelioration of morals can be achieved only through science and knowledge. In the deep darkness called life, it is only knowledge (*ilm*) that is the bright sun that shows man his humanity, his own self, his entire world.[70]

Yet following this quite standard association between learning and morality, the author argues that the most important virtue is having sound morals, as those with low morals can never be respected, "no matter how much knowledge and science they know." While there appears to be a contradiction between these two subsequent paragraphs, what is more significant is that learning, science, knowledge appear impossible to discuss without a reference to morality. It is not as much the exact nature of the connection between science and morality that matters as the basic idea that science and morality emerge as an indivisible couple in the 1880s and 1890s.

Perhaps what best illustrates this is another collection of readings for first-grade students. The titles of the first six readings by future novelist Ahmed Rasim are 1. Allah; 2. Science: Our Home, the Earth; 3. Prayer; 4. Science: The Sea; 5. Morals: Our Duties toward Our Fathers; and 6. Science: Agriculture. After this point, the texts alternate between "Science" and "Morals."[71] This is also a pattern we observe in the journal *Etfal*, mentioned above. The first piece it published in the first issue was a story entitled "Good Morals," followed by the "Science" section.[72]

But it is also important to note that statements that brought God, the sultan, and science together (as had always been the case in the nineteenth century) became strikingly common in this period. Consider the following speech by an imam at the opening of an elementary school.

God Almighty is ordering us to improve ourselves by learning the religious and the practical sciences. Our Sultan is spreading education even to the villages of his Well-Protected Domains. Our Prophet obliges us to go even to the farthest lands of the East in order to study the sciences. Our Glorious Master is gathering the sciences and knowledges of the entire world in His domains. . . . Thus, is there any other way for us to express our gratitude than educating our children and making them learned people?[73]

Similarly, even in the introduction of a book on the phonograph and the newly discovered methods of recording sound, the translator Ahmed

Rasim quotes a saying by Ali, the son-in-law of Muhammad and the fourth caliph of Islam, on the importance of knowledge and the dangers of ignorance. He also notes that, thanks to the sultan who was an advocate of progress, the Ottomans were going to reach new heights.[74]

Indeed, most textbooks from this period start with convoluted praises to the sultan. As previous chapters have shown, this was an established pattern in earlier periods as well, and the idea of science as a boon bestowed on the subjects by the sultan was already entrenched. So while it is hard to say that Abdülhamid's rule brought about a qualitative change, it remains the fact that these introductions proliferated due to the increase in the number of textbooks, and they were somewhat enriched with stronger references to the "glory of the sultan" in the 1880s and 1890s. Necib Asım's *Kıraat-ı Fenniye* ("Scientific Readings") starts with an introduction that refers to the sultan as "the greatest patron of knowledge and science" and prays that his reign is filled with new victories.[75] Ismail Cenabî, the translator of a general work on the physical sciences, states that Abdülhamid's reign was such an era of progress that everybody endeavored to bring about something to serve the general interest.[76] Ömer Subhi, who translated Archibald Geikie's *Physical Geography*, noted that the sultan had declared war on ignorance and opened many schools, as a result of which the only duty of "the faithful members of the nation and those loyal to the state and religion" was "to translate and write as many useful works as possible and produce sources of knowledge and learning."[77] Captain M. Münşi, in his *Bedraka-i Mühendisîn* ("Guide for Engineers"), refers to Abdülhamid as "the sun of learning" who "emit[ted] the light of knowledge."[78] Besim Ömer, a graduate of the Imperial School of Medicine, starts his work on dentistry with a two-page eulogy for the sultan, thanks to whose reign "the children of the nation were illuminated by the light of learning."[79]

The following poem published in the journal *Gülşen* in 1886 is also typical in this respect:

The greatest bliss for man
Is learning knowledges and sciences
You, zealot, the source of ignorance,
Your ignorance is like insanity! . . .
The reign of Sultan Hamid the Second,
Built schools all over the land.
Live forever, o magnificent ruler,
It's your graces that revived this universe![80]

C. The Result: Portraying the Ideal Man of Science

Ottoman literature of the 1880s and 1890s contains several examples of the ideal man of science, or even the man of science as *the* ideal man, based on the rather specific criteria outlined above.

A key character can be found in Mizancı Murad's famous didactic novel *Turfanda mı Turfa mı?*[81] Mansur, the protagonist, is a young man who studies medicine and political science in Paris and returns to Istanbul, hoping to use his useful knowledge in the service of his nation. On his return, he is extremely disappointed to realize the dominance of non-Muslims and Europeans in all affairs of the Ottoman Empire as well as the corruptness of the civil servants. As a man of science, he sets up a laboratory in his apartment, starts working at the Imperial School of Medicine, and, thanks to the family connections that he reluctantly takes advantage of, he is also appointed to a post in the foreign ministry. Yet the ignorance and indifference of many of the scribes and the lack of dignity he observes repulse him. When he receives a promotion, also due to family connections, he refuses to accept, stating that he "regard[s] service to the state and religion as holy duties" and that he will have to resign if he is rewarded for no real achievement.[82] He also cannot prevent himself from slapping an interpreter from a European embassy who shows a disrespectful attitude.[83] The novel contains many monologues in which Mansur advocates the spread of education throughout the empire and criticizes all nationalist movements that threaten the future of the Muslim world.

A physician is also the protagonist in Hüseyin Rahmi's *İffet* (Decency) of 1896. This work is particularly worth considering, as it clearly shows what kinds of figures the man of science (and a materialist one here, at that) is defined in contrast to, and how this makes him the ideal Ottoman Muslim. The protagonist "Mr. N.," a physician, first recounts to his friend, a litterateur, an incident from earlier that day:

In order to avoid contracting the patients' diseases, we put on "eau phenique," that is our lotion. I got on the trolley to go to Aksaray. Sitting next to me was a fop. And what a fop! A true snob indeed! . . . I didn't understand why he was looking at me with such disdain. As it turns out, the gentleman was disturbed by my smell. With a scornful face, he said, "You physicians smell like itinerant hospitals. This . . . strong smell will kill not just the germs but the people unlucky enough to sit next to you." . . . I just glared at him and said, "Delicate people like yourself should travel in private carriages, not in trolleys."[84]

The physician is not a delicate fop; he is not interested in his looks, truly understands how science works, and uses his knowledge to serve people. Note the emphasis on the manliness of the man of science as opposed to the effeminate fop. These portrayals are closely connected to those like Şinasi's (as discussed in chapter 4) that depict the Orient as a mature man who should resist the seductive whims of Europe. By underlining that the man of science is not delicate, Hüseyin Rahmi presents him as one who has passed the test, so to speak. The man of science is (that is to say, should be) a "real man" if he wishes to be accepted as an authentic Ottoman Muslim.

Moreover, in absolute contrast to a fop, the man of science knows exactly the condition common people live in; he is a true, sincere, loyal member of his society rather than a blind imitator of the West. As the following dialogue between the two friends indicates, this is the reason the physician is also unlike a litterateur who lives in a world of dreams and fiction. Note, finally, that as a physician, Mr. N. talks in the name of science, not simply medicine:

—I am one of those people whose work is serious. I am . . . infuriated by those poets who write gibberish . . . all day about clouds, seas, winds, nightingales, and think this is service to humanity.

—So in your opinion those who did not study medicine are worthless, huh?

—Not at all! I don't claim that the sciences taught in the civilized world amount to medicine only. . . . All sciences are essential for humanity. And humanity should be grateful to everybody who engages in science. Civilization relies on science, science relies on civilization. I appreciate sciences that deal with the material world. I don't appreciate metaphysical ones. And I regard our poetry and novel writing as nothing but metaphysics.[85]

Mr. N. then argues that nobody should consider writing a novel before reading serious scientific books on a particular topic. But he also implies that the man of science, because he both reads about the material world and, particularly in the case of the physician, encounters the world in its materiality, will always know more about the world than the poet or novelist. Furthermore, as a physician's everyday work involves dealing with real people, he knows his society better than the poet as well, and the actual story takes off when, in order to prove his point, the physician takes his friend to the home of one of his patients.

A final example comes from the work of Ahmed Midhat Efendi. Midhat's novel *Acaib-i Alem* (Wonders of the World) is about Suphi Bey, a young man ridiculed by people for his intense love for the "laws of nature." Suphi is a recluse who lives a modest life, in a house filled with books, curiosities, collections, and scientific instruments. In a conversation with his friend Hicabi, Suphi explains what his fascination is truly about: "I believe that if there is a purpose behind our coming to this world, it is admiring the might and mastery of God, the Creator of all, the Great Maker, through studying nature. This is what pleasure is all about."[86] Selling all his collections, Suphi, along with Hicabi, embarks on a journey to northern Russia in order to observe the effects of the earth's shape and rotation in areas closer to the poles. During the journey, he meets Miss Haft, an English woman, who admires the morals and manners of these Muslim Ottoman men. In a letter to her mother, she writes, "Suphi Bey and Hicabi Bey are fluent in French and German, respectively, and they are competent in European sciences. But they themselves have not become European. They are pure, true Ottomans."[87] In the novel's finale, Miss Haft proposes to Suphi. When her aunt hears the news, she exclaims, "Oriental lands and the Orientals are full of mysteries. What they know aren't like our sciences and arts. I am worried that this Turk confused you with some sort of magic!"[88] Miss Haft, smiling, tells her aunt that the magic involved is Suphi's *terbiye*—the word which, with denotations like education, decency, morals, and discipline, encapsulates the ideal Ottoman synthesis.

Conclusion

The official discourse of the last two decades of the nineteenth century combined the themes of the past decades and rendered science a topic that could not be discussed without overt references to morality. More specifically, it strictly separated scientific knowledge from the person who possessed it, and precisely by emphasizing that the former was both desirable and value-neutral, it brought to the fore the question of the moral qualities of the latter. Emphasizing that scientific knowledge was safe to import meant that science, in essence, had nothing to do with the values of the Europeans or Christianity. Thus, the undesirable arguments or lifestyle of a man of science could never be attributed to the scientific knowledge that he possessed, and the responsibility should be placed on the corrupted morality of the individual himself. The wedge between science as knowledge and the man of science worked as a disciplinary mechanism that incessantly compelled the representative of science to demonstrate the soundness of his morality.

Of course the morality in question was also defined within the official discourse. To be moral meant being a Muslim who obeyed God and the sultan and who worked diligently for the prosperity and honor of the nation. The emphasis on obedience and selfless hard work is clearly a response to the sentiments of a growing number of young men who learned in school not only the new sciences but also to see themselves as uniquely important for the future of the Ottoman state. The prestigious medical and military academies did indeed provide a kind of training that was rarely found in the other institutions of the empire. But the sense of empowerment and entitlement that this training gave rise to among the students was precisely the kind of threat that the centralizing policies of Abdülhamid II intended to eliminate. Hence the heightened emphasis on the specific definition of morality delineated above.

What is also crucial, however, is that there is more to the story than the scientism and the related "savior mentality" that these students' writings exhibit. As this chapter has demonstrated, even in the 1890s when the Young Turk movement became active, the output of many educated young Muslim Ottomans did not noticeably contradict the official discourse. In fact, while it is undeniable that an oppositional discourse emerged at the end of the nineteenth century, the official discourse remained hegemonic. Arguments about Muslims' contributions to science, portrayals of Islam as the only religion that truly promoted science, and the assertion that the man of science had to be a patriotic, modest, obedient Muslim—that is, a moral individual—are not confined to official documents. Similarly, the figures of the fop and the confused materialist continued to play particularly effective rhetorical roles by challenging the Muslimness, Ottomanness, and manliness of the Muslim Ottoman man of science. Portrayals of the young and superficially enthusiastic advocates of the new sciences as either ignorant parrots of the missionaries or misguided devotees of the Europeans not only ridiculed them but represented them as potential threats. The response, in turn, came in the form of equally simplistic representations of the ideal man of science as a real man, real Muslim, and real Ottoman: a lifeless and cartoonishly hollow embodiment of perfection.

Conclusion

When the Ottoman ambassador Yirmisekiz Mehmed Çelebi was sent to France in 1720, one of the instructions he received was to "gain mastery of the means of civilization and learning" of the French and evaluate their applicability in the Ottoman Empire.[1] Mehmed Çelebi did indeed visit numerous locations; he also met many people and discussed his impressions in his celebrated travelogue. His account makes it clear that his French hosts were intent on taking advantage of this opportunity to impress their guest and to present an image of France as the center of prosperity and intellectual activity. The Paris Observatory, the Gobelins Manufactory, the Hôtel des Invalides, the Versailles menagerie, and the Jardin du Roi were among the spots the Ottoman ambassador visited with his entourage, and several of his hosts also introduced him to their cabinets of curiosities.

Readers expecting to see paragraphs about the achievements and progress of European science in Mehmed Çelebi's travelogue might be disappointed, however. Not that the Ottoman diplomat was dismissive of or condescending toward what he had been shown. On the contrary, Mehmed Çelebi spares no words to express his fascination with the opulence and flamboyance that he witnessed in Paris. His text includes numerous quantitative estimates of the dimensions and worth of the awe-inspiring objects and buildings that he had seen and makes many positive remarks about the utility of the French novelties. Yet it is "wonders" that Mehmed Çelebi frequently talks about—"wonders" that amazed and, at times, confused him.[2] The things he saw were wondrous and intriguing, but not all of them were necessarily important or relevant. Similarly, there was not one simple, coherent, and comprehensive story (such as "the progress of civilization") that could be told about them, or one basic concept (such as "science") that they could all be reduced to. Finally, Mehmed Çelebi does not talk about the emergence of a thoroughly new, let alone superior, type of people in Europe in relation to the "curiosities" that he had observed. He was interested in and impressed with Jacques Cassini's observations at the Paris Observatory, the animals at the Versailles menagerie, the herbs in the Jardin du Roi, and the engineering behind the Canal du Midi, but he did not build on these impressions a comprehensive social commentary advocating radical social and political change in the Ottoman Empire. Undoubtedly, Mehmed Çelebi's text does exemplify the Ottoman elites' adoption of a more comparative perspective and, very importantly, their emphasis on the "new" in the early eighteenth century. But the outcome is not a narrative containing plainly delineated problems and solutions.

Fast-forward a century, and we find scholars like Şanizade and Mustafa Behçet who referred to new knowledge, especially in the field of medicine, but, once again, did not directly associate it with a new social or political order. The perception that the Europeans were producing new and impressively effective types of knowledge and that the apparent superiority of European militaries was attributable only to this fact was increasingly common. But for many educated Muslim Ottomans, it was possible to regard new knowledge as an extension of or, at least, as not necessarily incommensurable to inherited knowledge. In order to be truly knowledgeable one needed to be familiar with as many aspects of knowledge as possible, and it was certainly legitimate if the "knowers" occupied a higher place in the social hierarchy; but there was no need to define a new type of prestigious knowledge or to reimagine the nature and basis of social order.

Needless to note, perceptions and ideas are not self-contained, coherent entities that emerge from and exist in the ether. They become possible and make sense within discourse and are articulated by differently situated social actors. And at the turn of the nineteenth century, the actors who would construct a new discourse within which it would mean a rather different thing to talk about the new knowledges of the Europeans still constituted a relatively ineffective minority in the Ottoman Empire. The key theme of the new discourse that the members of this minority gradually built was the conceptualization of new knowledge not simply in terms of its "curious and wondrous" material consequences, but in terms of the qualities of its very "possessor."

The producers of this discourse were themselves new kinds of people who identified themselves with the new knowledge. The more unique and novel this knowledge, the more distinction could potentially accrue to one who was familiar with it, and the appeal of new knowledge for a new group that lacked prestige in a society where knowledgeability had long been associated with rank and worth is not surprising.[3] The young bureaucrats and diplomats of the Ottoman Empire and the students of the few but increasingly prestigious new Ottoman schools gradually transformed themselves into the dominant representatives of the new types of knowledge that the Europeans had been producing. As they were increasingly able to assume positions within the Ottoman state mechanism, they were also able to mold an official discourse on knowledge and ignorance—a discourse that was not simply about what it meant to possess or lack knowledge, but what individuals deemed knowledgeable or ignorant deserved. Ultimately, the discourse on knowledge and its boundaries was about social order and social boundaries.

In the official discourse of the 1820s and 1830s we observe a gradual association between new knowledge—now commonly referred to as *ilm*—and welfare in this world, while religious knowledge was associated only with bliss in the afterlife. This set of associations was, once again, inseparable from definitions regarding social boundaries, as the issue in question was establishing the criteria for evaluating Ottoman subjects. What kind of people were the Ottomans? What kind of people did the empire really need? As the sciences of the Europeans came to be defined as the beneficial and, increasingly commonly, the needed sciences, those who claimed familiarity with them also defined themselves as the types of individual who were essential for the salvation of the empire. The emphasis on the novelty and the unique benefits of these sciences rendered their representatives also uniquely valuable; those who were experts of inherited knowledge still deserved respect, but theirs was not the qualification needed to solve the empire's pressing problems.

But we should not assume that the discourse constructed by these young bureaucrats and students had a determinedly radical tone that rejected the authority and worth of inherited knowledge or expressed the need for the new sciences in an unusual, let alone revolutionary, style. Indeed, arguments that were uncompromisingly critical of or condescending toward the so-called older types of knowledge, especially those most overtly linked to the Islamic legacy, are strikingly hard to come by, at least until the end of the nineteenth century. Similarly, the case is that even texts that glorified the new sciences and their representatives were commonly written in a style that was barely atypical; the conceptual frame-work and the justifications used in these texts resembled those used in texts from earlier periods.

While references to holy war got increasingly uncommon during the nineteenth century, for instance, the principle of due reciprocity was con-sistently alluded to: the Ottomans were justified in adopting the ways of the Europeans in order to defend themselves. Second, the image of the sultan as the benevolent protector of learning was always present in texts on the benefits of the new sciences. Third, the concept *ilm* played an ex-ceptionally important part in the naturalization of the new European sci-ences. And fourth, particularly in the first three-quarters of the century, Ottoman authors regularly referred to the idea that the contemporary sciences were essentially based on—and, thus, categorically not that dif-ferent from—the sciences to which the Muslim scholars of the Golden Age of Islam had made momentous contributions. Yet each one of these examples of continuity is more complicated than at first seems, and their trajectories embody intriguing ironies.

The idea that the new sciences of the Europeans had to be imported in order to be able to defend the empire was rooted in the well-established Islamic principle of due reciprocity. That Europeans had access to a spe-cific type of knowledge that had enabled them to dominate the entire world justified Ottomans' adoption of this knowledge to be able to defend themselves and their religion in the new era. In this respect, arguments framed this way were additions to a long tradition of justificatory texts. While the principle was overtly referred to in the late eighteenth and early nineteenth centuries, however, the growing influence of the new elites in the construction of the official discourse resulted in the transforma-tion of the principle into a commonsense notion that no longer needed to be explicitly asserted. Repeated references to this argument in the offi-cial discourse—particularly within textbooks and decrees—rendered the principle behind it, so to speak, invisible.

But this gradual "backgrounding"—that is to say, the process through which it became no longer necessary to justify the statement "The adop-tion of the sciences of the Europeans is the only option"—cannot be ad-equately understood without a reference to the power relations within Ottoman society. Bourdieu's argument on such views that "go without saying" is pertinent in this context: these are the points of view of "those who dominate by dominating the state and who have constituted their point of view as universal by constituting the state."[4] Defining a specific type of skill and familiarity as essential, and linking these qualifications to political power, the new Ottoman elites constructed an official discourse that defined the "right" way of perceiving the world and the status of the empire—a way that would, by implication, portray them as the "right" type of individual as well. As their power increased, the principle of due reciprocity that was overtly mentioned in earlier periods gradually moved to the background. It was now "obvious" that the empire needed these new types of knowledge and skill.

The authority and prestige that the sultan's endorsement imparted to any institution or practice was also crucial for the legitimacy of the new knowledges. As early as the turn of the nineteenth century, we observe references to Selim III as the patron of the types of knowledge and skill developed by the Europeans and of the young Ottoman men who learned them. The efforts of Sultans Selim III and Mahmud II to construct an effective and stable central authority involved attempts to generate a well-regulated cadre, the members of which would devote their loyalty only to the sultan. The students of the new European-style military schools as well as the new Ottoman diplomats and bureaucrats were to form this cadre. But particularly in the early decades of the nineteenth century, these were men still without a significant amount of symbolic capital to rely on. Hence, the image of the sultan as the protector of the new knowledges and skills—and, consequently, the protector of the new "knowing class"— was ubiquitous in texts. Such portrayals not only reinforced the legitimacy of the new groups and the types of knowledge they represented but implied continuity with the status of the inherited types of knowledge. Once this initial stage was over, however, and as the bureaucracy achieved significant power during the Tanzimat Era, the patronage of the top bureaucrats themselves took center stage in many documents, especially textbook introductions. The sciences of the Europeans were endorsed by the state itself, represented by the names of cabinet ministers and other leading statesmen. Then, during the reign of Abdülhamid II, and as part of the comprehensive concern with the representations of power within the empire, the image of the benevolent sultan once again dominated the discourse on science.

While this omnipresence of the sultan thus established a continuity with earlier Ottoman texts, the implications of the way the support of the sultan (or of the state) was expressed is worth highlighting. Indeed, during the entire nineteenth century, the opening of new schools, translation of books, publication of journals, and establishment of institutions associated with the new sciences were, almost as a rule, praised with the frequent use of the gift metaphor. The sultan as well as the representatives of state power were represented as giving the Ottoman people the gift of science as a consequence of their wisdom and benevolence with the accompanying implication, and often the overt assertion, that this noble deed had to be reciprocated. In return for this gift of beneficial knowledge, Ottomans themselves were to become beneficial and virtuous individuals who served their state efficiently and humbly. In this sense, during the entire century the official discourse on science was about the relations and hierarchy between those who represented state power and those who could not.

The common use of the word *ilm* in discussions on the types of knowledge imported from Europe constitutes another area of continuity and irony. The new sciences were frequently referred to as *fen*, especially when the topic was a specific discipline. But the practicality and particularity suggested by this concept was, in a sense, counterbalanced by the common use of the word *ilm* as well. Once again, the way the concepts were used indicates a sense of continuity, or a sense that the imported was not categorically different from the familiar. What is more significant in this context is the adjectives that accompanied these words, such as "new," "beneficial," and "needed." Whether or not one referred to chemistry as a *fen* or an *ilm*, for instance, the reference was commonly surrounded with phrases such as the "needed *ilms*" or "beneficial *fens*."[5] Thus, what was being learned from the Europeans was placed in the same category as what already existed, but portrayed as a special and—at least for the present— more urgently needed version of it.

This observation is in harmony with the arguments of scholars who suggest that at least for part of the nineteenth century, Muslim elites perceived themselves as not alien to what they were dealing with, and they were able to place the new additions within the framework of the existing cultural repertoire and the institutional setting. The process did entail some syncretism and creativity, but it was not experienced as a radical and threatening change.[6] Indeed, many Muslim intellectuals were able to imagine themselves as contributing to the one "Civilization."[7] It is very important to be aware of these conceptual ingenuities and the creative attitudes of Muslim intellectuals, particularly in order to avoid simplistic talking points such as "Islam versus modern science." Yet what is also worth underlining are the social implications of the emphasis on the "beneficial" and "needed" characteristics of the new sciences. Even if various modes of knowledge production are placed in the same conceptual category, it matters significantly if some are deemed more urgently needed or overall more practically beneficial than others, as ultimately the question has to do with the status of the social actors who represent these types of knowledge. In the early nineteenth-century Ottoman Empire, the new *was* perceived in terms of the old, perhaps; but the representatives of the new increasingly portrayed themselves as the possessors of "(more) useful versions of the old" which, in turn, made them the types of individual the empire truly needed in the new world.

This emphasis on worldly benefits and practical urgency is also clearly related to the issue of science and state. What were the sciences of the Europeans beneficial for, after all? Around the mid-nineteenth century, many Ottoman elites started to answer this question in reference to the state. The new sciences made one a better member of the community who could perform duties more efficiently and effectively and, thus, serve the state more substantially. The old knowledge could be prestigious, but if its contributions were not as useful, it would not be too far-fetched to label its representatives as parasites who were entirely dependent on the state—an argument that would appear in implicit or explicit forms in Ottoman texts of the second half of the nineteenth century.

Finally, another line of continuity with an ironic twist was the emphasis on the Islamic origins of modern science. It was a common argument in European historiography that the contributions of the Muslim scholars of the so-called Golden Age of Islam (roughly the ninth through thirteenth centuries) constituted the link between the ancients and the European Renaissance. Scholars like Avicenna (Ibn Sina), Averroes (Ibn Rushd), Al-Hazen (Ibn al-Haytham), and Al-Kindi were frequently referred to in nineteenth-century histories of intellectual progress or civilization. Muslim observers such as al-Tahtawi and, in the Ottoman case, Mustafa Sami, were impressed with and proud of such representations, and the idea that Europeans had made so much progress thanks only to their adoption of the discoveries and inventions of Muslims gained popularity among literate Muslims in the early nineteenth century.

On the one hand, this portrayal contributed to the naturalization of the new sciences in the Muslim world in general and the Ottoman Empire in particular. It enabled many Ottoman intellectuals to imagine themselves as continuing the tradition of Muslim contributions to civilization. But, on the other hand, it was precisely this tendency to portray the sciences as native in essence that helped transform the Ottoman debate on science into a debate on community and belonging. The statement that the sciences imported from Europe were "already ours" implied that the representatives of these sciences in the Ottoman Empire should not fail to be proper, native representatives of the community themselves.

Furthermore, the enthusiasm to which the "Muslim contributions" argument gave rise was an uneasy one from the beginning. For one thing, these early Muslim scholars' works were not particularly familiar to the Ottoman elites of the nineteenth century, primarily because the *medrese* scholars and students of the period were not trained in a system that high-lighted these contributions. As a result, the tradition alluded to was more invented than easily recognized.[8] Second, as the nineteenth century was characterized by centralizing policies and efforts to construct a predict-able, standardized citizenry in the Ottoman Empire as elsewhere, commu-nication became a key concern. As public education expanded, phrases like "plain Turkish" and "language that anyone can understand" became commonplace. In such a context, the very fact that many of the works of the old masters had been written in Arabic—and in an Arabic that even contemporary *medrese* graduates did not necessarily understand—gradually led to an association between new knowledge and plain Turkish while Arabic was the language of old (that is, not so beneficial) knowledge. Hence, while many an Ottoman author made passionate assertions about early Muslim scholars' contributions, this line of argument was not par-ticularly sustainable in practice.

These ironies aside, the dominant early to mid-nineteenth-century Ottoman discourse remains one that does not categorically deny the rel-evance and importance of old knowledge, nor does it imply that a radical transformation should occur; indeed, it is replete with explicit and im-plicit signs of continuity. While this characteristic of the dominant dis-course did not go through much change in the later parts of the century, its ironies as well as social and political implications became much more pronounced. As official statements from the Tanzimat Era as well as the writings of men like Münif and Safvet Pashas suggest, the idea that the empire now needed a specific type of person became the dominant view after the 1840s. It was crucial for this new Ottoman to be familiar with a specific type of knowledge, as it was this knowledge that would enable him to understand the actions of the representatives of state power. Un-derstanding the state would, in turn, make him into a virtuous patriot who would not be overly critical of statesmen or pose a threat to social order. Some familiarity with new knowledge would make him an obedient and hardworking Ottoman.

The books published on the new sciences, particularly by the Ottoman Learned Academy; the initial attempt to found a university; and, finally, the Ottoman *Journal of Sciences* are particularly striking examples of how science was perceived and represented as a social and political matter in the mid-nineteenth-century Ottoman Empire. That the new sciences were not only endorsed but visibly represented by the high-ranking bu-reaucrats of the empire indicated that these sciences were now "state property," and their credibility and prestige were inseparable from the obedience that the holders of state power demanded from the subjects.

The sciences imported from Europe were to be taken seriously because it was the state that supported, promoted, and represented them. The contemporary enlightened bureaucrats of the Ottoman Empire were not unlike the great scholars of the past who were not only wise and knowledgeable but also illustrious statesmen as well.

The continuity thus alleged is another instance of the new being perceived and portrayed in terms of the old. The use of the word *ilm* commonly operated in this fashion as well, as it was used to refer to the new sciences such as geology and zoology while maintaining its special connotations in the Islamic tradition. But neither of these forms of continuity can be considered as an inevitable, natural transition; the transformation of new knowledge into much more than one with immediate practical benefits entailed the intended conferral of prestige to a specific elite group and modified the relations between the state and knowledge production. Even though the institution of old knowledge represented by the ulema had always been endorsed by the state and remained essential for the legitimation of state authority, it maintained some autonomy, and its prestige could by no means be seen as but a function of its association with the state. For new knowledge, on the other hand, the endorsement of the state was paramount, so much so that the representatives of state authority were at the same time representatives of the sciences imported from Europe.

Moreover, as the writings of Münif Pasha illustrate, in the dominant Ottoman discourse of the 1850s and the 1860s, *ilm* could be used as a term to refer exclusively to the new sciences. High-ranking bureaucrats as well as young litterateurs could comment on the virtues of knowledge in a general sense using the term *ilm*, but the examples they gave were entirely about the benefits of these new sciences. With this strategy, the religiously constructed associations between knowledgeability and virtuousness (expressed commonly with the formulation "Knowledge equips one with the ability to distinguish right from wrong") were imported into the discourse on science. Those familiar with the new sciences (that is, the new bureaucrats, but also litterateurs and students of European-style schools) could thus be represented also as men of virtue—a representation that not only complemented but also justified the power and status of the holders of state authority.

It is also important to note the parallel between this portrayal and the Ottomanist policies of the Tanzimat Era. In a context where ethnic and religious identities were expected to be transcended by an Ottoman identity that would bring together all subjects of the sultan (and later citizens of the empire), religiously grounded knowledge claims could potentially be divisive rather than uniting. Thus, institutions for the production and dissemination of such types of knowledge could be under the jurisdiction of specific religious communities—as the established Ottoman system of administration already sanctioned—while the Ottoman state itself could be represented by a type of knowledge that, allegedly, transcended ethnoreligious traditions. The idea that the sciences imported from Europe were neutral in this sense further accelerated its identification with the state in the heyday of Ottomanism. The makeup of the Ottoman Society of Science and the contributors to its *Journal of Sciences* show clearly the transcommunal status of the new sciences. One could be Greek, Armenian, Muslim, or Jewish and have a specific background rooted in religious tradition, but when it came to learning and writing about the new sciences, everybody could be simply an Ottoman.

But all these claims and processes—the identification of the state with the new sciences, the idea that a certain type of knowledge could transcend communal membership and tradition, the appropriation of the religiously grounded associations between knowledge, virtue, and prestige for the new types of knowledge and their representatives—could not go unchallenged in a period of social, political, and economic transformation that led groups within the Muslim community to feel left behind. Very importantly, these challenges were not in essence related to the contents of the new types of knowledge—and there was no reason for them to be. Knowledge was represented by actual social actors, and what mattered was the way the deeds and arguments of these actors were perceived by other social actors in a specific social context.

Disillusioned young bureaucrats became the speakers of the anxious and disgruntled sectors of the Muslim community in the 1860s, and they directed their criticisms at the high-ranking bureaucrats of the Tanzimat Era. Many Muslim Ottomans perceived the political and economic influence of the European powers over the Ottoman Empire as a consequence of the weakness if not the betrayal of the Westernized top bureaucrats. Capitalizing on this perception, the generation of the Young Ottomans constructed the character of the fop—an effeminate, superficial, arrogant, snobbish young man who was alien to the lives and traditions of the common people and who imitated the Europeans in every way. Not only did the fop try to look and talk like French urbanites, but he frequented the districts of Istanbul where non-Muslim Ottoman as well as European merchants resided. "The sciences of the Europeans" was ultimately but one of the many topics that this fictional character spoke enthusiastically about, while looking down on the traditions and life styles of his "own people."

The belief that the contemporary sciences of the Europeans were based on the discoveries and inventions of early Muslim scholars was already popular among the elite, but the generation of the Young Ottomans made it central to their discourse on science. They also reinvented Islam as *the* pro-science religion. While both of these contributions were effective in the appropriation of the new sciences, this process of "naturalization of the new" had another implication: as sciences themselves were not in conflict with "our ways," the representatives of these sciences should not be, either. In other words, if a spokesperson of the new sciences exhibited foppish characteristics, he could not argue that the lifestyle and attitudes he had adopted had to do with the sciences of which he was an admirer. Such portrayals of science as neutral toward, if not in harmony with, "our values" thus placed an additional burden on those who wished to speak on its behalf: there was no problem with science, so if a science enthusiast failed to prove that he was one of "us," he himself was to blame, not the sciences he praised so heartily.[9]

Needless to note, this line of argument produced an ambivalent attitude toward Ottomanism and the alleged universality of science. The Young Ottomans were sympathetic toward Ottomanist policies, yet they represented primarily the dissatisfied Muslim community of the empire. As a result, while they portrayed science as potentially universal, their emphasis on the proper characteristics of the representatives of science made "Muslimness" a central theme in their discourse. A man of science perhaps did possess knowledge that was universally valid, but he himself was a member of a specific community, and he had to demonstrate this.

The main contribution of the generation of the Young Ottomans to the debate was thus very similar to the dominant discourse of the earlier periods: science was a less controversial matter than the characteristics of the man of science. The representation of the man of science as inherently virtuous and patriotic was thus replaced with the assertion that a true man of science should *also* be a religious and thus moral individual. A Muslim Ottoman who wished to present himself as a spokesperson for science had to demonstrate that he was indeed a Muslim Ottoman.

The final decades of the empire witnessed the emergence of the new official discourse on science that borrowed elements from the dominant and alternative discourses of the past. It tied science very tightly to the state and presented a compelling idea of science, the authority of which could not be separated from that of its truest patron, the sultan. But as Islamic representations became ever more fundamental to the establishment of state authority itself, the ultimate outcome was what we can call an authority triangle—a triangle formed by the sultan, Islam, and science. Each of these components added to and legitimized the authority of the other two. The ideal Ottoman citizen was one who respected and obeyed all three, and the combination of these three authorities was, in a sense, the ultimate authority that could make possible the construction of a truly deferential citizenry. These citizens would know that science and Islam could not be in conflict and that science only proved that Islam was the true religion; Islam, in turn, would unite the Muslims of the world around the Ottoman sultan, and the sultan would be the benevolent protector of both the religious establishment and science.

Particularly among the students and graduates of the military and medical academies and the school of administration this triangle had less sway, however. The sultan's personal control over the key positions within the government and the limited range of opportunities for prestigious employment and political participation continued to cause dissatisfaction. This was particularly the case for young men who, thanks to the education that they had been told was uniquely prestigious and urgently needed, were disappointed with their inability to acquire what they saw as an entitlement.[10] But it is not particularly precise to refer to these students as products of secular education, as Islamic morality was part and parcel of the curricula to which they were exposed.[11] Similarly they came from diverse backgrounds, and many of them had at least some religious education as well.[12] As a result, it would be inaccurate to argue that the majority of these young men were radical devotees of some form of materialism who portrayed science as an anti-Islamic enterprise that could only be discussed in terms borrowed from European authors. The connection between Islam and the authority of the sultan remained quite effective during this period, and the need to raise moral youth was alluded to by most participants of the debate.

Nevertheless, it is certainly the case that scientism and vulgar materialism influenced many young men in this period and accompanied their sense of entitlement, enabling them to imagine themselves as the true saviors of the empire.[13] Moreover, while they did not constitute a majority, their ideas were particularly influential in shaping the intellectual debates of the late nineteenth century. When the writings of young men like Beşir Fuad or, later, the birth of the Young Turk movement indicated the vulnerability of the authority triangle so central to Abdülhamid's policy for establishing order, the result was the emergence of the figure of the misguided materialist—essentially a new version of the by then very familiar fop. Just like the fop, the misguided materialist was a parrot with no original thought of his own. His outfits and attitudes were perhaps more modest (and less effeminate) than that of the fop, but his views were based on those of the Europeans and, more specifically and dangerously, Christian missionaries within the Ottoman Empire. The tightly regulated Ottoman press and schoolbooks were filled with references to and condemnations of this character. But it was hardly the superficiality and simplicity of his views that were criticized. Rather, once again, misguided materialists themselves were defined as the problem: portrayed as respecting only the science corner of the authority triangle, the materialists were claimed to not only ignore but to betray their own community, and they became representatives of the enemies of the empire. They were "Franks in fezzes," as the journal *Kasa* expressed so succinctly.

That these misguided materialists had misgivings about the two corners of the authority triangle does not mean, however, that they were not interested in the question of social order and obedience. In fact, the adoration of science by its advocates in the Ottoman Empire was hardly ever dissociated from the question of order, and in this the difference between the Young Turks and the sultan virtually evaporates. The version of scientism that we observe in the nineteenth-century Ottoman Empire was a particularly moralistic one. Science was a type of knowledge that demonstrated to its possessors the proper place of everything and everyone, and it made them both diligent and deferential, that is, moral, as a result. There was a broad consensus that this type of citizen needed to be constructed, and that scientific knowledge would enable this. The core disagreement was on the proper place of the sultan and of the young graduates of the prestigious schools.

This study started with the naive question "What were Muslim Ottomans talking about when they talked about science?" The discussions in the preceding chapters have suggested a number of answers, and, indeed, the Ottomans were talking about numerous things when they talked about science. Among them were the meaning of needed and beneficial knowledge, the content and significance of tradition, the proper characteristics of the rulers and the ruled as well as the nature of the hierarchy between them, and the essence of religion. But ultimately, and very simply, Muslim Ottomans talked about *people* when they talked about science in the nineteenth century. The entire debate was about what kind of people the Ottomans were (and were not), and what kind of people they should (and should not) become. When Muslim Ottomans talked about science, they asked questions like "What does familiarity with the new sciences transform a person into?" "What does it mean to be an ignorant person?" and "What are the virtues associated with the possession of scientific knowledge?" They discussed the appropriate form of relationship between knowledgeable and ignorant people, the qualities that members of the ruling elite should acquire, and the type of person the commoner should become. They talked about virtue and vice, laziness and industriousness, dependence and self-sufficiency, modesty and arrogance, sincerity and hypocrisy, loyalty and treachery, and contempt and deference. The meaning and boundaries of science (and for that matter, religion) were important questions to ask, but the final answers had to do with people and their qualities.

This is an important insight. Granted, the case of the nineteenth-century Ottoman Empire is about a very specific historical, social, and cultural context—and it was references to this context on which the analysis presented in this book relied. Particularly after the 1830s, it was barely possible for many Ottomans to perceive the appropriation of the new sciences from Europe as if this process was independent of their political and social context. While the Ottoman Empire was never formally colonized, the ability of the Great Powers to influence Ottoman domestic policy, their domination over the Ottoman economy, and their protection of the non-Muslim communities of the Ottoman Empire concerned many Muslim Ottomans. In such a context, the resultant emphasis on the loyalty of the man of science is not particularly surprising.

But the broader finding—that the most central questions to shape the Ottoman debate on science had to do with ethicopolitical rather than epistemological matters—is consistent with the arguments of several recent studies, particularly on the public understanding of science and religion.[14] Public concerns or perceptions that are portrayed as "opposition to science" are found to be much less about people's beliefs about the way scientific research is conducted or what science is. Rather, they have to do with what people believe to be the moral implications of specific attitudes as represented by scientists.

Not that these are essentially unavoidable or undeniably accurate concerns or perceptions. Indeed, the portrayal of specific scientific attitudes and the linkage of these attitudes to moral concerns are acts of boundary work themselves. Social actors, such as the representatives of well-established educational, cultural, or religious institutions, or members of particular interest groups can portray specific aspects of the scientific life as potentially detrimental to morality while defining the sphere they themselves represent as the source of true morals. What matters is not the relevance, significance, or accuracy of the concerns in question. The simple fact is the resultant centrality of matters of virtue and vice, and the meaning of "the right way to live" to debates on science. How people perceive the relationship between science and morality (regardless of how they define these two concepts), and how they view the role of science in the "good life" do matter, and "human-free" analyses of science fail to capture these fundamental aspects of science as a social institution.[15] When we look at what it means for the public to have a debate on science, we see that public debates on science are ultimately about what makes people, institutions, activities, decisions, and lives good and right. Students as well as practitioners of science cannot wish this away—even if, as in the case this book analyzes, the good or the right are defined in restrictive and potentially repressive fashions. Instead, these are questions scientists and students of science themselves need to tackle head on.

These observations also indicate that the perception and representation of the figure of the scientist are also of paramount importance. Studies on the representations of the scientist in Europe and North America suggest that the variety of publicly recognized images (ranging from the reclusive genius to the selfless savior) that are available to the scientists enable them to renegotiate their identity in different settings and potentially to maintain or further their interests.[16] Note, however, that each such image is about the kind of person whom the individual scientist is, rather than about one way of doing science or another. In fact, delimiting the proper characteristics of scientists is a central, albeit not necessarily articulated, concern in debates on science, and stands as a proxy for deeper concerns. The conflict, if there is one, is not between different epistemologies but between different ways of defining and practicing the right and the good—not only as individual scientists but as members of moral communities.

––––––––

Returning to the specific context of the nineteenth-century Ottoman Empire, we do not detect the emergence of a range of possible acceptable images for the man of science. The elitist, Europhile, arrogant, snobbish materialist and the absolutely altruistic, demonstrably religious and patriotic, modest native emerged as the only two options at the end of the nineteenth century, and the former was not even considered a true man of science. To some degree, the images are still alive in contemporary Turkey, as the statements uttered to students by the Turkish prime minister about the "immoralities of the West" indicate.[17] There is a remarkably broad consensus among different sectors of society on the benefits, goals, and methods of scientific research. The real struggle takes place on the very body of the scientist; it is how a representative of science thinks, looks, and lives that matters.

This overall agreement has a particularly important component: the link between scientific knowledge and patriotism. Familiarity with the new sciences was defined from the 1830s onward as a way to become a good Ottoman subject/citizen—that is, a person who understood and appreciated the holders of state power as well as his own place in the world and who worked hard to serve the state. While different groups had different opinions on who should in fact hold state power, the view that education, particularly the dissemination of scientific knowledge, was necessary if not sufficient for the emergence of a population of learned patriots was commonly shared. Learning and morality (specifically obedience) remained connected concepts for groups that otherwise disagreed, especially in the late nineteenth century.

This is because the appeal of the authoritativeness of scientific knowledge had gained general acknowledgment by this period. While the Hamidian regime constructed on the legacy of the earlier discourses on science an authority triangle in which science, Islam, and the sultan complemented and justified the authority of one another, many leading members of the Young Turks and later the founders of the Turkish Republic referred to science as the ultimate authority, but continued to link this authority to their imaginations of the ideal political and social order. Instead of the sultan, the triangle they proposed included the state. Similarly, the well-established assertions about the harmony between Islam and science were popular not only during the Hamidian era and in the twentieth-century Ottoman Empire but remain very much alive in the Turkish Republic as well. Thus, while the specific content and significance of the three corners varied from one group to another, the triangle itself remained central to political ideology in Turkey. After all, authoritarian science is an idea few social and political movements find unappealing, as long as this authority can be represented as legitimizing a specific power structure.

Perhaps one of the most famous sayings of Kemal Atatürk, the founder of the Turkish Republic and a product and member of the Young Turk Era himself,[18] is from a speech he gave to a group of teachers: "For everything in the world, for civilization, for life, for success, the truest guide is knowledge and science (*ilm* and *fen*). Searching for other guides is unawareness, ignorance, misguidedness." This saying, found in many textbooks and school buildings in Turkey, is commonly seen as a sign of Atatürk's faith in science and of the scientism inherent to the founding ideology of the

Turkish Republic. Yet when studied with the approach I adopted in this study, the statement acquires additional significance, as in the rest of the same speech Atatürk notes that while education is fundamental for the upbringing of the new members of the new Turkish Republic, it is also essential that this education be neither cosmopolitan, nor religious, but "national."[19] The speech is indeed unequivocal in its denigration of religious education, and, in this respect, it is possible to argue that the third corner of the authority triangle referred to above was relatively insignificant in this new discourse. Yet it is also clear that the call is for raising new generations that will avoid cosmopolitanism—indeed, the so-called "second man" of the Republic, Ismet Inönü, also noted that a flaw of religious education was that it was simply another type of cosmopolitan education anyway, and the "national" was the needed element in the education of the new Turks.[20] Hence, science—however defined—*was* essential, but also essential was to produce educated representatives of the "national will."[21] Elsewhere, Atatürk described the characteristics of the needed individual rather clearly:

A nation hit by a catastrophe is a nation that is diseased. Hence, salvation entails examining and curing the disease that is in society. Health can be achieved if the treatment is based on knowledge and science. . . . Patriotism, purity of heart, and altruism are essential for those who want to save the nation. But in order to see and cure the social disease, and to progress this nation according to the requirements of the age, knowledge and science are also required. The home base of scientific activity is the school. . . . It is the school that teaches young minds respect for humanity, love for the nation and the country, the dignity of independence. . . . Those who want to save the country and the nation should [thus] also be honorable experts and people of knowledge (*ilm*).[22]

Hence, scientific knowledge is required for "curing" the society, but this knowledge needs to be combined with, and used in the name of, patriotism. A true person of science is thus a savior, a servant, and a representative of the essence of his nation—a characterization that both glorifies the possessor of scientific knowledge and rigidly restricts the range of acceptable ways of being one. Not surprisingly, while many religious critics censure Atatürk's characterization of science as the "truest guide," the same sayings have also been used to demonstrate that the teaching of Marxist thought in Turkish universities would be a betrayal of Atatürk's principles![23]

This once again shows how an emphasis on the superiority of scientific knowledge to any alternative type of knowledge—this is commonly the way scientism is defined in many studies—does not in itself mean much. Science is not an idea but a practice with real practitioners, and praise for science is not always praise, let alone carte blanche, for the representative of science. Indeed, as many scholars also note, scientism itself can appear in a variety of forms.[24] This is why it is not surprising to find, say, religious movements that also enthusiastically advocate scientific research. When findings can be demonstrated not to contradict religious doctrine—and as countless examples demonstrate, this is not particularly difficult in most cases—the only result is that the authority of science and the authority of religion reinforce one another. A typical example of this approach from the Islamic world is the remarkably popular Nur movement that originated in Turkey with the writings of the religious thinker and leader Said-i Nursi (1878–1960) whose followers have approached scientific study almost as the religious duty of a good Muslim.[25]

A practitioner of science of this type does not use different methods or conduct uncommon or revolutionary types of scientific research; it is not the science but the scientist who is different. In the Ottoman case, the persona of such of a scientist was built on the claim to be able to distinguish the beneficial sciences of the West from its fashionable perversions—that is, to remain an authentic Muslim (or Turk) while learning and practicing the sciences. This persona did not vanish in the Turkish Republic, where tests of authenticity remained central in the lives of scientists. In contexts such as these, the representatives of science, provided they pass these tests, can (and do) imagine themselves as equipped, at least to some extent, also with the authority of the state and/or religion as well—arguably a rather tempting status. The outcome is but a constellation that leads scientists to race to define themselves as *the* learned patriots.

Acknowledgments

This book is the product of a long and convoluted journey that has spanned several cities and involved many individuals and institutions. My interest in the formation and discursive representations of scientists stems all the way back from my years as a student at the Science High School in Ankara, Turkey. I thank my friends and teachers from those years for the origination of some of the questions I address in this book.

I owe a lot to my professors in the Sociology Department of Middle East Technical University, as it was they who instilled the sociological imagination in my engineer-to-be mind. I am grateful particularly to Ayşe Saktanber, who has remained a mentor for me over the years, and who also encouraged me to venture into historical research on this topic. I also thank my professors in Sabancı University, Istanbul, where I took courses in the doctoral program in history. I will remain forever indebted to Professor Şerif Mardin, who was an intellectual and personal mentor for me for many years. His endlessly inspiring work and wisdom, as well as his guidance and enthusiastic support, helped me immensely to find my intellectual pathway. It was also Professor Mardin who introduced me to the field of social studies of science, for which I am also very grateful.

At the University of California, San Diego, I benefited greatly from the faculty and my friends in the sociology and science studies programs. Steven Epstein was an outstanding mentor who helped me stay focused and avoid straying too far afield. His timely and thorough feedback was essential for the completion of the project on which this book is based. John Evans's suggestions have been nothing short of tremendous, from the beginning of my research to the preparation of the final manuscript. Without his always timely support and encouragement, this book would have hardly been possible. Andy Scull consistently gave me both wise advice and the reassurance to pursue historical inquiry as a sociologist. His mentorship especially in my early years in graduate school enabled me to focus on my research without delay. Hasan Kayalı not only offered much appreciated intellectual and personal support during my time at UCSD, but his feedback was essential for me to ground my work in the relevant literature. Robert Westman's interest in my research was a great source of motivation, and our discussions on religion and science in particular helped me formulate some of the arguments I make in this book. I also thank my friends at UCSD, and in particular Michael Evans, Brian Lindseth, and Anthony Rodriguez-Alcala for all the support, encouragement, and fun. In both the U.S. and Turkey stages of my research, Harun Küçük's friendship and enthusiasm as well as feedback on my work have been priceless. My colleagues at Ohio Wesleyan University and, in particular, Ted Cohen, Mary Howard, Jim Peoples, John Durst, and Paul Dean offered me an intellectually stimulating and supportive community, exactly what I needed during the process of writing and revising this book.

Some of the research on which this book is based has been possible thanks to the grants awarded by UCSD's graduate program in sociology, the Science Studies Program, and the Institute for International, Comparative, and Area Studies (IICAS).

I presented some of the findings of my research at conferences, including the meetings of the Society for the Social Studies of Science (4S), Middle East Studies Association, and History of Science Society, as well as the International Congress of History of Science, Technology, and Medicine, the Cambridge Symposium of Middle Eastern Studies, and the Science in Public Conference. I thank the commentators and participants of these meetings for their valuable feedback.

The research for this book was conducted in the National Library and the library of the Turkish Historical Society in Ankara, and the Süleymaniye Library, Beyazıt State Library, Center for Islamic Studies (ISAM) Library, and the Taksim Atatürk Library in Istanbul. I am grateful to the librarians in these institutions, as well as the librarians of the University of California, San Diego, the University of California, Davis, and Ohio Wesleyan University, thanks to whose efforts I was able to promptly access many different types of material.

I received much encouragement for the book project from Ronald

Numbers, for which I am very grateful. Benjamin Fortna offered very valuable feedback on sections of the book. I am particularly obligated to Carter Findley for his thorough review of the book and insightful comments. I also thank two anonymous reviewers for their suggestions that greatly improved the text. *Of course, the opinions expressed and errors made are all mine.* I am also indebted to Marwa Elshakry for allowing me to read her book manuscript before publication, and Kostas Tampakis and Eireni Mergoupi-Savaidou for bibliographical recommendations. Karen Darling and the staff at the University of Chicago Press have been very supportive and helpful during the publication of this book.

Last and most importantly, I thank my wife, Rebekah, and both our families. I am grateful to Rebekah for being a wonderful companion during this, at times challenging, journey. Without her spirit, sense of humor, patience, and love, none of this would have been possible. I can hardly express my gratitude to my parents who have supported me unconditionally throughout the years, even when they did not necessarily agree with my decisions. They have been great role models for me, and everything I have done is thanks to their love and understanding. I dedicate this book to Rebekah and my parents.

Notes

1. In this and many Ottoman texts the word "art" refers to crafts and industries. All translations in this book are mine unless otherwise noted.
2. On Ersoy's life and views, see Ismail Kara, *Türkiye'de İslâmcılık Düşüncesi: Metinler, Kişiler* (Istanbul: Risale, 1986), and Hasan Kayalı, "Islam in the Thought and Politics of Two Late Ottoman Intellectuals: Mehmed Akif and Said Halim," *Archivum Ottomanicum* 19 (2001): 307–33.
3. Mehmet Akif was not alone in perceiving and portraying Japan as exemplifying how a society could remain devoted to its traditions while learning the new sciences from Europeans. On Ottoman Muslim intellectuals' infatuation with the "Japanese model" and the way this ideological construct operated in the late nineteenth and early twentieth centuries, see Renée Worringer, "'Sick Man of Europe' or 'Japan of the Near East'?: Constructing Ottoman Modernity in the Hamidian and Young Turk Eras," *International Journal of Middle East Studies* 36 (2004): 207–30.
4. The word I translate as "depravity" is *maskaralık*—a word that means both "buffoonery" and "disgracefulness." In this respect, it defines the activities in question not only as outrageous but also as childish and immature. This formulation, in turn, presents the "non-Westerner" as the voice of wisdom and maturity—a theme that we will return to in the analysis of Ottoman texts in the following chapters.
5. Mehmet Akif Ersoy, *Safahat* (Ankara: TC Kültür Bakanlığı, 1989), 156. He makes the same argument in the lines following this introduction's epigraph: "To rip through all ages of

progress / Keep as guide your own moral character / For futile is the hope for salvation without it" (172).

6. "Erdoğan: Batının Ahlaksızlıklarını Aldık," *Milliyet,* January 24, 2008, http://www.milliyet.com.tr/2008/01/24/son/sonsiy18.asp (accessed September 22, 2013).

7. I should also state at the outset that I regularly talk about "Muslim men," "men of science," and commonly use the pronoun "he" when making generalizations about individuals in this book. It is not at all comfortable to use this sexist language, but I believe it is needed to characterize the social actors that this study focuses on accurately. While the number of Muslim women who participated in public debate increased toward the end of the nineteenth century in the Ottoman Empire, the groups that shaped the debate on science that I analyze were composed of men. Similarly, while the Turkish language does not differentiate between male or female pronouns, and thus it is not always directly clear if the author of a piece is talking about a man or a woman, it would be safe to assume that the real or imaginary "people of science" who were referred to were men, if one considers the social realities of the late Ottoman Empire.

8. That their abandonment of and subsequent inability to import science led to the "decline" of Muslim societies is a common theme of studies that are based on unilinear conceptions of modernization and monolithic portrayals of the West and the non-West, with the former serving as the ideal for the latter. Such studies present science as a term that requires no further explanation and its impact on "progress" as obvious. See, e.g., chapter 3 of Bernard Lewis's highly popular *What Went Wrong? The Clash between Islam and Modernity in the Middle East* (New York: Oxford University Press, 2002). For a more sophisticated discussion with a similar conclusion, see Toby Huff's *The Rise of Early Modern Science: Islam, China, and the West* (Cambridge: Cambridge University Press, 1993). A classic work on science in the Ottoman Empire with a similar narrative is Adnan Adıvar, *Osmanlı Türklerinde İlim,* 4th ed. (Istanbul: Remzi, 1982) (rev. Turkish translation of *La Science chez les Turcs Ottomans* (Paris: G. P. Maisonneuve, 1939).

9. The argument that "scientifically/technologically backward" societies, in a way, "deserve" to be dominated is one that not only was used frequently by colonizers but is still made, albeit in more subtle forms. For a study on this ideology, see Michael Adas, *Machines as the Measure of Men: Science, Technology, and Ideologies of Western Dominance* (Ithaca, NY: Cornell University Press, 1990). There is a vast literature on the place of science in the colonialist project. See, among others, Gyan Prakash, *Another Reason* (Princeton, NJ: Princeton University Press, 1999); Zaheer Baber, *The Science of Empire: Scientific Knowledge, Civilization, and Colonial Rule in India* (Albany: State University of New York Press, 1996); Patrick Carroll, *Science, Culture, and Modern State Formation* (Berkeley: University of California Press, 2006); Roy MacLeod, ed., "Nature and Empire: Science and the Colonial Enterprise," *Osiris,* 2nd

ser., 15 (2001); Paolo Palladino and Michael Worboys, "Science and Imperialism," *Isis* 84 (1993): 91–102; and Nathan Reingold and Marc Rothenberg, *Scientific Colonialism: A Cross-Cultural Comparison* (New York: Smithsonian, 1987).

10. In a sense, my approach to this assumption is similar to the arguments made by Steven Shapin ("Cordelia's Love: Credibility and the Social Studies of Science," *Perspectives on Science* 3 [1995]: 255–75) and by Barry Barnes and David Bloor ("Relativism, Rationalism, and the Sociology of Knowledge," in *Rationality and Relativism*, ed. Martin Hollis and Steven Lukes [Oxford: Blackwell, 1982], 21–47) that the fact that something is true does not explain why the belief is held. How some claims and claims-makers attain credibility while others do not is a sociological question par excellence.

11. Andrew Cunningham made this point in his "Getting the Game Right: Some Plain Words on the Identity and Invention of Science," *Studies in the History and Philosophy of Science* 19 (1988): 365–89. While to some extent critical of Cunningham's thesis, Peter Dear largely expressed overall agreement in his "Religion, Science, and Natural Philosophy: Thoughts on Cunningham's Thesis," *Studies in the History and Philosophy of Science* 32 (2001): 377–86. Also see Peter Harrison, "'Science' and 'Religion': Constructing the Boundaries," *Journal of Religion* 86 (2006): 81–106.

12. For a comparison of the English concept "science" with German *wissenschaft*, see Babette Babich, "Nietzsche's Critique of Scientific Reason and Scientific Culture: On 'Science as a Problem' and 'Nature as Chaos,'" in *Nietzsche and Science*, ed. Gregory M. Moore and Thomas Brobjer (Aldershot: Ashgate, 2004), 133–53. Comparative studies on institutionalized science in Europe include Robert Fox and George Weisz, eds., *The Organization of Science and Technology in France, 1808–1914* (Cambridge: Cambridge University Press, 1980), Andrew Jamison, "National Styles of Science and Technology: A Comparative Model," *Sociological Inquiry* 57 (1987): 144–58. Such comparisons were commonly made by European "men of science" themselves. See, e.g., Harry W. Paul, *The Sorcerer's Apprentice: The French Scientist's Image of German Science, 1840–1919* (Gainesville: University of Florida Press, 1972). On the making of the category "science," see Richard Yeo, *Defining Science: William Whewell, Natural Knowledge, and Public Debate in Early Victorian Britain* (Cambridge: Cambridge University Press, 1993). "Scientist" was also a nebulous category in the nineteenth century, as has been demonstrated by works such as Sydney Ross, "'Scientist': The Story of a Word," *Annals of Science* 18 (1962): 65–85, and Paul Lucier, "The Professional and the Scientist in Nineteenth-Century America," *Isis* 100 (2009): 699–732.

13. The most famous and influential contribution in this vein in science studies is the work of Thomas Gieryn, particularly his *Cultural Boundaries of Science: Credibility on the Line* (Chicago: University of Chicago Press, 1999), and "Boundary-Work and the Demarcation of Science from Non-science: Strains and Interests in Professional Ideologies of Scientists," *American Sociological*

Review 48 (1983): 781–95. But note that the cartographic metaphor has been used in a variety of ways by many sociologists and anthropologists such as Gregory Bateson and Pierre Bourdieu. What is also important to note is that the "map" metaphor should not imply that each culture provides one and only one such map and that cultures are fixed and coherent. As I further discuss below, societies are characterized by struggles among social groups regarding the dissection and representation of reality.

14. Gieryn, "Boundary-Work," 782.

15. For a few of the countless examples, see N. Mizrachi, J. T. Shuval, and S. Gross, "Boundary at Work: Alternative Medicine in Biomedical Settings," *Sociology of Health and Illness* 27 (2005): 20–43; Emanuel Gaziano, "Ecological Metaphors as Scientific Boundary Work: Innovation and Authority in Interwar Sociology and Biology," *American Journal of Sociology* 101 (1996): 874–907; Michael S. Evans, "Defining the Public, Defining Sociology: Hybrid Science—Public Relations and Boundary-Work in Early American Sociology," *Public Understanding of Science* 18 (2009): 5–22.

16. For the classic expression of this approach, see Pierre Bourdieu, *Distinction: A Social Critique of the Judgment of Taste* (Cambridge, MA: Harvard University Press, 1979).

17. See, e.g., Michèle Lamont, *Money, Morals and Manners: The Culture of the French and American Upper-Middle Class* (London: University of Chicago Press, 1992); Dale Southerton, "Boundaries of 'Us' and 'Them': Class, Mobility and Identification in a New Town," *Sociology* 36 (2002): 171–93; Monica Prasad et al., "The Undeserving Rich: 'Moral Values' and the White Working Class," *Sociological Forum* 24 (2009): 225–53. For a review of such approaches, see Michèle Lamont and Virag Molnar, "The Study of Boundaries in the Social Sciences," *Annual Review of Sociology* 28 (2002): 167–95.

18. Attributing interests to social actors is a contentious issue both in sociology and in science studies. While I do not rule out cases in which actors do indeed act on interests they consciously define, I do not see "acting on interests" as an explanation that presumes rational choice based on an awareness of one's interests. In a way similar to Bourdieu, I refer to interests as propensities to act in particular ways rather than others. The experiences one goes through due to a particular location within the social structure shape ways of perceiving, desiring, and acting, resulting in habituated responses to social stimuli. Consequently, one's improvised everyday actions are built on one's particular dispositions that are shared with people with similar experiences. It is individuals acting on this tacit principle that I refer to when I make an argument on interests rather than simply, or necessarily, actions based on consciously made calculations. Participation in a specific field such as the state is itself interested behavior, in another sense, as it involves investment in and struggle for its stakes.

19. For a study on American ways of talking about science, see Daniel P. Thurs, *Science Talk: Changing Notions of Science in American Popular Culture* (New

Brunswick, NJ: Rutgers University Press, 2007). Thurs's approach to discourse (see 7–11) is similar to the one used in this study.

20. William Sewell, "The Concept(s) of Culture," in *Beyond the Cultural Turn: New Directions in the Study of Society and Culture*, ed. Victoria E. Bonnell and Lynn Hunt (Berkeley: University of California Press, 1999), 56. Parallels can be drawn between this approach and Mikhail Bakhtin's concept of "authoritative discourse." What Michel Foucault refers to as the "author-function" also serves a similar purpose: the name of the author is used to impose unity and coherence on a set of texts, making possible "a limitation of the cancerous and dangerous proliferation of significations." Michel Foucault, "What Is An Author?" in *Aesthetics, Method, and Epistemology*, ed. James D. Faubion (New York: New Press, 1998), 221.

21. Pierre Bourdieu, "Rethinking the State: Genesis and Structure of the Bureaucratic Field," in *State/Culture: State-Formation after the Cultural Turn*, ed. George Steinmetz (Ithaca, NY: Cornell University Press, 1999), 56.

22. For discussions around these themes, see Pierre Bourdieu, *The State Nobility: Elite Schools in the Field of Power* (Stanford, CA: Stanford University Press, 1996); Pierre Bourdieu, "Social Space and Symbolic Power," *Sociological Theory* 7 (1989): 14–25; and Pierre Bourdieu and Jean-Claude Passeron, *Reproduction in Education, Society, and Culture* (London: Sage, 1977).

23. This was the case for the entire Middle East, as Timur Kuran demonstrates in his "The Economic Ascent of the Middle East's Religious Minorities: The Role of Islamic Legal Pluralism," *Journal of Legal Studies* 33 (2004): 475–515.

24. See Fatma Müge Göçek, *Rise of the Bourgeoisie, Demise of Empire: Ottoman Westernization and Social Change* (New York and Oxford: Oxford University Press, 1996). On the economic transformation of the empire and its social consequences, also see Çağlar Keyder, *State and Class in Turkey* (London: Verso, 1987); Çağlar Keyder, "Europe and the Ottoman Empire in the Mid-Nineteenth Century: Development of a Bourgeoisie in the European Mirror," in *East Meets West: Banking and Commerce in the Ottoman Empire*, ed. Philip Cottrell (Aldershot: Ashgate, 2008): 41–58; and Reşat Kasaba, *The Ottoman Empire and the World-Economy: The Nineteenth Century* (Albany: SUNY Press, 1988).

25. For a presentation of the latter argument and a broad evaluation of the various approaches, see Kuran, "Economic Ascent."

26. This disproportion was perpetuated by the foreign corporations within the empire. While the labor historian Donald Quataert argues that there were many instances of collaboration and solidarity among Muslim and non-Muslim members of the Ottoman working class, he also notes in this context that foreign merchants and corporations "routinely hired foreigners for the top jobs, foreigners and Ottoman Christians for the middle positions, and Muslims for the lower ranks." See Donald Quataert, "Ottoman Workers and the State," in *Workers and Working Classes in the Middle East: Struggles, Histories, Historiographies*, ed. Zachary Lockman (Albany: SUNY Press, 1994), 25.

27. Metin Heper, "Center and Periphery in the Ottoman Empire: With Special Reference to the Nineteenth Century," *International Political Science Review/Revue internationale de science politique* 1 (1980): 81–105; Christoph K. Neumann, "Whom Did Ahmed Cevdet Represent?" in *Late Ottoman Society: Intellectual Legacy*, ed. Elisabeth Özdalga (New York: Routledge, 2005), 117–34. The final two decades of the nineteenth century present a more complex picture, however, due to Abdülhamid II's attempts to reassert the authority of the sultan.

28. See Philip Abrams, "Notes on the Difficulty of Studying the State," *Journal of Historical Sociology* 1 (1977): 58–89, and the essays in George Steinmetz, ed., *State/Culture* (Ithaca, NY: Cornell University Press, 1999), which offer alternative, nuanced ways of employing the concept "the state." For a study that challenges the idea of a rigid boundary between society and state, see Steven Epstein, *Inclusion: The Politics of Difference in Medical Research* (Chicago: University of Chicago Press, 2007). For a critical approach to the usage of "the state" in Ottoman studies, see Gabriel Piterberg, *An Ottoman Tragedy: History and Historiography at Play* (Berkeley: University of California Press, 2003).

29. Inspired by Bourdieu's "field theory," we can refer to this outcome as the resultant vector when a number of forces are combined within a field. But while this dynamism should always be acknowledged, it is also the case that the outcomes of struggles may be rendered more durable through institutionalization.

30. For discussions about the types of capital, see Pierre Bourdieu, "The Forms of Capital," in *Handbook of Theory and Research for the Sociology of Education*, ed. John Richardson (New York: Greenwood, 1986), 241–58; Alejandro Portes, "Social Capital: Its Origins and Applications in Modern Sociology," *Annual Review of Sociology* 24 (1998): 1–24; Michele Lamont and Annette Lareau, "Cultural Capital: Allusions, Gaps, and Glissandos in Recent Theoretical Developments," *Sociological Theory* 6 (1988): 153–68.

31. See esp. chap. 7 of Pierre Bourdieu, *The Logic of Practice* (Stanford, CA: Stanford University Press, 1990).

32. Bourdieu, "Rethinking the State," 58.

33. Once again, I should emphasize that this approach does not imply that a particular and homogenous culture is imposed by force and reproduced flawlessly and continuously. While the process in question may indeed lead to the emergence and maintenance of a dominant set of cultural categories and definitions, alternatives do not disappear. The durability of these dominant definitions is also never to be taken for granted, especially if we remember that the state itself is a field of constant struggle, and the very "holding" of statist capital needs to be recognized and deemed legitimate by other social actors to be effective. Nevertheless, the approach outlined here is important as it allows us to ask why particular materials are included in the—using the terminology of Ann Swidler—cultural "toolkits" of certain individuals living

in a certain territory at a certain time, and why other items are not. Toolkits, repertoires, narratives are not coherent and uniform; but they do include and exclude, as well as differentiate and hierarchize, and all these are the outcomes of particular histories.

34. Christoph Neumann, "Political and Diplomatic Developments," in *The Cambridge History of Turkey: The Later Ottoman Empire, 1603–1839*, ed. Suraiya Faroqhi (Cambridge: Cambridge University Press, 2006), 54.

35. The authoritative works on the transformation in question are Carter Findley, *Bureaucratic Reform in the Ottoman Empire: The Sublime Porte, 1789–1922* (Princeton, NJ: Princeton University Press, 1980), and Carter Findley, *Ottoman Civil Officialdom: A Social History* (Princeton, NJ: Princeton University Press, 1989). For a case study regarding the eighteenth-century transformation, see Virginia Aksan, *An Ottoman Statesman in War and Peace: Ahmed Resmi Efendi, 1700–1783* (Leiden: E. J. Brill, 1995).

36. While most approaches in science and technology studies emphasize the variety of interactions among a variety of groups in their analyses of how scientific knowledge is generated and stabilized, the activities of the scientist are commonly prioritized, as in the work of Bruno Latour. For criticisms of this approach, see Emily Martin, *Flexible Bodies: Tracking Immunity in American Culture: From the Days of Polio to the Age of AIDS* (Boston: Beacon Press, 1994), 6–7; Stefan Timmermans and Marc Berg, "Standardization in Action: Achieving Local Universality through Medical Protocols," *Social Studies of Science* 27 (1997): 273–305; and Tereza Stöckelová, "Immutable Mobiles Derailed: STS, Geopolitics, and Research Assessment," *Science, Technology, and Human Values* 37 (2012): 286–311.

37. As Gieryn points out, in episodes of boundary work, "disputes over nature are settled in and through disputes about culture." See Gieryn, *Cultural Boundaries of Science*, 29.

38. Many examples attest to the strength of this approach. See David Thurs, *Science Talk*; Epstein, *Inclusion*; Steven Epstein, *Impure Science: AIDS, Activism, and the Politics of Knowledge* (Berkeley: University of California Press, 1996); Aileen Fyfe and Bernard Lightman, eds., *Science in the Marketplace: Nineteenth-Century Sites and Experiences* (Chicago: Chicago University Press, 2007); Kelly Moore, *Disrupting Science: Social Movements, American Scientists, and the Politics of the Military, 1945–1975* (Princeton, NJ: Princeton University Press, 2011); David Hess, *Science in the New Age: The Paranormal, Its Defenders and Debunkers, and American Culture* (Madison: University of Wisconsin Press, 1993).

39. It is not a coincidence that some of the most influential actors we will focus on were products of an office within the Ottoman state: the Translation Bureau.

40. Talal Asad, "The Concept of Cultural Translation in British Social Anthropology," in *Writing Culture: The Poetics and Politics of Ethnography*, ed. James Clifford and George Marcus (Berkeley: University of California Press, 1986), 151.

41. D. B. Macdonald, "'*Ilm*," *Encyclopaedia of Islam*, http://referenceworks.brillonline.com; Paul E. Walker, "Knowledge and Learning," *Encyclopaedia of the Qur'ān*, ed. Jane Dammen McAuliffe, http://www.encquran.brill.nl. On the Ottoman case, see Ekmeleddin İhsanoğlu, "Institutionalisation of Science in the Medreses of Pre-Ottoman and Ottoman Turkey," in *Turkish Studies in the Philosophy and History of Science*, ed. Gürol Irzık and G. Güzeldere (Dordrecht: Springer, 2005), 265–84. I use Turkish orthographic conventions in this book.

42. See Franz Rosenthal's classic *Knowledge Triumphant: The Concept of Knowledge in Medieval Islam* (Leiden: Brill, 1970).

43. Niyazi Berkes, *The Development of Secularism in Turkey* (New York: Routledge, 1988 [1964]).

44. See Ismail Kara, "Modernleşme dönemi Türkiyesi'nde 'ulûm,' 'fünûn' ve 'sanat' kavramlarının algılanışı üzerine birkaç not," in *Din ve Modernleşme Arasında: Çağdaş Türk Düşüncesinin Meseleleri* (Istanbul: Dergah, 2003), 126–97.

45. Şükrü Hanioğlu, "Blueprints for a Future Society: Late Ottoman Materialists on Science, Religion, and Art," in *Late Ottoman Society: The Intellectual Legacy*, ed. Elisabeth Özdalga (London: Routledge, 2005), 33.

46. The classic definition is in Susan Leigh Star and J. R. Griesemer, "Institutional Ecology, 'Translations,' and Boundary Objects: Amateurs and Professionals in Berkeley's Museum of Vertebrate Zoology, 1907–39," *Social Studies of Science* 19 (1989): 387–420.

47. This also constitutes the limit to the applicability of the concept "boundary object." *Ilm*, a concept used very commonly to claim social superiority, could not connect competing elite groups for long. On the connection between *ilm* and social hierarchy in Muslim societies, see Brinkley Messick, *The Calligraphic State: Textual Domination and History in a Muslim Society* (Berkeley: University of California Press, 1993), esp. 154–60, and Louise Marlow, *Hierarchy and Egalitarianism in Islamic Thought* (Cambridge: Cambridge University Press, 1997).

48. This basic observation should inform any analysis of science, as "humanfree" accounts of science are not only inaccurate as depictions of a thoroughly human activity but have serious implications. This is because such accounts consider the very consequences of the activities that are categorized as "science" as external to science, and thus a secondary, if not negligible, matter. On this, see Sandra Harding, *Science and Social Inequality: Feminist and Postcolonial Issues* (Urbana: University of Illinois Press, 2006); Zygmunt Bauman, *Modernity and the Holocaust* (Cambridge: Polity, 1989); Zygmunt Bauman, *Postmodern Ethics* (Cambridge: Polity, 1993).

49. Steven Shapin, *A Social History of Truth: Civility and Science in Seventeenth-Century England* (Chicago: University of Chicago Press, 1994), and Steven Shapin, *The Scientific Life: A Moral History of a Late Modern Vocation* (Chicago: University of Chicago Press, 2008). On the scientist's persona, also

see Charles Thorpe, *Oppenheimer: The Tragic Intellect* (Chicago: University of Chicago Press, 2006).

50. See Alan Irwin and Brian Wynne, eds., *Misunderstanding Science? The Public Reconstruction of Science and Technology* (Cambridge: Cambridge University Press, 1996); Moore, *Disrupting Science.*

51. John H. Evans, "Epistemological and Moral Conflict between Religion and Science," *Journal for the Scientific Study of Religion* 50 (2011): 707–27; A. L. Roth, J. Dunsby, and L. A. Bero, "Framing Processes in Public Commentary on US Federal Tobacco Control Regulation," *Social Studies of Science* 33 (2003): 7–44; Kristina Petkova and Pepka Boyadjieva, "The Image of the Scientist and Its Functions," *Public Understanding of Science* 3 (1994): 215–24.

52. The establishment of specific characterizations about how nature works has to do with the construction of networks of trust and hence is built on socially accepted principles about reliability and trustworthiness. In this respect, solutions to the question of natural order are inseparable from solutions to the question of social order, as science studies scholars have convincingly argued. See Steven Shapin, "Of Gods and Kings: Natural Philosophy and Politics in the Leibniz-Clarke Disputes," *Isis* 72 (1981): 187–215; Steven Shapin and Simon Schaffer, *Leviathan and the Air-Pump: Hobbes, Boyle, and the Experimental Life* (Princeton, NJ: Princeton University Press, 1985); and Yaron Ezrahi, *The Descent of Icarus: Science and the Transformation of Contemporary Democracy* (Cambridge, MA: Harvard University Press, 1992). On portrayals of science and related expectations of obedience, see Steven Shapin and Barry Barnes "Science, Nature, and Control: Interpreting Mechanics' Institutes," *Social Studies of Science* 7 (1977): 31–74; Jonathan Topham, "Science and Popular Education in the 1830s: The Role of the Bridgewater Treatises," *British Journal for the History of Science* 25 (1992): 397–430.

53. Using Bourdieu's perspective, we can argue that when activities are misrecognized as disinterested rather than seen as guided by the underlying interests of specific social actors within specific fields, it is possible to refer to those activities as possessing symbolic power. Symbolic capital accrues to social actors who benefit from this misrecognition and can present themselves as disinterested. As we shall see, it was rather common for students of the new elite schools of the Ottoman Empire to portray themselves as altruistic enlighteners; their critics, on the other hand, insisted on representing them as simply self-interested snobs. Hence, the debate can be seen as one on whose activities are to be deemed disinterested. On this also see David Swartz, "Bridging the Study of Culture and Religion: Pierre Bourdieu's Political Economy of Symbolic Power," *Sociology of Religion* 57 (1996): 71–85. As they involve disputes on which types of actions should be seen as "irreducible to interest," examples of the so-called conflict between science and religion can be seen partly as conflicts over symbolic capital. For a relevant example, see Frank M. Turner, "The Victorian Conflict between Science and Religion: A Professional Dimension," *Isis* 69 (1978): 356–76.

CHAPTER ONE

1. Helmuth von Moltke, *Türkiye'deki Durum ve Olaylar Üzerine Mektuplar*, trans. Hayrullah Örs (Ankara: Türk Tarih Kurumu, 1960 [1841]), 324; translation based on Moltke's observations as discussed in "Moltke's Campaign against the Egyptians," *Macmillan's Magazine* 46 (1882): 473–81. The historian Abdurrahman Şeref recounts the story as one about Mehmed Emin Âli Pasha, one of the most prominent statesmen of the reform era. The mistake is significant as a hint about the way members of the new group of elites who emerged in the early nineteenth century were perceived. See Abdurrahman Şeref, *Tarih Musahebeleri* (Istanbul: Matbaa-i Amire, 1340/1921), 84.

2. This, incidentally, was the school Louis Pasteur would attend eight years later.

3. Many of Hüsrev's protégés occupied key positions within the Ottoman state machinery and, thanks to these networks, he remained an influential figure in Ottoman politics for decades, perhaps representing the apogee of patrimonial authority in the Ottoman Empire. Max Weber's typology of authority structures contrasts "legal-rational" authority to patrimonialism and traditional authority by defining the former as characterized by bureaucratic posts based on educational credentials and skills rather than personal connections and patronage, fixed salaries, and a well-organized hierarchy with specific regulations about mobility. While the nineteenth-century Ottoman Empire witnessed a transition toward this type of authority, personal ties maintained their importance to a significant extent. The frustration that this caused among young, educated, low-ranking bureaucrats of the future decades will be a theme that we will pick up several times in the following chapters. This theme is developed effectively in Carter Findley, *Bureaucratic Reform in the Ottoman Empire: The Sublime Porte, 1789–1922* (Princeton, NJ: Princeton University Press, 1980). On the factional politics of the era, see Carter Findley, *Ottoman Civil Officialdom: A Social History* (Princeton, NJ: Princeton University Press, 1989), 70–80.

4. Quoted in Edhem Eldem, "Fransa'ya Eğitime Gönderilen Sadrazam Ibrahim Edhem," *Popüler Tarih* 2 (2006): 52.

5. On these students, see Adnan Şişman, *Tanzimat Döneminde Fransa'ya Gönderilen Osmanlı Öğrencileri (1839–1876)* (Ankara: Türk Tarih Kurumu, 2004). Reşid "the Spectacled" himself was not one of the first four—he was sent two years after them. He would become the marshal of the Imperial Arsenal and then the governor of Baghdad in 1851. See Mehmed Süreyya, *Sicill-i Osmanî* (Istanbul: Tarih Vakfı Yurt Yayınları, 1996), 5:1382.

6. It is worth noting at the outset, however, that in contrast to the way they would be characterized by critics throughout the nineteenth century, the attitudes of the members of this group—including those educated in Europe—did not necessarily involve a complete enthusiasm for anything European. When he was the governor of Baghdad, Reşid Bey told a Persian traveler that

the aim of the Ottomans was to imitate and later surpass the Europeans in military matters, and he recommended that the Persians do the same. Reşid's "desire at England's annihilation" was evident according to the traveler (Adib al-Molk, quoted in Anja Pistor-Hatam, "Iran and the Reform Movement in the Ottoman Empire: Persian Travellers, Exiles, and Newsmen under the Impact of the Tanzimat," in *Proceedings of the Second European Conference of Iranian Studies* (Rome: Is.M.E.O, 1995): 566.

7.   On these and earlier educational institutions in the Ottoman Empire, see the essays in Ekmeleddin İhsanoğlu, *Science, Technology, and Learning in the Ottoman Empire: Western Influence, Local Institutions, and the Transfer of Knowledge* (Aldershot: Ashgate, 2004).

8.   For a meticulous study on this institution, see Kemal Beydilli, *Türk bilim ve matbaacılık tarihinde Mühendishâne, Mühendishâne Matbaası ve Kütüphânesi, 1776–1826* (Istanbul: Eren, 1995).

9.   An important study that analyzes the period with a similar approach to mine is Berrak Burçak, "Science, a Remedy for All Ills: Healing 'the Sick Man of Europe': A Case for Ottoman Scientism" (PhD diss., Princeton University, 2005). For a detailed discussion of Mustafa's *Diatribe*, see Berrak Burçak, "Modernization, Science, and Engineering in the Early Nineteenth-Century Ottoman Empire," *Middle Eastern Studies* 44 (2008): 69–83. Burçak portrays Seyyid Mustafa as a representative of a new social type, the engineer, who sees himself as the savior of the empire. We owe the pioneering analyses of Seyyid's work to Adnan Adıvar and Niyazi Berkes, who, in their classic works, portrayed the *Diatribe* as evidence of the enthusiasm of Ottoman youth for modern science. See Adnan Adıvar, *Osmanlı Türklerinde İlim*, 4th ed. (Istanbul: Remzi, 1982) (rev. Turkish translation of *La Science chez les Turcs Ottomans* (Paris: G. P. Maisonneuve, 1939), 208–9, and Niyazi Berkes, *The Development of Secularism in Turkey* (Montreal: McGill University Press, 1964), 78–80. Kahraman Şakul, on the other hand, analyzes this work in parallel to the works of other members of the close circle around Selim III, and argues convincingly that it was an effort to invite European powers to see the empire in a new light. See Kahraman Şakul, "Nizam-ı Cedid Düşüncesinde Batılılaşma ve İslami Modernleşme," *Divan İlmi Araştırmalar Dergisi* 19 (2005): 117–50. This comparison also indicates that while Seyyid Mustafa's *Diatribe* did emphasize science more than the others, it was not entirely different in terms of its emphasis on "the new" from similar treatises written by other members of the same circle who were not engineers or graduates of the new schools. In this respect, it is more apt to consider Seyyid and his work as not fundamentally a case for science, but a case for the new in general. This is all the more appropriate if we remember also that the set of reforms that were introduced during the reign of Selim III was called the New Order (*Nizam-ı Cedîd*).

10.   The contemporary connotations of the word "diatribe" suggest that the book was essentially written to attack the enemies of Selim III and his enlightened

followers. This is indeed an important aspect of the book, but *la diatribe* had a somewhat different meaning in early nineteenth-century French. The definition provided by Abel Boyer's *Dictionnaire Français-Anglais* (24th ed., 1816), for instance, is "tedious pedantick [*sic*] dissertation"!

11. Seyyid Mustafa, *Diatribe de l'Ingénieur Séid Moustapha sur l'état actuel de l'art militaire, du génie et des sciences, à Constantinople*, ed. L. Langles (Paris: Ferra, 1810 [1803]), 16.
12. I will come back to this issue and its significance later.
13. Mustafa, *Diatribe*, 18.
14. Ibid., 20.
15. Ibid., 21.
16. Ibid., 32–33.
17. Ibid., 36–37.
18. Ibid., 52.
19. On this principle (*mukabele-i bi'l-misl*) and the way it was utilized in this period, see Kahraman Şakul, "Nizam-ı Cedid Düşüncesinde Batılılaşma ve İslami Modernleşme," *Divan İlmi Araştırmalar Dergisi* 19 (2005): 117–50.
20. While many sources indicate that Seyyid Mustafa himself was also killed in this period, there is evidence to the contrary. See Kemal Beydilli, *İlk mühendislerimizden Seyyid Mustafa ve Nizâm-i Cedîd'e dair risâlesi* (Istanbul: Edebiyat Fakültesi, 1983–87).
21. See Kemal Beydilli, *Mahmud Râif Efendi ve Nizâm-i cedîd'e dâir eseri* (Ankara: Türk Tarih Kurumu, 2001), and Şakul, "Nizam-ı Cedid."
22. Johann Strauss, "The Millets and the Ottoman Language: The Contribution of Ottoman Greeks to Ottoman Letters (19th–20th Centuries)," *Die Welt des Islams* 35 (1995): 196.
23. Mahmud Raif, *Journal du Voyage de Mahmoud Raif Efendi en Angleterre, écrit par lui même*, reproduced in Vahdettin Engin, "Mahmud Raif Efendi Tarafından Kaleme Alınmış İngiltere Seyahati Gözlemleri" in *Prof. Dr. Ismail Aka Armağanı* (İzmir: Beta, 1999), 157. For a detailed study on Raif's appointment in England and an English translation of his report, see Alaeddin Yalçınkaya, "Mahmud Raif Efendi as the Chief Secretary of Yusuf Agah Efendi, The First Permanent Ottoman-Turkish Ambassador to London (1793–1797)," *OTAM* 5 (1994): 385–434.
24. This work was translated into Ottoman Turkish by Yakovaki Efendi (Iakovos Argyropoulos), a Greek subject of the sultan who was at the time the Ottoman envoy in Vienna and would later become the dragoman of the fleet.
25. Strauss, "The Millets," 198–99.
26. Mahmud Raif, *Osmanlı İmparatorluğu'nda Yeni Nizamların Cedveli*, trans. Arslan Terzioğlu and Hüsrev Hatemi (Istanbul: Turing, 1988).
27. For more on this work and Raif, see Kemal Beydilli, *Mahmud Râif Efendi* and Şakul, "Nizam-ı Cedid."
28. Karen Barkey, *Empire of Difference: The Ottomans in Comparative Perspective* (Cambridge: Cambridge University Press, 2008), 274.

29. Being aware of the meaning, aims, and more generally, the domestic context of these policies and the accompanying developments is important in order not to reduce the eighteenth- and nineteenth-century transformations of the Ottoman Empire simply to Westernization, or to seeing them as projects that the Ottomans were forced to undertake and followed blindly due to the irresistible influence of the European powers.

30. *Takvim-i Vekayi* (25 Cemaziyelevvel 1247/November 1, 1831), 1.

31. Orhan Koloğlu, *Takvim-i Vekâyi: Türk Basınında 150 Yıl* (Ankara: ABS, 1982), 131.

32. For an analysis of the attitudes of the ulema, see Uriel Heyd, "The Ottoman Ulema and Westernization in the Time of Selim III and Mahmud II," in *The Modern Middle East: A Reader*, ed. Albert Hourani et al. (Berkeley: University of California Press, 1993), 29–59.

33. Ahmed Lütfi Efendi, *Vak'anüvis Ahmed Lütfi Efendi Tarihi*, vol. 9, ed. Münir Aktepe (Istanbul: Edebiyat Fakültesi, 1984), 156.

34. Mahmud Cevad İbnü'ş-şeyh Nâfi, *Maarif-i Umumiye Nezareti Tarihçe-i Teşkilat ve İcraatı—XIX. Asır Osmanlı Maarif Tarihi*, ed.Taceddin Kayaoğlu (Ankara: Yeni Türkiye, 2001), 3.

35. See Selçuk Akşin Somel, *The Modernization of Public Education in the Ottoman Empire* (Leiden: Brill, 2001), 27.

36. Quoted in Osman Nuri Ergin, *İstanbul Mektepleri ve İlim, Terbiye ve San'at Müesseseleri Dolayisiyle Türkiye Maarif Tarihi*, 2nd ed. (Istanbul: Eser, 1977), 1–2:336–37.

37. In the field of medicine the old/new distinction started to emerge in the eighteenth century thanks to both translations and the works and practices of non-Muslim Ottoman physicians educated in Europe. For a brilliant analysis of this emphasis on "the new" in this period, see Harun Küçük, "Early Enlightenment in Istanbul" (PhD diss., University of California, San Diego, 2012). On medicine also see Cevat Izgi, *Osmanlı Medreselerinde İlim* (Istanbul: İz, 1997), 2:40–41. Note that Behçet's statement involves a comparison between the expertise of Muslim and non-Muslim Ottoman physicians as well. The emerging perception that non-Muslim Ottomans were much more familiar with the new knowledges than Muslim subjects and that this put the latter in a particularly disadvantaged position gradually turned into a theme that would be central to the Ottoman debate on science, even though the argument was not always made overtly. We will come back to this topic in the following chapters.

38. James Ellsworth de Kay, *Sketches of Turkey in 1831 and 1832 by an American* (New York: J. and J. Harper, 1833), 419.

39. De Kay provides a summary of the treatise in *Sketches of Turkey*, 518–20. For the Tunisian case, see Nancy E. Gallagher, *Medicine and Power in Tunisia, 1780–1900* (Cambridge: Cambridge University Press, 1983), 58–59. On Behçet's life and works, see Feridun Nafiz Uzluk, *Hekimbaşı Mustafa Behçet: Zâtı, Eserleri Üstüne Bir Araştırma* (Ankara: Ankara Üniversitesi Tıp Tarihi Enstitüsü, 1954).

40. Most interpretations are based on the sarcastic remarks Adnan Adıvar makes about this book in his *Osmanlı Türklerinde İlim* (see 217–19). The examples Adıvar mentions include "mystery number 61"—that having a convict eat the tongue of a quail would cause him to confess his crime—and "mystery number 295"—that the cure for hiccup is to hold a goose's beak close to the mouth.

41. *Hezar Esrar* (Istanbul: Muhib, 1285), 4–7. As the book was published posthumously by Behçet's nephew, the physician and intellectual Hayrullah Efendi, who also wrote a foreword to the book, the introduction, too, may at least in part be attributed to Hayrullah.

42. Don Bates, "Why Not Call Modern Medicine 'Alternative'?" *Perspectives in Biology and Medicine* 43 (2000): 502–18.

43. See Robert Darnton, *Mesmerism and the End of the Enlightenment in France* (Cambridge, MA: Harvard University Press, 1968), and Patricia Fara, "An Attractive Therapy: Animal Magnetism in Eighteenth-Century England," *History of Science* 33 (1995): 127–77. William Gregory, professor of chemistry at Edinburgh University, published his popular *Letters to a Candid Enquirer on Animal Magnetism* in 1851. On this also see Martin Willis, "George Eliot's *The Lifted Veil* and the Cultural Politics of Clairvoyance," in *Victorian Literary Mesmerism*, ed. Martin Willis and Catherine Wynne (Amsterdam: Rodopi, 2006): 145–62.

44. Heyd, "Ottoman Ulema."

45. The work in question is Anton von Störck's *Medicinisch-praktischer Unterricht für die Feld und Landwundärzte der österreichischen Staaten*, first published in 1776. On this translation and Şanizade, see Aykut Kazancıgil, *XIX. Yuzyılda Osmanlı İmparatorluğunda Anatomi* (Istanbul: Özel, 1991).

46. Şanizade Mehmed Ataullah, *Miyarü'l-Etıbbâ* (Istanbul: Tabhane-i Şahane, 1820), 3; the wording is *garâib-i fünûn*.

47. Ibid., 7; the wording is *tabib-i kâmil*.

48. See Thomas X. Bianchi, "Miroir des corps, ecrit en turc par Chani-zadéh," *Revue Encyclopédique* 10 (1821): 297.

49. "State of the Turkish Press at Constantinople," *Oriental Herald and Journal of General Literature* 17 (1828): 309.

50. Thomas X. Bianchi, *Vocabulaire Français-Turc* (Paris: Everat, 1831), v–vi.

51. I thank Harun Küçük for suggesting this interpretation.

52. See Heyd, "Ottoman Ulema."

53. As Heyd argues, however, the fact that they were of lower rank makes it rather difficult to study the opinions of these critics. It was almost impossible for them to get their views published even if they dared to do it. See ibid., 34–36.

54. Reproduced in the best study on this topic, Ekmeleddin İhsanoğlu, *Başhoca İshak Efendi: Türkiye'de Modern Bilimin Öncüsü* (Ankara: Kültür Bakanlığı, 1989), 20–21; the wording is *ulûm ve fünûn-ı lâzımeye âşina*.

55. Ibid., 51.

56. İshak Efendi, *Mecmua-i Ulûm-ı Riyaziye* (Cairo: Bulak Matbaası, 1841–45 [1831–34]) 1:2.

57. Ibid., 2–3.

58. İhsanoğlu, *Başhoca İshak Efendi*, 47.

59. De Kay, *Sketches of Turkey*, 138–42. The American zoologist and the composer of the *Zoological Report of New York State* (1842), James Ellsworth De Kay (1792–1851) came to the Ottoman Empire in 1831, along with his father-in-law, the Scottish-American shipbuilder Henry Eckford, who would become the superintendent of Ottoman shipyards. De Kay left the next year, following Eckford's death.

60. Ibid., 141.

61. The Ottoman ambassador to Russia at the time was Halil Rifat Pasha, yet another protégé of Hüsrev Pasha. On his return to Istanbul he would become chief commander of the navy. He is best known for his warning to the sultan: "If we do not become like the Europeans, we will have to retreat to Asia." Nezahat Nureddin Ege, *Prens Sabahaddin: Hayatı ve İlmî Müdafaaları* (Istanbul: Güneş, 1977), 4.

62. Edward Thomason, *Sir Edward Thomason's Memoirs during Half a Century* (London: Longman, Brown, Green and Longmans, 1845), 298. In a letter dated February 18, 1835, Edward Thomason, inventor, businessman, and vice-council of the Ottoman Empire in Birmingham, describes Namık Pasha as a man "well-informed on arts and sciences" and "on subjects of mechanism." On his return to Istanbul, Namık Pasha presented to the sultan a manuscript on his observations. Recounting his visits to Cambridge University in detail, he argued that the Ottomans should establish similar schools throughout the empire. As a gesture that epitomized his dreams about the empire, he drew pictures of the balloons that he had seen in England, attaching to them Ottoman flags. See Şükrü Hanioğlu, *The Young Turks in Opposition* (New York; Oxford: Oxford University Press, 1995), 9. Also see Ahmet Nuri Sinaplı, *Devlete, Millete Beş Padişah Devrinde Kıymetli Hizmetlerde Bulunan Şeyhülvüzera, Serasker Mehmed Namık Paşa* (Istanbul: Yenilik, 1987), 85–90.

63. Sinaplı, *Mehmed Namık Paşa*, 67.

64. Julia Pardoe, *The City of the Sultan and Domestic Manners of the Turks in 1836* (London: Henry Colburn, 1838), 188.

65. Ibid., 193–94.

66. Ibid., 197.

67. Ibid., 199–200.

68. Alphonse Royer, "Sultan Mahmoud II," *Revue de Paris* 53 (1837): 234 (italics mine).

69. The French orientalist L.-M. Langlès, who wrote the foreword to the French publications of Seyyid Mustafa's *Diatribe*, presents Selim III and Seyyid Mustafa as the representatives of the Enlightenment in the Ottoman Empire. Also see Langlès's letter to A. L. Millin, the publisher of the *Magasin Encyclopedique*, who published the *Diatribe* in his magazine ("Lettre de M. L. Langlès

à M. Millin, sur un ouvrage imprimé en français, en 1803, dans l'imprimerie de scutari," *Magasin Encyclopedique* 5 (1809): 5–11). The French officer Jules Planat assumes a more condescending tone and refers to Seyyid as a person whose "half-knowledge" made him a sage in the Ottoman Empire. Nevertheless, in a patronizing way, Planat also appreciates Mustafa's "enlightenment." See Jules Planat, *Histoire de la régénération de l'Egypte* (Paris: J. Barbezat, 1830), 2, 6.

70. Reşat Kaynar, *Mustafa Reşit Paşa ve Tanzimat* (Ankara: TTK, 1954), 91–92.

71. See Bruno Belhoste, Amy Dahan-Dalmédico, and Antoine Picon, eds., *La formation polytechnicienne: 1794–1994* (Paris: Dunod, 1994); François Georgeon, "La formation des élites à la fin de l'Empire ottoman: le cas de Galatasaray," *Revue du monde musulman et de la Méditerranée* 72 (1994): 15–25.

72. "Commercial Relations with Turkey," *British and Foreign Review: or, European Quarterly Journal* 5 (1837): 468–69. This article also congratulates the Ottoman sultan Mahmud II for sending students to England to learn "the literature, arts, and sciences of Europe" without completely imitating the Europeans (503–4). The author may be David Urquhart, whose views on Anglo-Ottoman trade shaped the 1838 treaty that opened the Ottoman markets to British imports and had a significant impact on the Ottoman economy. I discuss this development in more detail below.

73. See Harun Küçük, "Early Enlightenment," and Shirine Hamadeh, "Ottoman Expressions of Early Modernity and the 'Inevitable' Question of Westernization," *Journal of the Society of Architectural Historians* 63 (2004): 32–51.

CHAPTER TWO

1. As Rojo and van Dijk argue, official discourses justify official actions, establish certain interpretations as true, and delegitimate alternative discourses. See L. M. Rojo and T. van Dijk, "'There Was a Problem, and It Was Solved!': Legitimating the Expulsion of 'Illegal' Migrants in Spanish Parliamentary Discourse," *Discourse and Society* 8 (1997): 523–66. To this we should add Bourdieu's reminder that what comes to be "official" at a given time is a result of struggles among actors with specific interests and resources within the state field. See Pierre Bourdieu, "Social Space and Symbolic Power," *Sociological Theory* 7 (1989): 14–25, and Pierre Bourdieu, "Rethinking the State: Genesis and Structure of the Bureaucratic Field," in *State/Culture: State-Formation after the Cultural Turn*, ed. George Steinmetz (Ithaca, NY: Cornell University Press, 1999).

2. Document reproduced in Osman Nuri Ergin, *İstanbul Mektepleri ve İlim, Terbiye ve San'at Müesseseleri Dolayisiyle Türkiye Maarif Tarihi*, 2nd ed. (Istanbul: Eser, 1977), 1–2:397.

3. Before the Translation Bureau was fully established, it was rather hard to find Muslims who could speak European languages. Until 1821, the translator to the Imperial Council was a Greek subject of the sultan; after that, Muslim

converts Yahya Efendi and Ishak Efendi (of the School for Military Engineering) occupied the post. Tellingly, while the physician and historian Şanizade Ataullah Efendi knew French, he was not allowed to become a translator as he was a member of the ulema. See Carter Findley, *Bureaucratic Reform in the Ottoman Empire: The Sublime Porte, 1789–1922* (Princeton, NJ: Princeton University Press, 1980), 133.

4.　The text of the treaty may be found in Lewis Hertslet, ed., *A Complete Collection of the Treaties and Conventions and Reciprocal Regulations at Present Subsisting between Britain and Foreign Powers* (London: Henry Butterworth, 1840) 5:506–18.

5.　It should be emphasized, however, that not all sectors within manufacturing were equally adversely impacted, and a period of growth in manufactures started in the 1870s. See Halil Inalcık and Donald Quaetert, eds., *An Economic and Social History of the Ottoman Empire, 1300–1914* (Cambridge: Cambridge University Press, 1994): 888–933, and Carter Findley, *Turkey, Islam, Nationalism, and Modernity* (New Haven, CT: Yale University Press, 2010):111–15.

6.　See Çağlar Keyder, State and Class in Turkey (London: Verso, 1987); Çağlar Keyder, "Europe and the Ottoman Empire in the Mid-Nineteenth Century: Development of a Bourgeoisie in the European Mirror," in *East Meets West: Banking and Commerce in the Ottoman Empire*, ed. Philip Cottrell (Aldershot: Ashgate, 2008); and Fatma Müge Göçek, Rise of the Bourgeoisie, *Demise of Empire: Ottoman Westernization and Social Change* (New York: Oxford University Press, 1996). Note that an Ottoman Muslim who spoke a European language was extremely hard to come by, even in the capital. Christian communities, however, had been in close contact with Europe for centuries. Wealthy Greek families in Istanbul, for instance, had started sending their sons to Italian universities as early as the seventeenth century.

7.　The English translation of the decree may be found in Halil Inalcık, "The Hatt-i Sherif of Gülhane, 3 November 1839," in *The Middle East and North Africa in World Politics: A Documentary Record*, ed. J. C. Hurewitz (New Haven, CT: Yale University Press, 1975), 270. The equality of all citizens would be stated even more unequivocally in the Reform Decree of 1856.

8.　Hasan Kayalı, *Arabs and Young Turks: Ottomanism, Arabism, and Islamism in the Ottoman Empire, 1908–1918* (Berkeley and Los Angeles: University of California Press, 1997), 30. Probably the best-known example regarding the common Turkish perception of the new order involves some words spoken by a police captain. When a Turk was brought to the police station by a Christian whom he had called a gâvur (infidel), the policeman addressed the Turk: "O my son, didn't we explain? Now there is the Tanzimat, a gâvur is no longer to be called a gâvur!" See Abdurrahman Şeref, *Tarih Musahebeleri* (Istanbul: Matbaa-i Amire, 1340/1921), 73.

9.　Reproduced in Reşat Kaynar, *Mustafa Reşit Paşa ve Tanzimat* (Ankara: TTK, 1954), 91–92.

10.　Findley, *Bureaucratic Reform*, 137.

11. It should be noted here that the decree was not considerably different in tone or even in most of its content than traditional Sunni Islamic texts that portrayed justice and law as the source of strength and prosperity for the state. Reşid himself had some religious education, his mentor Pertev Pasha was an orthodox Sunni, and the orthodox Naqshbandi-Mujaddidi order was particularly popular in Istanbul at the time, influencing the new sultan Abdülmecid II (reigned 1839–61) as well. Hence, combining ideas that would appease the European powers with Islamic political theory, the document identifies the negligence of the sharia as the reason for the problems Ottomans faced. While it indicates the importance of justice and basic rights, it is written entirely from a state-centered perspective. See Butrus Abu-Manneh, "The Islamic Roots of the Gülhane Rescript," *Die Welt des Islams* 34 (1994): 173–203.

12. Quoted in Şerif Mardin, *The Genesis of Young Ottoman Thought: A Study in the Modernization of Turkish Political Ideas* (Princeton, NJ: Princeton University Press, 1962), 110.

13. Frederick Millingen, *La Turquie sous le Règne d'Abdul-Aziz (1862–1867)* (Paris: Librarie Internationale, 1868), 256–58.

14. Edwin de Leon, "The Old Ottoman and the Young Turk," *Harper's Magazine* 44 (1872): 610. In a similar vein, Julia Pardoe refers to Reşid and men like him as individuals "who prove to Europe what the Turks already are, and what they are capable of becoming." See Julia Pardoe, *The City of the Sultan and Domestic Manners of the Turks in 1836* (London: Henry Colburn, 1838), 268.

15. Also see the July 13, 1863, edition for a description of a feast given by Fuad Pasha where the author states: "Not longer than 10 or 15 years ago any Turkish Minister of an eccentric turn who would have had the temerity to throw his house open for the reception of the European community of both sexes, who would have tolerated under his roof the amusements of an European ball-room would have been branded as a giaour (*infidel*). . . . In the present day how different seems to be the public feeling!" Fuad Pasha was a graduate of the Imperial School of Medicine. He worked at the Translation Bureau and numerous Ottoman embassies in Europe before assuming the posts of foreign minister and prime minister several times. In the latter two posts, he alternated with Âli Pasha, another top Tanzimat bureaucrat with similar experience.

16. "Proceedings of the American Oriental Society," *Journal of the American Oriental Society* 11 (1884): cxc.

17. Dated 25 Receb 1291/September 7, 1874; reproduced in Mehmed Zeki Pakalın, *Safvet Paşa* (Istanbul: Ahmet Sait, 1943), 101. See also pp. 102–14 for more examples regarding his appreciation of Europe as well as his bon vivant lifestyle.

18. Quoted in Roderic Davison, "Halil Şerif Paşa: The Influence of Paris and the West on an Ottoman Diplomat," in *Nineteenth-Century Ottoman Diplomacy and Reforms* (Istanbul: Isis, 1999 [1986]), 83.

19. Ibid. Halil Şerif, who was independently wealthy, later settled in Paris on his own. He became a flamboyant figure in French high society as well as a patron of the arts, which made him particularly well-known. He commissioned from Gustave Courbet a set of paintings including the infamous *L'Origine du Monde* (1866), a most candid nude that would end up in Jacques Lacan's collection in 1955. On Halil Şerif, see Deniz Türker, "The Oriental Flâneur: Khalil Bey and the Cosmopolitan Experience" (master's thesis, MIT, 2007).

20. Fatma Aliye, *Ahmet Cevdet Paşa ve Zamanı* (Istanbul: Bedir, 1995 [1913]), 79–82; Göçek, *Rise of the Bourgeoisie*, 81; Kemal Karpat, *Ottoman Population, 1830–1914: Demographic and Social Characteristics* (Madison: University of Wisconsin Press, 1985), 93.

21. The number of Europeans particularly in Istanbul steadily increased in the early nineteenth century. Most common among these were merchants and middlemen; among the foreign residents of Istanbul were also orientalists, artists, travelers, and authors. Many members of the literate elite interacted with them, but as I will show in the next chapter, the Europeanized manners of these individuals—including those who had not been abroad—were commonly condemned in literature and the press.

22. Details of his biography are based on Fatih Andı's introduction in Mustafa Sami, "Avrupa Risalesi," in *Bir Osmanlı Bürokratının Avrupa İzlenimleri: Mustafa Sami Efendi ve Avrupa Risalesi*, ed. Fatih Andı (Istanbul: Kitabevi, 1996).

23. The Ottoman ambassador at the time was Ahmed Fethi Pasha (1802–58). Raised in the palace school and protected by Hüsrev Pasha, he became a military officer and then Ottoman ambassador to Vienna. After his return to Istanbul, Fethi Pasha married the sultan's daughter. He not only occupied many high-ranking posts during his career, but, inspired by his observations and experiences in Europe, he also founded several factories and became the chief organizer of the first Ottoman Museum, where the imperial collections were exhibited. On this creation, see Wendy M. Shaw, *Possessors and Possessed: Museums, Archaeology, and the Visualization of History in the late Ottoman Empire* (Berkeley: University of California Press, 2003).

24. Mustafa Sami, "Avrupa Risalesi," 54.

25. As we will see below, the 1838 memorandum of the Council on Public Affairs is built on a similar theme.

26. Mustafa Sami, "Avrupa Risalesi," 71–72.

27. Ibid., 73–75.

28. Ibid., 76–78.

29. Ibid., 78.

30. The comparison between "our poetry" and "their science" is a recurrent theme in Ottoman debates on science, which will be taken up in the next chapters.

31. Mustafa Sami, "Avrupa Risalesi," 79–80.

32. Published in English as *An Imam in Paris: Al-Tahtawi's Visit to France (1826–31)*, trans. and ed. David Newman (London: Saqi Books, 2002).

33. Benjamin A. Elman, "'Universal Science' Versus 'Chinese Science': The Changing Identity of Natural Studies in China, 1850–1930," *Historiography East and West* 1 (2003): 68–116.
34. Mustafa Sami, "Avrupa Risalesi," 80–81.
35. Ibid., 79.
36. In Bourdieu's words, any form of capital transforms into symbolic capital "when it is perceived by social agents endowed with categories of perception which cause them to know it and to recognize it, give it value." For many sectors of Ottoman society, but particularly the traditional elites and the disgruntled members of the bureaucracy, the rising bureaucrats' knowledge and expertise remained irrelevant and failed to translate into prestige and status. See Bourdieu, "Rethinking the State," 47.
37. Ibnülemin M. K. Inal, *Son Asır Türk Şairleri* (Istanbul: Devlet Matbaası, 1939), 1: 1647.
38. Ibid.
39. On Fethi Pasha see note 22. In February 1841, Lebib would lose his post entirely and be sent to Izmir as the tax collector (*muhassıl*) by Fethi Pasha. For related archival documents, see Prime Ministry's Ottoman Archives in Istanbul (BOA) Irade Mesail-i Mühimme 89/2559, 3 Zilkade 1255/January 8, 1840, and 90/2569, 7 Muharrem 1256/March 11, 1840; Hatt-ı Hümayûn 1423/58216, 29 Zilhicce 1256/February 21, 1841.
40. Inal, *Son Asır Türk Şairleri*, 9:1593.
41. Sami's comments in his treatise clarify his situation. Fethi was unable to speak French when he became ambassador to Paris. Prince de Joinville, *Memoirs (Vieux Souvenirs) of Prince de Joinville*, trans. Mary Loyd (New York: Macmillan, 1895), 230–31. The author presents him as quite an awkward person to be an ambassador. He did speak some French according to Charles MacFarlane, who met him ten years later, but knew next to nothing about issues he was interested in, like agriculture and manufactures. Charles Mac-Farlane, *Turkey and Its Destiny* (London: John Murray, 1850), 2:161–64.
42. For his biography and examples of his poetry, see Inal, *Son Asır Türk Şairleri*, 3:481–86.
43. Ibid., 9:1598–99. Rather significantly, we see the names of Lebib Efendi and İbrahim Hakkı among the members of the Council of Poets (Encümen-i Şuara), an informal group of classicist poets formed in 1861. Meeting regularly to discuss and recite poetry, all members of this council were civil servants with ties to dervish lodges or religious orders. Despite certain novelties they embraced, their main inspiration was classical Ottoman poetry. M. Korkut Çeçen, "Encümen-i Şuara'nın Tanzimat Birinci Dönem Sanatçılarına Etkisi," *Çukurova Üniversitesi Sosyal Bilimler Enstitüsü Dergisi* 15 (2006): 133–52. Among the members also were three young men by the names of Namık Kemal, Ziya, and Refik—founders of the Young Ottoman movement that will be discussed in chapter 4. Suffice it to say here that these young and fierce critics of the Tanzimat elite were, similarly, disillusioned lower-ranking bureaucrats.

NOTES TO PAGES 53–57

44. Ahmed Lütfi, *Vak'anüvis Ahmed Lütfi Efendi Tarihi* (Istanbul: YKY), 1:8.
45. Ahmet Hamdi Tanpınar, *XIX. Asır Türk Edebiyatı Tarihi* (Istanbul: İbrahim Horoz, 1956), 125.
46. For examples, see Sadık Rifat Pasha, *Müntehabat-ı Asar* (Istanbul: Divitciyan, 1873–76), 9:11–14. For more on Rifat's political views, see Mardin, *Genesis*, 175–90.
47. Sadık Rifat, *Müntehabat-ı Asar*, 1:7.
48. The wording here is *ulûm-ı lâzıme*.
49. Sadık Rifat, *Müntehabat-ı Asar*, 6:45.
50. Mardin, *Genesis*, 186–87.
51. BOA Irade Hariciye 45/2092, 4 Cemaziyelevvel 1264/April 8, 1848.
52. Wanda, *Souvenirs Anecdotiques sur la Turquie (1820–1870)* (Paris: Firmin-Didot, 1884), 265.
53. Mardin, *Genesis*, 111.
54. The word "civilization" was one of the first European words imported directly into Ottoman Turkish and it remained in use until a Turkish word was coined. Sadık Rifat Pasha was the first person to refer to European *sivilizasyon*.
55. Selçuk Akşin Somel, *The Modernization of Public Education in the Ottoman Empire* (Leiden: Brill, 2001), 62. For Somel's analysis of Rifat's work, see 61–64.
56. Sadık Rifat, *Müntehabat*, 9:60.
57. Ibid.
58. Sadık Rifat Pasha, *Zeyl-i Risale-i Ahlak* (Istanbul: Darü't-Tabaatü'l-Amire, 1273/1857), 2.
59. Ibid., 15.
60. Ibid., 16.
61. Ibid.
62. As Hanioğlu notes, this text can be seen as comparable to a "cold-blooded strain" in European thought where analyses do not refer to moral values. But the use of a concept like ilm with potential moral associations itself makes it possible to read the text also as one that makes a moral case for the adoption of the new sciences. See Şükrü Hanioğlu, *The Young Turks in Opposition* (New York: Oxford University Press, 1995), 12.
63. Ibid., 17.
64. Ibid., 18.
65. Ibid., 3.
66. Document reproduced in Mahmud Cevad İbnü'ş-şeyh Nâfi, *Maarif-i Umumiye Nezareti Tarihçe-i Teşkilat ve İcraatı—XIX. Asır Osmanlı Maarif Tarihi*, ed. Taceddin Kayaoğlu (Ankara: Yeni Türkiye, 2001), 7–11.
67. That the report was clear about what it meant by "other sciences" (*fünûn*) is important in that the words *fen* or *fünûn* (pl.) continued in this period to refer also to "arts" in a broader sense. The school for public servants, for instance, was intended to teach, among others, "scribal arts" (*fünûn-ı kalemiyye*) (Mahmud Cevad, *Maarif-i Umumiye*, 23). This report, however, uses

fünûn in a more prestigious sense, treating them as equivalent to religious sciences.

68. Note, however, that this discussion is accompanied by the proposal that the teachers in elementary schools should be inspected by public officials. The established idea that elementary education should essentially be religious training was not challenged in this memorandum, but the aim appears to have been the standardization of training and evaluation as well as increasing the authority of civil bureaucracy in the organization of elementary education which was up till then under the control of the ulema.

69. For this approach see Somel, *Modernization*.

70. It was probably Reşid Pasha who encouraged the sultan to issue this edict (ibid., 37).

71. The wording here is *menba-i ulûm ve fünûn ve mehaz-ı sanayi-i maarif-nümûn*.

72. The chair, Melekpaşazade Abdülkadir Efendi, was a regular at the meetings of the Beşiktaş Scientific Society (Beşiktaş Cemiyet-i İlmiyyesi): resembling a salon, this was a regularly meeting group of well-educated ulema with mysticist leanings, such as Şanizade Ataullah and Kethüdazade Arif. They discussed the scientific developments in Europe in addition to subjects like literature and philosophy. Kethüdazade is known to have argued: "A thousand and so years ago the ancients wrote so many books . . . [that were forgotten afterwards]. Then the Franks studied these books and furthered science. All the factories, steam engines, machines, wheels are thanks to science. The Franks both study the sciences and implement it practically. Their states aid those who perform experiments. . . . With us, the practice is ignored even if the knowledge is studied." Emin, *Menakıb-ı Kethüdazade el-Hac Mehmed Arif Efendi* (Istanbul, 1877), 219. For the Beşiktaş Scientific Society, see Ekmeleddin İhsanoğlu, "19. yy. Başında Kültür Hayatı ve Beşiktaş Cemiyet-i İlmiyesi," *Belleten* 51 (1987): 801–20. Another member, Esad Efendi, was the court historian in the 1820s and the chief editor of the official gazette at its establishment. The remaining two members, Arif Hikmet and Mehmed Arif Efendis, were future *şeyhülislam*s. The Permanent Council founded the following year would be dominated by bureaucrats, however. On these developments the best sources to consult are Ekmeleddin İhsanoğlu, *Darülfünûn: Osmanlı'da Kültürel Modernleşmenin Odağı* (Istanbul: IRCICA, 2010) and Somel, *Modernization*.

73. Reproduced in Yahya Akyüz, *Türk Eğitim Tarihi (Başlangıçtan 2001'e)* (Istanbul: Alfa, 2001), 28.

74. Ibid., 43.

75. According to *Kamus-ı Türki*, *kemâlât* was commonly used to refer to "being perfect with respect to knowledge and morality" ("*kemal*," 1182).

76. Ahmed Cevdet Paşa, *Tezakir*, ed. Cavid Baysun (Ankara: Türk Tarih Kurumu, 1967), 4:47.

77. The wording here is *âfitâb-ı ulûm ü fünûn memalik-i şarkiyede pertev-endaz olup, . . . ol vakte göre lâzım olan fünûn ü maarif*.

78. Ahmed Cevdet, *Tezakir* 4:48. According to *Kamus-ı Türkî*, the phrase "Encümen-i Dâniş," which literally means "Council of Learning," was intended as the equivalent of "Academy" (*"dâniş,"* 600). Some sources translate the name as "Academy of Science(s)" into English (Şükrü Hanioğlu, *A Brief History of the Late Ottoman Empire* [Princeton, NJ: Princeton University Press, 2008]; Göçek, *Rise of the Bourgeoisie*) or as "Bilim Akademisi" into modern Turkish (Cahit Bilim, "İlk Türk Bilim Akademisi: Encümen-i Daniş," *Hacettepe Üniversitesi Edebiyat Fakültesi Dergisi* 3 [1985]: 81–104) but the current meanings of these words do not exactly capture the range of fields the Academy was expected to present. Conspicuously, foreign authors referred to *Encümen-i Dâniş* as "Academy of Science and the Belles Lettres" (James Porter, *Turkey: Its History and Progress*, 2 vols. [London: Hurst and Blackett, 1854], Xavier Heuschling, *L'Empire de Turquie* [Bruxelles: H. Tarlier; Paris: Guillemin, 1860]) or "Society for Advancement of Turkish Literature" (Smithsonian Institution, *List of Foreign Correspondents of the Smithsonian Institution* [Washington, DC: Smithsonian, 1872]). References to science in this text can also be seen in this more comprehensive sense, as "learning" in general.
79. I will discuss this further in the following chapters. The version of the memorandum published in the official gazette emphasized that works on arts and sciences should be in a language that "common people" (*amme-i nas*) could understand and thus make use of. It is also much clearer in terms of its description of the transformation of knowledge: "Science and knowledge changes with the passing of time and grows with the amalgamation of ideas. In each era the producers of ideas demonstrate different skills and arts, and the most precious fabric of science and knowledge is embellished with a different pattern each day. Hence, in each period, it is required to spread the needed sciences and knowledges [of that period]." Kenan Akyüz, *Encümen-i Dâniş* (Ankara: Ankara Üniversitesi Basımevi, 1975), 50.
80. Cevdet, *Tezakir* 4:56.
81. Ibid., 4:55.
82. Reproduced in Akyüz, *Encümen-i Dâniş*, 54.
83. BOA Irade Hariciye 118/5800, 20 Cemaziyelahir 1271/March 10, 1855 and 184/10237, 4 Zilkade 1277/May 14, 1861; Amedi Kalemi 75/13, 1273/1856–57, and 79/12,1273/1856–57. I contacted the Smithsonian Institute as well but unfortunately the staff was not able to locate any documents regarding the correspondence with the Ottoman Academy
84. "Proceedings of the American Oriental Society, Prepared from the Records, 1849–50," *Journal of the American Oriental Society* 2 (1851): xxviii. The new bureaucrats certainly embraced the "universality" of science: When he was minister of education in 1874 Safvet Pasha promptly had the pieces of a meteor that fell in Vidin brought to Istanbul and then sent to France to be examined at École des mines. See BOA Maarif Mektubi 18/124, 16 Cemaziyelevvel 1291/July 1, 1874 and 19/117, 14 Receb 1291/August 27, 1874. Also see G. A.

Daubrée, "Note sur un météorite tombée le 20 Mai, 1874, en Turquie, á Virba près Vidin," *Comptes Rendus* 79 (1874): 276–77.

85. Akyüz, *Encümen-i Dâniş*, 29.

86. Tanpınar, *XIX. Asır*, 115.

87. A copy of *Kavaid* was sent by Fuad to the American Oriental Society as well.

88. The wording here is *fenn-i hatîr*.

89. Mehmed Ali Fethi, Rusçuklu *İlm-i Tabakât-ı Arz* (Istanbul: Darü't-Tibaatü'l-Amire, 1269/1853), 2.

90. Ibid., 3.

91. Ibid., 6.

92. Ibid., 7.

93. Cevdet, *Tezakir* 1:13. Among the chief adversaries, according to Cevdet, was Ahmed Fethi Paşa, who had been a close ally of Reşid until the establishment of the Academy.

94. Akyüz, *Encümen-i Dâniş*, 28.

95. See Feza Günergun, "Derviş Mehmed Emin pacha (1817–1879), serviteur de la science et de l'État ottoman," in *Medecins et Ingénieurs Ottomans à l'âge des Nationalismes*, ed. Méropi Anastassiadou-Dumont (Paris: Maisonneuve et Larose, 2002): 171–83.

96. Ahmed Lütfi Efendi, *Vak'anüvis Ahmed Lütfi Efendi Tarihi*, ed. Münir Aktepe (Istanbul: Edebiyat Fakültesi, 1984), 9:157. Note that Lütfi refers to ambassadorship as an art, using the word *fen*. The same word was used by Âli Paşa to refer to geology—see n. 91.

97. Derviş Mehmed Emin Paşa, *Usûl-i Kimya* (Istanbul: Darü't-tıbaatü'l-Amire, 1264/1847–48), 3.

98. Ibid., 5.

99. Derviş's autograph, where he spells his name in the French way as Dervisch, reads "À son excellence Edhem Pacha, souvenir de l'auteur." Edhem Pasha, a renowned statesman of the second half of the nineteenth century, was a member of the first group of Ottoman youths sent to Paris for education. The copy is at the library of the Turkish Historical Association in Ankara.

CHAPTER THREE

1. Details on the Ottoman students in France are from Adnan Şişman, *Tanzimat Döneminde Fransa'ya Gönderilen Osmanlı Öğrencileri (1839–1876)* (Ankara: Türk Tarih Kurumu, 2004), unless otherwise noted.

2. Report reproduced in ibid., 165.

3. Ibid., 167.

4. Mahmud Cevad İbnü'ş-şeyh Nâfi, *Maarif-i Umumiye Nezareti Tarihçe-i Teşkilat ve İcraatı—XIX. Asır Osmanlı Maarif Tarihi*, ed.Taceddin Kayaoğlu (Ankara: Yeni Türkiye, 2001), 58.

5. "Natural sciences" according to Selçuk Akşin Somel, *The Modernization of Public Education in the Ottoman Empire* (Leiden: Brill, 2001), 170; "political

economy" according to French archival material that Şişman discusses in *Tanzimat Döneminde*, 35.

6. Şişman, *Tanzimat Döneminde*, 35. Tahsin is also famous for his couplet stating that one who has never seen Paris cannot claim to have truly lived. On Tahsin's life, with anecdotes about his experiences in Paris, see Ibnülemin M. K. Inal, *Son Asır Türk Şairleri* (Istanbul: Devlet Matbaası, 1940), 4:1870–82. Tahsin's career at the university and his works will be referred to later.

7. I was unable to locate the yearly reports about their studies in Paris as specified in the instructions they had received.

8. Şerif Mardin, *The Genesis of Young Ottoman Thought: A Study in the Modernization of Turkish Political Ideas* (Princeton, NJ: Princeton University Press, 1962), 222.

9. See Thomas Gieryn, "Cultural Boundaries: Settled and Unsettled," in *Clashes of Knowledge: Orthodoxies and Heterodoxies in Science and Religion*, ed. Peter Meusburger et al. (Berlin: Springer, 2008), 91–100.

10. Charles MacFarlane, *Turkey and Its Destiny* (London: John Murray, 1850), 2:179. For more on this widely cited but little-known observer of the mid-nineteenth-century Ottoman Empire, see the introduction to Charles MacFarlane, *Reminiscences of a Literary Life* (London: John Murray, 1917), and David Hill Radcliffe, "Charles Macfarlane: Reminiscences of a Literary Life," http://lordbyron.cath.lib.vt.edu/contents.php?doc=ChMacfa.1917.Contents (accessed August 22, 2013).

11. BOA Sadaret Mühimme Kalemi Evrakı 204/18, 14 Cemaziyelahir 1277/December 28, 1860.

12. Osman Nuri Ergin, *İstanbul Mektepleri ve İlim, Terbiye ve San'at Müesseseleri dolayisiyle Türkiye Maarif Tarihi*, 2nd ed. (Istanbul: Eser, 1977), 1–2:550. As in most topics pertaining to science in the Ottoman Empire, the most authoritative work on the Darülfünûn is by İhsanoğlu. See Ekmeleddin İhsanoğlu, *Darülfünûn: Osmanlı'da Kültürel Modernleşmenin Odağı* (Istanbul: IRCICA, 2010) for a detailed description of the history of the institution.

13. "Darülfünûn'da Ders-i 'âmm Vuku-ı Küşadı," *Mecmua-i Fünûn* 1, no. 6 (Cemaziyelahir 1279/December 1862): 259.

14. For Derviş, see chap. 2.

15. "Darülfünûn'da Ders-i 'âmm Vuku-ı Küşadı," *Mecmua-i Fünûn* 1, no. 7 (Receb 1279/January 1863): 301–2.

16. *Takvim-i Vekayi* 742 (26 Şevval 1280/April 4, 1864): 2.

17. "Darülfünûn'da Ders-i 'âmm Vuku-ı Küşadı," *Mecmua-i Fünûn* 1, no. 7 (Receb 1279/January 1863): 302–3.

18. A graduate of the Imperial Military School and a divisional general, not to be confused with the future minister of education Safvet Pasha.

19. Osman Saip was one of the earliest teachers and principals of the School of Medicine. He translated medical works from French in addition to sections from Italian geographer Adriano Balbi's *Abrégé de Géographie* (1832) in 1841.

20. Mahmud Cevad, *Maarif-i Umumiye*, 73–74.

21. Harry Collins, "Public Experiments and Displays of Virtuosity: The Core-Set Revisited," *Social Studies of Science* 18 (1988): 725–48.

22. On the importance of "trained senses" in science, see Lorraine Daston and Peter Galison, *Objectivity* (Cambridge, MA: MIT Press, 2007).

23. Simon Schaffer, "Natural Philosophy and Public Spectacle in the Eighteenth Century," *History of Science* 21 (1983): 1–41; Simon Schaffer, "The Consuming Flame: Electrical Showmen and Tory Mystics in the World of Goods," in *Consumption and the World of Goods*, ed. J. Brewer and R. Porter (London: Routledge, 1993): 489–526; Iwan Rhys Morus, *Frankenstein's Children: Electricity, Exhibition, and Experiment in Early Nineteenth-Century London* (Princeton, NJ: Princeton University Press, 1998).

24. Note the identification of specific types of knowledge with "knowledge as such." This issue will be discussed in more detail in the following sections.

25. "Darülfünûn dersleri," *Mecmua-i Fünûn* 1, no. 8 (Şaban 1279/February 1863): 331–32.

26. It may be useful here to remember Lütfi Efendi's criticism of Derviş Pasha's appointment to St. Petersburg as the Ottoman ambassador; see chap. 2.

27. The wording here is *tarik-i ilmiyye*, i.e., hierarchy within the ulema class.

28. "Darülfünûn dersleri," *Tasvir-i Efkâr* (9 Şaban 1279/January 30, 1863): 1. For more on Şinasi, see chap. 4.

29. Steven Shapin and Simon Schaffer, *Leviathan and the Air-Pump: Hobbes, Boyle, and the Experimental Life* (Princeton, NJ: Princeton University Press, 1985), 59.

30. See Donna Haraway, *Modest_Witness@Second_Millennium. FemaleMan©_Meets_OncoMouse™: Feminism and Technoscience* (New York and London: Routledge, 1997), 23–45.

31. Mehmed Ali Aynî, *Darülfünûn Tarihi*, ed. Aykut Kazancıgil (Istanbul: Kitabevi, 2007 [1927]): 16–17.

32. "Darülfünûn dersleri," *Mecmua-i Fünûn* 1, no. 8 (Şaban 1279/February 1863): 330–31.

33. Quoted in Aynî, *Darülfünûn*, 18.

34. My discussion here is based on Maryam Ekhtiar, "Nasir al-Din Shah and the Dar al-Funun: The Evolution of an Institution," *Iranian Studies* 34 (2001): 153–63, and Kamran Arjomand, "The Emergence of Scientific Modernity in Iran: Controversies Surrounding Astrology and Modern Astronomy in the Mid-Nineteenth Century," *Iranian Studies* 30 (1997): 5–24.

35. Osman Nuri Ergin, *İstanbul Mektepleri ve İlim, Terbiye ve San'at Müesseseleri dolayisiyle Türkiye Maarif Tarihi*, 2nd ed. (Istanbul: Eser, 1977), 1–2:482.

36. Mahmud Cevad, *Maarif-i Umumiye*, 93. On this document, also see Somel, *Modernization*, 86.

37. Mahmud Cevad, *Maarif-i Umumiye*, 94.

38. Ibid., 97. Note once again the growing identification of the new sciences with the Turkish language.

39. Ibid., 98. Finding it "a remarkable document," the *American Journal of Educa-*

tion published all articles of the act, without the introduction. See "Law on Public Instruction in Turkey," *American Journal of Education* 4 (1870): 17–31.

40. See Somel, *Modernization*, and Selçuk Akşin Somel, "Kırım Savaşı, Islahat Fermanı ve Eğitim," http://research.sabanciuniv.edu/5529/1/Kirim_Savasi,_ Islahat_Fermani_ve_Egitim.pdf (accessed September 9, 2013).

41. The petition is reproduced in Ahmet Karaçavuş, "Tanzimat Dönemi Osmanlı Bilim Cemiyetleri" (PhD diss., Ankara University, 2006), 198.

42. Marcel Mauss, *The Gift* (Oxon: Routledge, 2002).

43. Halil Pasha (1822–79), later Halil Şerif Pasha. Coming from an established Egyptian family, he went to the École Militaire Egyptienne and also studied political science in Paris before becoming an Ottoman diplomat; see also chap. 2.

44. Münif Pasha (1830?–1910) got his initial training at a *medrese* in Damascus, learned French in addition to Arabic and Persian, then joined the Translation Bureau. He was then employed at the Berlin Embassy and followed lectures in Berlin University. He would become a Pasha in 1880, but I will refer to him as Münif Pasha throughout the text to facilitate reading.

45. Said Pasha (1831?–96) studied mathematics at the University of Edinburgh, then got additional training at the Woolwich Military Academy, Enfield Rifle Factory, Waltham Gunpowder Mills, and the Greenwich Observatory.

46. Mehmed Kadri (1832–84) was the son of the governor of Cyprus. After basic Islamic training, he held bureaucratic posts in the provinces. Later he joined the Translation Bureau and became the translator-in-chief of the Supreme Council of Judicial Ordinances.

47. Sadullah (1838–91), later Sadullah Pasha, was a graduate of Darülmaarif, a prestigious new school that was intended to prepare students for the Darülfünûn. Later he joined the Translation Bureau, held high official posts, and became ambassador to Berlin. He was also famous for his poem "The Nineteenth Century" which I will refer to later.

48. Karabet, an Armenian, was secretary at the Council of the Treasury.

49. Andreas David Mordtmann (1811–79) was a German orientalist. He moved to Istanbul in 1860 and became a judge at the Commercial Court.

50. Istefan, an Armenian, was a translator.

51. Ekmeleddin İhsanoğlu, "Cemiyet-i Ilmiye-i Osmaniye'nin Kuruluş ve Faaliyetleri," in *Osmanlı İlmî ve Meslekî Cemiyetleri*, ed. Ekmeleddin İhsanoğlu (Istanbul: İÜ Edebiyat Fakültesi, 1987), 209.

52. The letter of acceptance is reproduced in *Mecmua-i Fünûn* 1, no.1 (Muharrem 1279/June 1862): 17.

53. The very first periodical in Turkish is generally regarded as the bulletin of the School of Medicine, which published a total of twenty-eight issues in 1849 and 1850. See Hıfzı Topuz II, *Mahmut'tan holdinglere Türk basın tarihi* (Istanbul: Remzi, 2003).

54. "Cemiyet-i İlmiye-i Osmaniye Nizamnamesidir," *Mecmua-i Fünûn* 1, no.1 (Muharrem 1279/June 1862): 2–10.

55. George A. Vassiadis, *The Syllogos Movement of Constantinople and Ottoman Greek Education, 1861–1923* (Athens: Centre for Asia Minor Studies, 2007), 67.

56. Ibid., 57.

57. Johann Strauss, "The Greek Connection in Nineteenth-Century Ottoman Intellectual History," in *Greece and the Balkans: Identities, Perceptions, and Cultural Encounters since the Enlightenment*, ed. Dimitris Tziovas (Aldershot: Ashgate, 2003): 53.

58. "Mukaddime," *Mecmua-i Fünûn* 1, no. 1 (Muharrem 1279/June 1862): 18.

59. "Mukayese-i İlm ü Cehl," *Mecmua-i Fünûn* 1, no. 1 (Muharrem 1279/June 1862): 21.

60. Ibid., 26.

61. Ibid., 35 (italics mine).

62. Ibid., 28.

63. Ibid., 33–34.

64. See Franz Rosenthal, *Knowledge Triumphant: The Concept of Knowledge in Medieval Islam* (Leiden: Brill, 1970).

65. See articles "Ilm," "Djahilliya," "Ulama" in the *Encyclopedia of Islam*, 2nd ed. For many examples on the association between knowledgeability and virtuousness, see Taşköprülüzade's mid-sixteenth-century collection of the biographies of Ottoman scholars (*eş-Şakâiku'n Nu'maniyye fî 'ulemai'd-Devlet-i Osmaniyye*; published as *Osmanlı Bilginleri*, trans. Muharrem Tan [Istanbul: Iz, 2007]), and Fahri Unan's meticulous analysis of this work, "Taşköprülü-zâde'nin Kaleminden XVI. Yüzyılın 'Ilim ve Âlim' Anlayışı," *Osmanlı Araştırmaları* 17 (1997): 149–264.

66. See Patrick Petitjean, "Science and the "Civilizing Mission": France and the Colonial Enterprise," in *Science across the European Empires, 1800–1950*, ed. Benedikt Stutchey (Oxford: Oxford University Press, 2005): 107–28, and Michael Adas, *Machines as the Measure of Men: Science, Technology, and Ideologies of Western Dominance* (Ithaca, NY: Cornell University Press, 1990).

67. Münif Pasha, "Mukaddime-i Ulûm-ı Jeoloji" *Mecmua-i Fünûn* 1, no. 2 (Safer 1279/July 1862): 65.

68. Âli Pasha, "Iltifatname," *Mecmua-i Fünûn* 1, no. 2 (Safer 1279/July 1862): 52. The word Âli uses here is *terbiye*, which signifies both instruction and the instilling of proper morals. The emphasis is on being well-mannered, and Şemseddin Sami defines the word in his *Kamus-ı Türki* as "the teaching of knowledge and manners." In this respect, the concept has more to do with discipline than education per se, as Somel, *Modernization*, also suggests. I elaborate on the importance of this concept in chap. 7.

69. Âli Pasha, "Iltifatname," 54.

70. Hyde Clarke, "Public Instruction in Turkey," *Journal of the Statistical Society of London* 30 (1867): 513.

71. Steven Shapin and Barry Barnes, "Science, Nature, and Control: Interpreting Mechanics' Institutes," *Social Studies of Science* 7 (1977): 42.

72. Inal, *Son Asır Türk Şairleri*, 3:573–76.

73. Refik, like Âli Pasha, uses the word *terbiye*.

74. "Esas-ı Medeniyyet," *Mir'at* 1 (1279/1863): 3.
75. Ibid., 4.
76. See "Buhara dair mâlumat," *Mir'at* 3 (1279/1863): 42.
77. Münif Pasha, "Hudus-ı Mecmua-ı İber-i İntibah," *Mecmua-i Fünûn* 1, no. 8 (Şaban 1279/January 1863): 353–55.
78. Münif Pasha, "Zuhur-ı Mir'at" *Mecmua-i Fünûn* 1, no. 9: (Ramazan 1279/ March 1863): 399.
79. "Vecibe," *Mir'at* 1, no. 3 (Zilkade 1279/April 1863): 47–48.
80. *Bab-ı Âli*, the metonymy for the Ottoman government.
81. See chap. 2.
82. Mustafa Reşid Pasha's son, ambassador to Paris.
83. Members of the Translation Bureau probably used books owned by the bureau library. Library records were recently uncovered and examined by Sezai Balcı, "Osmanlı Devleti'nde Tercümanlık ve Bab-ı Âli Tercüme Odası" (PhD diss., Ankara University, 2006). We learn from these records, e.g., that Mehmed Şevki, the author of a series of essays on the political and military history of Europe, had checked out Paoli-Chagny's *Histoire de la politique des puissances de l'Europe* (1817), a book entitled *Travers d'Espagne*, as well as *Histoire de la République de Venise*, probably by Pierre Daru (1819). See Balcı, 142.
84. For more on the Greek contribution to Ottoman intellectual life, see Johann Strauss, "The Millets and the Ottoman Language: The Contribution of Ottoman Greeks to Ottoman Letters (19th–20th Centuries)," *Die Welt des Islams* 35 (1995), and Strauss, "The Greek Connection in Ottoman Intellectual History."
85. "Mahiyet ve Aksam-ı Ulûm," *Mecmua-i Fünûn* 2, no. 13 (Muharrem 1280/June 1863): 9.
86. Ibid., 2. At the time Münif's writing, and indeed until its collapse, the Ottoman court had a post of chief astrologer. The traditional duty of the astrologer was to prepare horoscopes as well as calendars. While the use of horoscopes to determine the proper time (*eşref saat*) for important events was common procedure, there appears to have been a gradual decline in horoscope preparation after the 1830s. Several documents in the Ottoman Archives regarding the use of horoscopes from the period between 1840 (1255) and 1908 (1325) are from April 1846, June 1857, September 1891. See BOA Irade Dahiliye 118/6015, 14 Rebiyülevvel 1262/March 12, 1846; Sadaret Nezaret ve Devair 227/15, 2 Zilkade 1273/June 24, 1857; Yıldız Esas Evrakı 58/17, 26 Muharrem 1309/September 1, 1891. For a thorough study of the institution, see Selim Aydüz, "Osmanlı İmparatorluğunda Müneccimbaşılık," in *Osmanlı Bilimi Araştırmaları*, ed. Feza Günergun (Istanbul: IU Edebiyat Fakültesi, 1995): 159–207.
87. The wording here is *ulûm/fünûn-ı nâfia*.
88. Münif Pasha, "Mukaddime," *Ruzname-i Ceride-i Havadis* (16 Rebiyülahir 1277/ November 1, 1860): 1.
89. Münif Pasha, "Tarih-i Devlet-i Osmaniye Dersi," Mecmua-i Fünûn 3, no. 28 (Rebiyülahir 1281/September 1864): 157–58. For other references to useful

sciences see, e.g., "Ehemmiyet-i Terbiye-i Sıbyan," *Mecmua-i Fünûn* 1, no. 5 (Cemaziyelevvel 1279/November 1862): 176–85.

90. Ergin, *Maarif Tarihi* 1–2:464.

91. Mahmud Cevad, *Maarif-i Umumiye* 74 and 104, respectively.

92. Jonathan Topham, "Science and Popular Education in the 1830s: The Role of the Bridgewater Treatises," *British Journal for the History of Science* 25 (1992): 419.

93. Ibid., 406; also see Shapin and Barnes, "Science, Nature, and Control."

94. "Beyim, ilm olıcak nâfi gerektir / Hevâ vü nefs dâfi gerektir." Yaşar Yücel, *Osmanlı devlet teşkilâtına dair kaynaklar* (Ankara: Türk Tarih Kurumu Basımevi, 1988).

95. See Fahri Unan, *Kuruluşundan Günümüze Fatih Külliyesi* (Ankara: Türk Tarih Kurumu, 2003): 380–92, particularly 385–86.

96. Quoted in Enver Ziya Karal, "Osmanlı Tarihinde Türk Dili Sorunu," in *Bilim, Kültür ve Öğretim Dili Olarak Türkçe* (Ankara: Türk Tarih Kurumu, 1978), 54. For other examples, see Hasan Akgündüz, Klasik Dönem Osmanlı Medrese Sistemi: Amaç, Yapı, İşleyiş (Istanbul: Ulusal, 1997): 218, 270, 282. In the earlier periods when the "rational/intellectual sciences" were part of medrese curricula, they were also regarded as "useful sciences," as the foundation deed (vakfiye) of the medrese of Mehmed II (fifteenth century) suggests; see 270.

97. Muhammad Q. Zaman, *The Ulama in Contemporary Islam: Custodians of Change* (Princeton, NJ: Princeton University Press, 2002): 65–66.

98. Quoted in Anne DeWitt, *Moral Authority, Men of Science, and the Victorian Novel* (Cambridge: Cambridge University Press, 2013), 1. Also see Ruth Barton, "Just before Nature: The Purposes of Science and the Purposes of Popularization in Some English Popular Science Journals of the 1860s," *Annals of Science* 55 (1998): 1–33.

99. Quoted in Konstantinos Tampakis, "Onwards Facing Backwards: The Rhetoric of Science in Nineteenth-Century Greece," *British Journal for the History of Science* (2013): 13, doi: 10.1017/S000708741300040X (accessed September, 23 2013).

CHAPTER FOUR

1. My account of the lives and arguments of the members of the movement is based primarily on Şerif Mardin, *The Genesis of Young Ottoman Thought: A Study in the Modernization of Turkish Political Ideas* (Princeton, NJ: Princeton University Press, 1962). But my emphasis in this chapter is not on just the representatives of the movement, as I argue that their formulations were shared by many critics of the Tanzimat regime, even if they were not involved with the political objectives of the Young Ottomans. What matters for our purposes is the *generation* of the Young Ottomans.

2. On the perceptions of the reforms by Ottoman communities, and particu-

larly the Muslims, see Cevdet Paşa, *Tezakir* 1 (Ankara: Türk Tarih Kurumu, 1953): 67–89.

3. For examples, see Mardin's discussions on how the Young Ottomans borrowed concepts from the Islamic tradition (such as *biat*, *şu'ra*, and *meşveret*) and reinterpreted them as indications that Islam stipulates the participation of a wider portion of the entitled groups in government.

4. In the words of Eldem, trade between Europe and the Ottoman Empire assumed the shape of an "effective domination or influence of Western economic actors over Ottoman markets, production and consumption" in the nineteenth century, thus rendering the relations "quasi-colonial." See Edhem Eldem, "Capitulations and Western Trade," in *The Cambridge History of Turkey*, vol. 3: *The Later Ottoman Empire, 1603–1839* (New York: Cambridge University Press, 2006), 284.

5. On the question of "super-westernization," see Şerif Mardin, "Super-Westernization in Urban Life in the Last Quarter of the Nineteenth Century," in *Turkey: Geographical and Social Perspectives*, ed. Peter Benedict et al. (Leiden: Brill, 1974), 403–46.

6. While I refer to the fop as an invented character in many sections of this book, it was most likely built on already existing stereotypes in oral and written traditions (I thank Carter Findley for this comment). *Hacivat*, one of the two main figures in Ottoman shadow theater, for instance, is to some extent reminiscent of the fop, as he is a pedantic character with an erudite-sounding but awkward style of speech who represents the establishment.

7. It is worth remembering Mehmet Akif's stanza (quoted in the introduction) at this point. While Akif argued that science was neutral, he warned the Muslim reader of the "depravities" it tended to travel with. The so-called depravities had much to do with the social, political, and economic transformations of the Ottoman Empire in the global context of the nineteenth century, but they tended to be reduced to the inappropriate lifestyles of the fops, as in the case of Akif's poem and the examples that will be discussed in this chapter.

8. Muslim Ottomans commonly referred to Europeans as "Franks," and European manners and styles as "alla franca."

9. Quoted in Niyazi Akı, *XIX. Yüzyıl Türk Tiyatrosunda Devrin Hayat ve İnsanı: Sosyopsikolojik Deneme* (Erzurum: Atatürk Üniversitesi, 1974), 92.

10. Plato was of course known and respected by Ottomans, as his philosophy was a major inspiration for Islamic philosophy as well. But the novelty Şinasi introduces, much in accordance with the newly emerging European historiography of science, is the presentation of Newton as a "great man" comparable to, or in the same category as, Plato.

11. Şinasi is mimicking the calls of itinerant junk buyers common on the streets of Istanbul.

12. On this subject that needs more analysis, see Ismail Kara, "Tarih ve Hurafe," in *Din ve Modernleşme Arasında: Çağdaş Türk Düşüncesinin Meseleleri* (Istanbul: Dergah, 2003), 90–91.

13. Şinasi, "Mukaddime," *Tercüman-ı Ahval* (6 Rebiülahir 1277/October 22, 1860): 1.
14. Şinasi, "Mukaddime," *Tasvir-i Efkâr* (30 Zilhicce 1278/June 18, 1862): 1.
15. Şinasi, "İstanbul Sokaklarının Tenvir ve Tathiri," *Tasvir-i Efkâr* (28 Zilkade 1280/May 5, 1864): 2.
16. Zeynep Çelik, *Displaying the Orient: Architecture of Islam at Nineteenth-Century World's Fairs* (Berkeley: University of California Press, 1992), 12.
17. Mardin, *Genesis*, 268.
18. Hayreddin, "Medeniyet ve Türkistan," *Terakki* (5 Muharrem 1286/April 17, 1869): 3.
19. "Muteber imzasile aldığımız varakadır" *Basiret* (23 Cemaziyelevvel 1288/August 10, 1871): 2.
20. BOA Sadaret Nezaret ve Devair 347/39 (20 Ramazan 1277/April 1, 1861)
21. Ali Efendi, "Şehir Mektubu," *Basiret* (26 Rebiülevvel 1290/May 24, 1873): 2. Reproduced in Basiretçi Ali Efendi, *Istanbul Mektupları*, ed. Nuri Sağlam (Istanbul: Kitabevi, 2001), 147–48.
22. "Islamiyet ve Medeniyet," *Basiret* (6 Rebiyülahir 1289/June 13, 1872): 2.
23. Muallim Naci, "Terkib-i Bend," in *Terkib-i Bend-i Muallim Naci* ([Ruse, Bulgaria]: Tuna Vilayeti Matbaası, 18??), 4. The wording is *Tahsil-i fünûn ile serâir bilinir mi / Sırdan o haberdar olur kim geçe serden*.
24. Quoted in Ali Budak, *Batılılaşma Sürecinde Çok Yönlü Bir Osmanlı Aydını: Münif Paşa* (Istanbul: Kitabevi, 2004), 272.
25. See Ahmet Hamdi Tanpınar, *XIX. Asır Türk Edebiyatı Tarihi* (Istanbul: İbrahim Horoz, 1956), 152, and Mardin, *Genesis*, 240.
26. See *Tasvir-i Efkâr*, nos. 402–15 (22 Safer–12 Rebiyülahir 1283/July 6–August 24, 1866). The editors and writers of the *Tasvir-i Efkâr* were associated with the Young Ottoman movement. Mehmed Mansur's articles were published as a book in 1883, with two additional chapters. See Mehmed Mansur, *Meşhur İskenderiye Kütüphanesine Dair Risaledir* (Istanbul: Ceride-i Askeriye Matbaası, 1300/1882–83). My citations are from this version.
27. The burning of the Library of Alexandria, and the role of Christians or Muslims in this incident was, and remains, a topic of heated debate, particularly among apologists from different groups. For a study on the history of the topic in Europe, see Jon Thiem, "The Great Library of Alexandria Burnt: Towards the History of a Symbol," *Journal of the History of Ideas* 40 (1979): 507–26. For analyses of the debate, see Mostafa El-Abbadi and Omnia Mounir Fathallah, eds., *What Happened to the Ancient Library of Alexandria?* (Leiden: Brill, 2008).
28. Aleksandr (Karatheodori), "Taharri-i Musannefat-ı Atika," *Mecmua-i Fünûn* 2, no. 14 (Safer 1280/July 1863): 23.
29. Mansur, *Iskenderiye*, 66–67.
30. Ali ibn Embasevî, "İslam Kütüphaneleri," *Mecmua-i Fünûn* 5, no. 44 (Muharrem 1284/May 1867): 25–29.
31. Mansur, *Iskenderiye*, 70–71.
32. *Tasvir-i Efkâr* (September 4, 1866/23 Rebiyülahir 1283): 1.

33. Note that Mansur was Namık Kemal's French teacher at the Translation Bureau.

34. Ali Suavi, "Encümen-i Daniş-i Şarkî," in *Yeni Türk Edebiyatı Antolojisi*, ed. Mehmet Kaplan et al. (Istanbul: Istanbul Üniversitesi Edebiyat Fakültesi Yayınları, 1978), 540. According to Mordtmann, who was a member of the Ottoman Society for Sciences (see chap. 3), Münif was regarded by many as an atheist. See Andreas David Mordtmann, *Istanbul ve Yeni Osmanlılar (Stambul und das Moderne Türkentum)*, trans. Gertraude Habermann Songu (Istanbul: Pera, 1999 [1877]). This claim, supported by Ali Süavi's remarks, points to a reaction that probably emerged as early as 1859, when Münif Pasha published his first work: an anthology of dialogues he selected and translated from the works of Voltaire, Fenelon, and Fontenelle (*Philosophical Dialogues/Muhaverat-ı Hikemiye*, 1859). On Ali Suavi, see Mardin, *Genesis*, and Hüseyin Çelik, *Ali Süavi ve Dönemi* (Istanbul: İletişim, 1994). Çelik also refers to Süavi's condemnations of Münif; so does Budak, *Batılılaşma Sürecinde*.

35. "Havadisat-ı Dahiliyye," *Tasvir-i Efkâr* (12 Receb 1283/November 20, 1866): 1.

36. Namık Kemal, "Hubbü'l-vatan mine'l-iman," *Hürriyet* (8 Rebiyülevvel 1285/ June 29, 1868): 1–2.

37. Münif Paşa, "Mahiyet ve Aksam-ı Ulûm," *Mecmua-i Fünûn* 2, no. 13 (Muharrem 1280/June 1863): 9.

38. For a rather harsh criticism of the translation, see Mükrimin Halil Yinanç, "Tanzimattan sonra bizde tarihçilik," in *Tanzimat I: Yüzüncü Yılı Münasebetile* (Ankara: Türk Tarih Kurumu, 1940), 584. Ayvazoğlu underlines that while Ottoman authors must have certainly been aware of the intellectual achievements of the Muslims of Spain, they rarely referred to Andalusia before the nineteenth century. See Beşir Ayvazoğlu, "Edebiyatımızda Endülüs," in *Endülüs'ten İspanya'ya* (Ankara: Türkiye Diyanet Vakfı, 1996), 79–80. On the theme of Andalusia in Ottoman literature, also see İnci Enginün, "Edebiyatımızda Endülüs," in *Araştırmalar ve Belgeler* (Dergah: Istanbul, 2000), 32–41.

39. [Untitled], *Hürriyet* 11 (20 Cemaziyelevvel 1285/September 7, 1868): 8.

40. Namık Kemal, "Âzâr-ı mevhume," *Hürriyet* 35 (10 Zilkade 1285/February 22, 1869): 8.

41. Namık Kemal, "Nutk-ı hümayun," *Hürriyet* 50 (26 safer 1286/June 7, 1869): 4.

42. Kemal and the other Young Ottomans referred to the "glorious science of fiqh (*fenn-i celîl-i fıkh*)" in many of their writings. For examples and a relevant discussion, see Ihsan Sungu, "Tanzimat ve Yeni Osmanlılar," in *Tanzimat* 1 (Istanbul: Maarif, 1940): 777–857. Mardin, on the other hand, shows that Namık Kemal's understanding of Islamic law was not as puristic as he made it out to sound; see Mardin, *Genesis*, 313–19.

43. Kemal, "Âzâr-ı mevhume," 7.

44. Ziya Paşa, "Terkib-i Bend," in Kenan Akyüz, *Batı Te'sirinde Türk Şiiri Antolojisi*, 3rd ed. (Ankara: Doğuş, 1970), 48.

45. Ziya Paşa, "Meşrut-ı Ahval-i Şâ'iri," in *Ziya Paşa'nın Hayatı, Eserleri, Edebi*

Şahsiyeti ve Bütün Şiirleri, ed. Önder Göçgün (Ankara: Kültür ve Turizm Bakanlığı, 1987), 73–75.

46. Quoted in Kazım Yetiş, "Yeni bir Edebiyat Anlayışı," in *Dönemler ve Problemler Aynasında Türk Edebiyatı* (Istanbul: Kitabevi, 2007), 35.

47. Ali Süavi, "Türk," in *Türk Edebiyatı,* ed. Kaplan et al., 501. Original publication in *Ulûm* (1 Rebiülahir 1286/July 11, 1869), 1–17.

48. Ali Süavi, "Türk," 504.

49. Ahmed Midhat, "Velâdet," *Dağarcık* no. 2 (1288/1872): 49–52.

50. Ahmed Midhat, "Duvardan Bir Sada," *Dağarcık* no. 4 (1288/1872): 99–100.

51. Mikulas Teich, "Circulation, Transformation, Conservation of Matter and the Balancing of the Biological World in the Eighteenth Century," *Ambix* 29 (1982): 17–28. On Liebig and his influence, see Frederick Gregory, *Scientific Materialism in Nineteenth-Century Germany* (Dordrecht: D. Reidel, 1977). Şükrü Hanioğlu, "Blueprints for a Future Society: Late Ottoman Materialists on Science, Religion, and Art," in *Late Ottoman Society: The Intellectual Legacy,* ed. Elisabeth Özdalga (London: Routledge, 2005), analyzes the impact of these ideas on Ottoman thought.

52. Ahmed Midhat, "Duvardan," 102.

53. Ahmed Midhat, "Bir Mülahaza-ı Dîniyye," *Dağarcık* no. 4 (1288/1872): 103.

54. Ahmed Midhat, "İnsan- Dünyada İnsanın Zuhuru," *Dağarcık* no. 4 (1288/1872): 109–17.

55. [Harputlu İshak Hoca],"Mevaliden bir zat tarafından matbaamıza vürud eden varakadır," *Basiret* (4 Muharrem 1290/March 4, 1873): 1.

56. The Islamic term used to refer to Muslims, Christians, and Jews.

57. İshak, "Mevaliden," 2. While the Young Ottomans did not advocate revolutionary political and social change by any means, the upheavals in Europe did give them inspiration, and their newspapers occasionally published accounts sympathetic to the Commune. For a cursory study on this topic, see Serol Teber, *Paris Komünü'nde Üç Yurtsever Türk: Mehmet, Reşat ve Nuri Beyler* (Istanbul: De, 1986).

58. Harputlu İshak Hoca, "Varaka-i Cevabiyye," *Basiret* (9 Muharrem 1290/ March 9, 1873): 2.

59. Pierre Bourdieu, *The Logic of Practice* (Cambridge: Cambridge University Press, 1990).

60. İshak, "Varaka-i Cevabiyye," 2.

61. Ibid.

62. Ahmed Midhat, "Redd-i Itiraz ve Izah-i Hakikat," *Dağarcık* no. 8 (1288/1873) 234–50.

63. The wording here is *ulûm-ı efrenciyye.*

64. Midhat, "Redd-i Itiraz ve Izah-i Hakikat," 238.

65. Namık Kemal, "Efkâr-ı Cedide," *Hadika* (26 Ramazan 1289/November 27, 1872): 1–3.

66. Abdülhak Hamid Tarhan, *Abdülhak Hamid'in Tiyatroları,* vol. 1, ed. İnci Enginün (Istanbul: Dergah, 1998), 53–54.

67. "Türkistan'ın Esbab-ı Tedennisi," *Hürriyet* (6 Rebiyülahir 1285/July 27, 1868): 1. The unsigned essay is commonly attributed to Namık Kemal. Türköne claims that the two essays published in the fifth and sixth issues of the *Hürriyet* that put the blame on ignorance rather than on the despotism of Tanzimat bureaucrats, the usual target of the Young Ottomans, were actually commissioned by the Ottoman government itself. The rebellious intellectuals published these pieces half-heartedly in return for financial aid. See Mümtaz'er Türköne, *Siyasi Düşünce Olarak İslamcılığın Doğuşu* (Istanbul: İletişim, 1991), 98.

68. "Türkistan'ın Esbab-ı Tedennisi," 3.

69. "Hüda Kâdirdir, Eyler Seng-i Hârâdan Güher Peydâ," *Hürriyet* (July 5, 1869/25 Rebiyülevvel 1285): 3.

70. Kemal, "Âzâr-ı mevhume," 8.

71. *Basiret* (27 Safer 1287/May 29, 1870): 4.

72. Ebuzziya Tevfik, *Sirac* 5 (21 Muharrem 1290/March 21, 1873): 1. The "backward" state of education was indeed an issue of debate within the Ottoman Jewish community, and a Jewish educational reform did start in the 1860s, with the help of the Alliance Israélite Universelle. See Aron Rodrigue, *French Jews, Turkish Jews: The Alliance Israélite Universelle and the Politics of Jewish Schooling in Turkey, 1860–1925* (Bloomington: Indiana University Press, 1990).

73. "Bilmek ve Bilmemek," *Mümeyyiz* (12 Şevval 1286/January 15, 1870): 3.

74. *Mümeyyiz* further clarified its approach to knowledge by publishing in its forty-eighth issue an abridged version of Münif Pasha's "Comparison of Knowledge and Ignorance," which identified ignorance with lack of knowledge on the new sciences. It is also important that in a text on "people to interact and be friends with," the newspaper recommends "people of knowledge and skill" as the primary group, whereas "people of good morals" comes second. See "İhtilat ve Ünsiyet Olunacak Tavaif," *Mümeyyiz* (19 Safer 1287/May 21, 1870): 3.

75. Ahmed Remzi, *Müfredat-ı Tıb ve İlm-i Tedavi* (Istanbul, 1288/1871–72), 2.

76. [Untitled], *Hürriyet* (19 Safer 1286/May 31, 1869): 4.

CHAPTER FIVE

1. The establishment of a constitutional monarchy in 1876 is commonly regarded as the end of the Tanzimat era in Ottoman historiography. Sultan Abdülhamid II would suspend the parliament indefinitely and start building an autocratic rule only two years later.

2. Ali Efendi, "Şehir Mektubu," *Basiret* (17 Muharrem 1288/April 1, 1871): 1–2. Reproduced in Basiretçi Ali Efendi, *Şehir Mektupları*, ed. Nuri Sağlam (Istanbul: Kitabevi, 2001), 10–12.

3. Niyazi Akı indicates that similar introductions were very common in plays published in this period as well. See Niyazi Akı, *Türk Tiyatro Edebiyatı Tarihi*, vol. 1 (Istanbul: Dergah, 1989), 6.

4. Ahmed Midhat, "Kürre-i Arzın Fen Nazarında Keyfiyet-i Teşekkülü," *Dağarcık* no. 9 (1289/1873–74): 271.

5. Ibid., 273–74.

6. "Mukaddime," *Revnak* no. 1 (1290/1874–75): 2.

7. "Mukayese-i İlm ü Cehl," *Sandık* no. 1 (1290/1874–75): 13–14.

8. "Mukaddime," *Afitab-ı Maarif* no. 1 (1291/1874–75): 2. The translation is intended to mimic the verbosity of the original text.

9. Ibid., 4. The word translated as "virtue" is *kemâlât*, the word that indicates maturity and the attainment of a high level of virtuousness and wisdom (see chap. 2, n. 75). Once again, note the inseparability of new knowledge and virtue. Unfortunately I was unable to find any information on Mehmed, Vassaf, Nazım, and Recai, the editors of the journal and probably the authors of this essay. But from the tone of the text, it appears reasonable to conjecture that they were students or graduates of one of the highest-ranking new schools of the empire, the Imperial Military Academy or the School of Medicine.

10. Talat, who was twenty-five at the time, was a graduate of one of the less prestigious new schools of the empire, the *Mahrec-i Aklam*: an intermediate-level school for providing basic skills and general knowledge on a broad range of subjects for future civil servants. He occupied a number of lower-level posts in judicial institutions and also wrote a play (*Feryad* [*Cry for Help*]) criticizing arranged marriages. The protagonist of the play was a "well-educated but chaste" young girl.

11. "Şark Usulü: Afitab-ı Maarifin Münderecat-ı Mühimmesi / Mukaddime," *Kasa* no. 3 (1291/1874–75): 43.

12. Ibid., 44. The third name appears to be "Nikola" rather than Tycho; I leave it as Tycho here as the article in the *Afitab* that this piece mocks refers to Tycho Brahe.

13. "Tiyatro ve Ahlak," *Kasa* no. 3 (1291/1874–75): 53–55.

14. Nuri, "Batıl Zehab," *İbret* (20 Rebiyülahir 1289/June 27, 1872): 1.

15. Ebuzziya Tevfik, "Mektebsizlikten görülen bela ve mekteblerin vücub-ı ıslahı," *İbret* (27 Rebiyülahir 1289/July 4, 1872): 3.

16. "Havadis-i Hariciyye," *Tasvir-i Efkâr* (15 Muharrem 1279/July 13, 1862): 3.

17. See "Halepli Ahmed Efendi" (20 Cemaziyelevvel 1284/September 19, 1867): 4; "Hoca Abdi Efendi" (17 Ramazan 1285/January 1, 1869): 3. These were people believed to heal the ill with their breath.

18. Namık Kemal, "Usûl-ı meşrûta dair geçenki nüshada yazılan mektubun ikincisi," *Hürriyet* (4 Cemaziyelahir 1285/September 21, 1868): 7.

19. Namık Kemal, "Maarif," *İbret* (28 Rebiülahir 1289/July 4, 1872), 1.

20. Namık Kemal, *Namık Kemal'in Hususi Mektupları*, ed. Fevziye A. Tansel (Ankara: Türk Tarih Kurumu, 1969): 2, 192.

21. Ibid., 237.

22. Typical examples of this strikingly dull character include Rakım Efendi, from *Felatun Bey ile Rakım Efendi* (1875), and Nasuh, from *Paris'te Bir Türk* (1876), two of the many otherwise colorful novels of Ahmed Midhat.

23. "Terakki Gazetesine," *Terakki* (14 Muharrem 1286/April 26, 1869): 3.

24. "Gece Yolcuları," *Kasa* no. 2 (1290/1873–74): 30. The translation in question was from the popular French mystery novelist, Pierre Alexis Ponson du Terrail. It is striking that the author ignores the fact that the author of the original text itself was a European.

25. Cited in Cemal Kutay, *Sultan Abdülaziz'in Avrupa Seyahati* (Istanbul: Boğaziçi, 1991), 121.

26. "ve lekad keremna," *Terakki* (27 Safer 1286/June 7, 1869), 2–4.

27. [Untitled], *Basiret* (2 Zilkade 1289/January 1, 1873): 2.

28. Excerpt from the poem "Garam" by Abdulhak Hamid, quoted in Akı, *XIX. Yüzyıl*, 36.

29. "Türkistan'ın Esbab-ı Tedennisi," *Hürriyet* (6 Rebiyülahir 1285/July 27, 1868): 3.

30. Ibid.

31. Ali Suavi, "Maarif," *Muhbir* 8 (21 Ramazan 1283/January 27, 1867): 2. Also in *Yeni Türk Edebiyatı Antolojisi*, ed. Mehmet Kaplan et al. (Istanbul: Istanbul Üniversitesi Edebiyat Fakültesi Yayınları, 1978), 555–56.

32. Hayreddin, "Muvazene-i Din ve Akıl," *Basiret* (21 Şevval 1287/January 14, 1871): 3.

33. "Vaiz Efendiler," *Basiret* (10 Şaban 1290/October 3, 1873): 1.

34. *Basiret* (1 Zilhicce 1289/January 30, 1873): 2–3.

35. "Varaka," *Basiret* (14 Zilhicce 1289/February 12, 1873): 2

36. "Teşekkür," *Basiret* (12 Muharrem 1290/March 12, 1873): 1–2.

37. See Eyüp Baş, *Dil-Tarih İlişkisi Bağlamında Osmanlı Türklerinde Tarih Yazıcılığı: XVI. ve XVII. Yüzyıl Örnekleriyle* (Ankara: Türkiye Diyanet Vakfı, 2006): 79–95; Hüseyin Atay, *Osmanlılarda Yüksek Din Eğitimi: Medrese Programları, İcazetnameler, Islahat Hareketleri* (Istanbul: Dergah, 1983).

38. Mehmed Mansur, *Meşhur İskenderiye Kütüphanesine Dair Risaledir* (Istanbul: Ceride-i Askeriye Matbaası, 1300/1882–83), 157–58.

39. Münif Pasha, "Islâh-ı Resm-i Hatta Dair Bazı Tasavvurat," *Mecmua-i Fünûn* 2, no. 14 (Safer 1280/July 1863): 69–77. Azeri/Iranian litterateurs and politicians played an important part in this debate; see Hamid Algar, *Mīrzā Malkum Khān: A Study in the History of Iranian Modernism* (Berkeley: University of California Press, 1973), esp. 82–93.

40. Hayreddin, "Maarif-i Umumiye," *Terakki* (21 Rebiülahir 1286/July 31,1869): 3.

41. [Ebuzziya Tevfik] "Şûra-yı Devlet Mülazımlarından Tevfik Bey Tarafından Vârid Olan Varakadır," *Terakki* (23 Rebiülahir 1286/August 2, 1869): 3–4. Ebuzziya's response continued in the following two issues. For a similar comment, see Namık Kemal, "Usul-i Tahsilin Islahına Dair," *Tasvir-i Efkâr* (26 Safer 1283/July 10, 1866), 1–2.

42. Ali Suavi, "Lisan ve Hatt-ı Türkî," *Ulûm* 3 (1286/1869–70).

43. Ali Efendi, "Şehir Mektubu," *Basiret* (23 Safer 1291/April 11, 1874): 2. Reproduced in Basiretçi Ali Efendi, *Şehir Mektupları*, 275–76.

44. "Tahsil-i Maaş," *Basiret* (23 Şevval 1286/January 26, 1870): 1.

45. "Mekteb-i Sanayi," *Basiret* (16 Zilkade 1286/February 17, 1870): 5.

46. Quoted in Şerif Mardin, *The Genesis of Young Ottoman Thought: A Study in the Modernization of Turkish Political Ideas* (Princeton, NJ: Princeton University Press, 1962), 128.

47. "Bir Varaka," *Basiret* (23 Zilkade 1286/February 24, 1870): 4. Also see the letters published in issues 30 and 36.

48. "Mukaddime," *Hadika* (17 Zilkade 1287/February 8, 1871): 2. For a similar text, see "Şehir Mektubu," *Basiret* (9 Zilkade 1291/September 18, 1874), 2. Reproduced in Basiretçi Ali Efendi, *Şehir Mektupları*, 358.

49. [Untitled], *Basiret* (2 Zilkade 1289/January 1, 1873): 2.

50. "Maarife Çalışalım," *Basiret* (4 Rebiülevvel 1290/May 2, 1873): 2. *Efendi* is a term used commonly to address educated persons who held public office or were members of an esteemed profession.

51. Ali Suavi, "Maarif," *Muhbir* (21 Ramazan 1283/January 27, 1867): 3. Şemseddin Sami would make essentially the same remarks almost fifteen years later, in his "Semerat-ı Ilim," *Hafta* 19 (4 Safer 1299/December 26, 1881): 290.

52. Mahmud Cevad İbnü'ş-şeyh Nâfi, *Maarif-i Umumiye Nezareti Tarihçe-i Teşkilat ve İcraatı—XIX. Asır Osmanlı Maarif Tarihi*, ed.Taceddin Kayaoğlu (Ankara: Yeni Türkiye, 2001), 80. Also see Ekmeleddin İhsanoğlu, *Darülfünûn: Osmanlı'da Kültürel Modernleşmenin Odağı* (Istanbul: IRCICA, 2010).

53. Ekmeleddin İhsanoğlu, "Tanzimat Döneminde İstanbul'da Darülfünûn Kurma Teşebbüsleri," in *150. Yılında Tanzimat*, ed. Hakkı D. Yıldız (Ankara: Türk Tarih Kurumu, 1992), 413.

54. Ahmed Lütfi, *Vak'anüvis Ahmed Lütfi Efendi Tarihi*, vol. 10, ed. Münir Aktepe (Ankara: Türk Tarih Kurumu, 1988): 132.

55. Ahmed Lütfi, *Vak'anüvis Ahmed Lütfi Efendi Tarihi*, vol. 12, ed. Münir Aktepe (Ankara: Türk Tarih Kurumu, 1989), 85.

56. Ibid.

57. Ioannes Aristoklis (1831/32–1891/92) was a graduate of the Greek schools of Istanbul; he taught ethnography in the School for Civil Servants and at the university. See Ali Çankaya, *Yeni Mülkiye Tarihi ve Mülkiyeliler*, vol. 2 (Ankara: SBF, 1969), 950–51.

58. On Afghani, see Nikki R. Keddie, *Sayyid Jamāl ad-Dīn 'al-Afghānī': A Political Biography* (Berkeley: University of California Press, 1973).

59. Mehmed Zeki Pakalın, *Safvet Paşa* (Istanbul: Ahmet Sait, 1943), 131–33.

60. Ibid., 134–35.

61. Aynî, *Darülfünûn*, 22.

62. Ibid., 30.

63. *Meclis-i Mebusan Zabıt Ceridesi*, vol. 1, ed. Hakkı Tarık Us (Istanbul: Vakit, 1939): 11.

64. The wording here is *ulûm ve fünûn-ı nafıa*; ibid., 14.

65. Ibid., 45.

66. *Meclis-i Mebusan Zabıt Ceridesi*, vol. 2, ed. Hakkı Tarık Us (Istanbul: Vakit, 1954), 145.

67. Ibid., 207.
68. The commonly used Ottoman phrase to refer to the empire.
69. *Meclis-i Mebusan Zabıt Ceridesi*, vol. 2, ed. Hakkı Tarık Us (Istanbul: Vakit, 1954), 208.
70. Ibid., 210. On the career of this future member of the Young Turk movement, see Hasan Kayalı, *Arabs and Young Turks: Ottomanism, Arabism, and Islamism in the Ottoman Empire, 1908–1918* (Berkeley: University of California Press, 1997), chap. 1.
71. Ibid., 266.
72. For examples of Coumbary's work in Istanbul, see his letters published in *Comptes Rendus* (May 29, 1865): 1114–15 and *Nature* (March 24, 1870): 538.
73. See, e.g., Ekmeleddin İhsanoğlu, "Ottoman Science: The Last Episode in Islamic Scientific Tradition and the Beginning of European Scientific Tradition," in *Science, Technology, and Industry in the Ottoman World. Proceedings of the XXth International Congress of History of Science*, vol. 6, ed. E. İhsanoğlu et al. (Turnhout: Brepols, 2000), 40–41.
74. Muammer Dizer, "Rasathane-i Amire," *Bilim Tarihi* 2 (1993): 3–10.
75. Mithat Cemal Kuntay, *Namık Kemal, Devrinin İnsanları ve Olayları Arasında*, vol. 2–1 (Istanbul: Milli Eğitim Basımevi, 1949), 132. That Coumbary is referred to as an Ottoman Greek rather than a Frenchman even in several scholarly sources (see n. 74 and Roderic Davison, "The Advent of the Electric Telegraph in the Ottoman Empire," in *Essays in Ottoman and Turkish History, 1774–1923: The Impact of the West* [Austin: University of Texas Press, 1990], 150) seems indicative of how he was perceived at the time by Muslim Ottomans themselves. It was probably assumed that non-Muslim Ottomans were the natural intermediaries between the empire and European centers in scientific matters, just as in commercial ones.
76. Ali Efendi, "Şehir Mektubu," *Basiret* (14 Receb 1288/September 29, 1871): 1–2. Reproduced in Basiretçi Ali Efendi, *Şehir Mektupları*, 61–62.
77. Ali Efendi, "Şehir Mektubu," *Basiret* (13 Şaban 1288/October 28, 1871): 2. Reproduced in Basiretçi Ali Efendi, *Şehir Mektupları*, 69–70.

CHAPTER SIX

1. See Selim Deringil, "Legitimacy Structures in the Ottoman State: The Reign of Abdulhamid II, 1876–1909," *International Journal of Middle Eastern Studies* 23 (1991): 346. To this end, Abdülhamid revived the title of the caliph. Similarly, extravagant ceremonies were designed to celebrate the arrival of recently "discovered" holy relics in Istanbul and the departure of the sultan's gifts to the holy cities of Mecca and Medina. For additional examples, see Selim Deringil, *The Well-Protected Domains: Ideology and the Legitimation of Power in the Ottoman Empire, 1876–1909* (London: Tauris, 1998), and Kemal Karpat, *The Politicization of Islam: Reconstructing Identity, State, Faith, and Community in the Late Ottoman State* (New York: Oxford University Press, 2001).
2. See Dale Eickelman, "Mass Higher Education and the Religious Imagination

in Contemporary Arab Societies," *American Ethnologist* 19 (1992): 643–55. Benjamin Fortna makes a similar observation in his *Imperial Classroom: Islam, the State, and Education in the Late Ottoman Empire* (Oxford: Oxford University Press, 2002), 23.

3. Benjamin Fortna, "Islamic Morality in Late Ottoman 'Secular' Schools," *International Journal of Middle Eastern Studies* 32 (2000): 375.

4. Fortna, *Imperial Classroom*, 215.

5. Ibid., 219.

6. Report reproduced in Atilla Çetin, "Maarif Nazırı Ahmed Zühdü Paşa'nın Osmanlı Imparatorluğundaki Yabancı Okullar Hakkındaki Raporu," *Güneydoğu Avrupa Araştırmaları Dergisi* 10 (1981), 196.

7. Carter Findley, "An Ottoman Occidentalist in Europe: Ahmed Midhat Meets Madame Gülnar, 1889," *American Historical Review* 103 (1998): 21.

8. Ahmed Midhat Efendi, "Islamiyet ve Fünûn," *Tercüman-ı Hakikat* (23 Receb 1300/May 30, 1883): 2.

9. On al-Jisr, see Marwa Elshakry, *Reading Darwin in Arabic, 1860–1950* (Chicago: University of Chicago Press, 2014), and Adel Ziadat, *Western Science in the Arab World: The Impact of Darwinism, 1860–1930* (London: Macmillan, 1986), 95. For an analysis of Midhat's work on Draper, see M. Alper Yalçınkaya, "Science as an Ally of Religion: A Muslim Appropriation of the 'Conflict Thesis,'" *British Journal for the History of Science* 44 (2011): 161–81.

10. See Marwa Elshakry, "Exegesis of Science in 20th Century Arabic Interpretations," in *Nature and Scripture in the Abrahamic Religions: 1700–Present*, vol. 2, ed. Jitse van der Meer and Scott Mandelbrote (Leiden: Brill, 2008), 500. Also see Elshakry, *Reading Darwin in Arabic*.

11. Ayşe Osmanoğlu, *Babam Abdülhamid* (Istanbul: Güven, 1960), 22.

12. *Tercüman-ı Hakikat* (11 Zilkade 1296/October 27, 1879).

13. "Leh ve Aleyh," *Vakit* (13 Zilkade 1296/October 29, 1879): 3.

14. "Araplarda neler var imiş?" *Tercüman-ı Hakikat* (14 Zilkade 1296/October 30, 1879): 4.

15. "İlerliyor muyuz, geriliyor muyuz?" *Tercüman-ı Hakikat* (19 Zilkade 1296/November 4, 1879): 3.

16. Şemseddin Sami, *Medeniyyet-i İslamiyye* (Istanbul: Mihran, 1296/1879), 14.

17. See Ahmed Midhat, *Felsefe Metinleri*, ed. Erdoğan Erbay and Ali Utku (Erzurum: Babil, 2002), 289.

18. Şemseddin Sami, *Medeniyyet*, 13.

19. Şemseddin Sami, "Medeniyet," *Hafta* (19 Zilkade 1298/October 13, 1881): 129–32; (26 Zilkade 1298/October 20, 1881): 145–49; (3 Zilhicce 1298/October 27, 1881): 161–65; "Medeniyet-i Cedidenin Ümem-i İslamiye'ye Nakli" *Güneş* 1, no. 4 (1301/1883–84): 179–85. These articles were recently transcribed into the Latin alphabet and published. See Zeynep Süslü and Ismail Kara, "Şemseddin Sami'nin 'Medeniyet'e Dair Dört Makalesi," *Kutadgu Bilig* 4 (2003): 259–83. The English version of the last one, translated by Şükrü Hanioğlu, can be found under the title "Transferring the New Civilization

to the Islamic Peoples," in *Modernist Islam, 1840–1940*, ed. Charles Kurzman (New York: Oxford University Press, 2002), 149–51.

20. Şemseddin Sami, "Medeniyet-i Cedidenin," 173.

21. Ibid., 179–81. Translation based on Hanioğlu's in *Modernist Islam*, ed. Kurzman.

22. Şemseddin Sami, "Medeniyet-i Cedidenin," 183.

23. Ahmed Midhat, "Osmanlı Kütüb-i Atikasını Karıştırdıkça Frenklere Gülmemek Elimden Gelmiyor," *Dağarcık* no. 8 (1288/1873): 256.

24. Ahmed Rasim, "Ifade-i Mahsusa" (Preface), *Fonoğraf: Sadayı Tahrir ve Iade Eden Alet*, trans. Ahmed Rasim (Istanbul: K. Bağdadlıyan, 1302/1884–85), [i–ii].

25. For an analysis of the literary and linguistic aspects of this debate, see Kazım Yetiş, "Cevdet Paşa'nın Belâgât-ı Osmaniyyesi ve Uyandırdığı Akisler," in *Belagatten Retoriğe* (Istanbul: Kitabevi, 2006), 246–99. Hacı İbrahim's life and writings are discussed in detail in Musa Aksoy, *Moderniyete Karşı Geleneğin Direnişi: Hacı İbrahim Efendi* (Ankara: Akçağ, 2005).

26. Ahmed Midhat, "Belâgât-ı Osmâniye," *Tercüman-ı Hakikat* (8 Rebiülahir 1299/ February 27, 1882): 3.

27. Abdurrahman Süreyya, "Belâgât-ı Osmâniye," *Tercüman-ı Hakikat* (16 Rebiülahir 1299/March 7, 1882): 2–3.

28. Ibrahim's response started in the "Mevadd-ı Lisâniyye" (Linguistic Matters) section of *Tercüman-ı Hakikat* (1 Mart 1298/March 13, 1882): 3–4. For this specific remark, see the third part of the response, *Tercüman-ı Hakikat* (4 Mart 1298/March 16, 1882): 3.

29. Ahmed Midhat, "Belâgât-ı Osmâniye: Ibrahim Efendi Hazretlerine Mukabele" *Tercüman-ı Hakikat* (12 Mart 1298/24 March 1882): 2–3. A highly relevant issue in this context involves the translation of scientific terms into Ottoman Turkish. While some Ottoman men of science saw Arabic as a resource from which they could draw words and phrases in order to construct scientific terms—like the use of Latin in the language of European science—the success of this initiative remained controversial. On this important topic that requires a monograph in itself, see Feza Günergun, "Ondokuzuncu Yüzyıl Türkiye'sinde Kimyada Adlandırma," *Osmanlı Bilimi Araştırmaları* 1 (2003): 1–33; Nil Sarı, "Cemiyet-i Tıbbiyye-i Osmaniyye ve Tıp Dilinin Türkçeleşmesi Akımı," in *Osmanlı İlmî ve Meslekî Cemiyetleri*, ed. Ekmeleddin İhsanoğlu (Istanbul: Edebiyat Fakültesi, 1987), 121–42.

30. Hacı İbrahim, "Mevadd-ı Lisâniyye," *Tercüman-ı Hakikat* (20 Mart 1298/ April 1, 1882): 3.

31. Hacı İbrahim, "Teşekkür ve temenni," *Tercüman-ı Hakikat* (31 Mart 1298/ April 12, 1882): 3.

32. On this debate see chap. 4.

33. Muallim Naci, *Musa bin Ebu'l Gazan, yahud Hamiyyet* (Istanbul: Matbaa-i Ebuzziya, 1299/1881–82), 7–8.

34. Ebuzziya Tevfik, "Mütenevvia," *Tercüman-ı Hakikat* (28 Ramazan 1299/ August 13, 1882): 3.

35. Musa Akyiğitzade, *Avrupa Medeniyetinin Esasına Bir Nazar* (Istanbul: Cemal Efendi Matbaası, 1315/1897–98), 7.

36. Halid Eyüb Yenişehirlizade, *Islam ve Fünûn* (Istanbul: Cemal Efendi Matbaası, 1315/1897–98).

37. Mehmed Tahir, *Türklerin Ulûm ve Fünûna Hizmetleri* (Istanbul: Ikdam, 1314/1896–97).

38. Bahai [Veled Çelebi], "Arabdan Pek Çok Istifade Edeceğimiz Ulûm Var," *Tarik* (3 Teşrinisani 1314/November 15, 1898): 3.

39. Hüseyin Cahid, "Arabdan Istifade Edeceğimiz Ulûm," *Tarik* (27 Teşrinisani 1314/December 9, 1898): 4.

40. See Benjamin A. Elman, "'Universal Science' Versus 'Chinese Science': The Changing Identity of Natural Studies in China, 1850–1930," *Historiography East and West* 1 (2003): 68–116, esp. 92–93, and Hongming Ma, *The Images of Science through Cultural Lenses: A Chinese Study on the Nature of Science* (Rotterdam, Netherlands: Sense, 2012), 29–30.

41. Gyan Prakash, *Another Reason: Science and the Imagination of Modern India* (Princeton, NJ: Princeton University Press, 1999): 64–71.

42. Mustafa Sabri, "Cür'etli Bir Dekadan," *Malûmat* (3 Kanunıevvel 1314/December 15, 1898): 264–65.

43. "Gayret-i Ciddiye," *Âfak* no. 2 (1 Muharrem 1300/November 12, 1882): 61–62.

44. Nef'i (1572–1635) and Nedim (1681–1730) are two prominent representatives of classical Ottoman poetry.

45. Namık Kemal, *Namık Kemal'in Türk Dili ve Edebiyatı Üzerine Görüşleri ve Yazıları*, ed. Kazım Yetiş (Istanbul: İÜ Edebiyat Fakültesi, 1989), 4.

46. Cyrus Schayegh, *Who Is Knowledgeable Is Strong: Science, Class, and the Formation of Modern Iranian Society, 1900–1950* (Berkeley: University of California Press, 2009), 33–35.

47. Necib, "Mevadd-ı Fenniye," *Tercüman-ı Hakikat* (18 Cemaziyelevvel 1300/March 27, 1883): 3.

48. "Edebiyat," *Saadet* (5 Zilhicce 1302/September 14, 1885): 3.

49. "Teşrîh-i Eş'âr," *Saadet* (7[8] Zilhicce 1302/September 17, 1885): 2.

50. Ibid., 3.

51. "Teşrîh-i Eş'ar," *Saadet* (1 Muharrem 1303/October 10, 1885): 3. For a similar remark, see "Teşrih-i Eş'ar," *Saadet* (25 Zilhicce 1302/October 5, 1885): 3.

52. "Harputlu Hayri Efendi," *Saadet* (20 Zilhicce 1302/October 30, 1885): 3.

53. "Teşrîh-i Eş'ar," *Saadet* (4 Muharrem 1303/October 13, 1885): 3.

54. "Teşrîh-i Eş'ar," *Saadet* (5 Muharrem 1303/October 14, 1885): 3.

55. "Aynen Varaka," *Saadet* (7 Muharrem 1303/October 16, 1885): 2–3.

56. "Latife-i Edebiyye: Şiir-i Fennî," *Saadet* (24 Safer 1303/December 2, 1885): 3.

57. Faik Hilmi, [untitled], *Saadet* (24 Safer 1303/December 2, 1885): 3. Faik Hilmi published short articles on new scientific developments in the journal *Armağan Dağarcığı* and is also the author of a short novel named *Cep Defteri*,

"The Pocket Book." In this novel, the protagonist, who, as a child, had learned the new chemistry, talks condescendingly about his father who spent his time with mysticists, thought chemistry was the same as alchemy, and that Ibn Sina (Avicenna) was an alchemist. See Faik Hilmi, *Cep Defteri* (Istanbul: Kitapçı Tahir Efendi, 1305/1887–88), 7–13.

58. Mehmed Celal, "Aynen," *Saadet* (27 Safer 1303/December 5, 1885): 3.

59. Beşir Fuad, "Victor Hugo," in *Şiir ve Hakikat,* ed. Handan Inci (Istanbul: YKY, 1999), 145. This compilation includes many of the relevant writings of Fuad and the other participants of this dispute.

60. Tahir and Fuad were the coeditors of an earlier scientific journal, the *Haver.* Their collaboration on this short-lived journal most likely ended over a disagreement on whether or not there should be room for religious sciences in this journal of science.

61. M. M. Tahir, "Viktor Hugo," *Gayret* no. 6 (7 Şubat 1301/February 19, 1886): 24; also in Beşir Fuad, *Şiir ve Hakikat,* 169–70.

62. Beşir Fuad, "'Hugo' Unvanlı Makale-i İntikadiyeye Mukabeleden Mabâd," *Saadet* (4 Zilkade 1303/August 4, 1886): 2–3; also in Beşir Fuad, *Şiir ve Hakikat,* 187–88.

63. Tahir elaborated on this in two essays, "Beşir Fuad Beyefendi'nin Victor Hugo Unvanlı Eserine Dair Yazdığım Makaleye Mukabil Saadet Gazetesiyle Neşreyledikleri Varakaya Cevaptır," *Gayret* no. 30 (22 Agustos 1302/September 3, 1886): 118–20, and *Gayret* no. 31 (5 Eylül 1302/September 17, 1886): 122–23; also in Beşir Fuad, *Şiir ve Hakikat,* 198–202.

64. Beşir Fuad, "Menemenlizade Tahir Beyefendi'nin *Gayret*'in 29, 30, 31, 33 Numrolu Nüshalarındaki Makale-i Cevabiyeye Cevab," *Saadet* (18 Teşrinisani 1302/November 30, 1886): 3; also in Beşir Fuad, *Şiir ve Hakikat,* 227.

65. M. C., "Bir Mütefenninle Bir Şair," *Gayret* no. 26 (27 Haziran 1302/July 9, 1886): 102; also in Beşir Fuad, *Şiir ve Hakikat,* 243–44.

66. Beşir Fuad, "Yetmiş Bin Beyitli Bir Hicviye," *Saadet* (22 Zilkade 1303/August 22, 1886): 2–3, and *Saadet* (23 Zilkade 1303/August 23, 1886): 2–3; also in Beşir Fuad, *Şiir ve Hakikat,* 246.

67. "Kudret-i Fatıra"—a term used to refer to God in Islamic texts.

68. Beşir Fuad, "Çevir Kazı Yanmasın," *Saadet* (15 Zilhicce 1303/September 14, 1886): 3; Also in Beşir Fuad, *Şiir ve Hakikat,* 266.

69. Ahmed Midhat, "Fen ve Şiir, ve Şiir-i Fenni," *Tercüman-ı Hakikat* (2 Zilhicce 1303/September 1, 1886): 3.

70. Tahir Kenan, "Vukuat-ı Elektrikiyye ve Ahval-i Bedriyye," *Tercüman-ı Hakikat* (3 Zilhicce 1303/September 2, 1886): 3.

71. Namık Kemal, "Ebuzziya Tevfik Bey Biraderime," *Mecmua-i Ebuzziya* 52 (10 Muharrem 1304/October 9, 1886): 1636; also in Beşir Fuad, *Şiir ve Hakikat,* 312.

72. Salahi, "Nazire," *Saadet* (22 Rebiülahir 1304/January 18, 1887); Beşir Fuad, "Gazel," *Tercüman-ı Hakikat* (4 Cemaziyelevvel 1304/January 29, 1887): 3; also in Beşir Fuad, *Şiir ve Hakikat,* 334–35.

73. Zülfikar, "Aynen Varaka," *Saadet* (February 2, 1887); also in Beşir Fuad, *Şiir ve Hakikat*, 338.

74. "Terbiye—Tahsil," *Tercüman-ı Hakikat* (11 Cemaziyelahir 1304/March 7, 1887): 2. The word *terbiye* indicates a good upbringing that provides sound morals and discipline in addition to mere education. I have translated it as "disciplining," as in Somel's *Modernization*, but with an emphasis on moral education.

75. For similar examples that criticized Fuad more directly, see Orhan Okay, *Beşir Fuad: İlk Türk Pozitivist ve Natüralisti* (Istanbul: Hareket, 1969), 96–100.

76. For an example, see Okay, *Beşir Fuad*. For a sophisticated analysis of the intellectual influences of Fuad that examines the impact of scientism and vulgar materialism on his thought, see Şükrü Hanioğlu, "Blueprints for a Future Society: Late Ottoman Materialists on Science, Religion, and Art," in *Late Ottoman Society: The Intellectual Legacy*, ed. Elisabeth Özdalga (London: Routledge, 2005).

77. Ahmed Midhat Efendi, *Beşir Fuad* (Istanbul: Oğlak, 1996 [1305/1888]), 14–15.

78. Ibid., 23. Compare this to Midhat's own confession to none other than Fuad himself—see n. 17. It is worth noting that while Beşir Fuad devoted several pages to the "conflict" between Christianity and progress, he also condemned "Christian slanders" that Muslims burned down the Library of Alexandria, possibly to avoid accusations of atheism. See Beşir Fuad, *Şiir ve Hakikat*, 90–92.

79. Ahmed Midhat, *Beşir Fuad*, 69.

80. Okay, *Beşir Fuad*, 99.

81. BOA İrade Dahiliye 1278/100584, 13 Zilkade 1309/June 9, 1892. Documents expressing such concerns abound in the Ottoman archives. For examples, see BOA Dahiliye Mektubi, 13 Zilkade1310/May 29, 1893; Maarif Mektubi 272/44, 15 Muharrem 1313/July 8, 1895; Yıldız Hususi Maruzat 334/103, 29 Safer 1313/August 21, 1895.

82. See, e.g., a document about a book on religion and science that was not to be imported, whereas another book entitled *The Harmony between Religion and Science* could be allowed (BOA Maarif Mektubi 190/100, 13 Cemaziyelahir 1311/December 22, 1893.

83. Nabizade Nazım, "Şairiyyet," *Manzara* no. 4 (15 Nisan 1303/April 27, 1887): 44. A divan is a collection of a poet's poems written in different forms.

84. Ibid.

85. Fuad's main opponent, Menemenlizade Tahir, on the other hand, reiterated his position in the introduction to his collection of poems, and noted that while Ottoman literature should change, it did not mean that it would be nothing but "scientific." See Menemenlizade Tahir, *Yad-ı Mazi* (Istanbul: Cemal Efendi, 1304/1888), 3.

86. Nabizade Nazım, "Islamiyet ve Fünûn," *Manzara* no. 10 (15 Temmuz 1303/27 July 1887), 109.

87. Mehmed Celal, "Edebiyatta Fen," *Maarif* (13 Şubat 1307/February 25, 1892):

1–2; Mehmed Celal, "Fende Edebiyat," *Maarif* (27 Şubat 1307/March 10, 1892): 33–34. Note that Celal's suggestions to men of science are built on the well-established idea of a man of science essentially as a man who learns and teaches, not as a man who invents, investigates, or builds.

88. Mehmed Celal, "Roman Mütalaası," *Maarif* (17 Teşrinisani 1310/November 29, 1894): 18–20. That science was the future and literature the past was clearly a popular view in this period. We learn from the memoirs of the twentieth-century politician Ali Münif that in the 1890s his father, even though he himself was a poet, discouraged him from writing poetry, saying, "You should occupy yourself with issues that have to do with the means of progress. Leaving science aside and busying yourself with literature is like using a flintlock rifle when there exist modern weapons." See Ali Münif, *Ali Münif Efendi'nin Hatıraları*, ed. Taha Toros (Istanbul: Isis, 1996), 14.

89. [Faik] Reşad and Ibrahim [Aşki], *Kıraat—Beşinci Seneye Mahsus* (Istanbul: Karabet, 1895), 67–70.

90. Şerafeddin Mağmumi, *Vücud-ı Beşer* (Istanbul: Nişan Berberyan, 1310/1892–93), 10. For an analysis of the scientism of Mağmumi, see Hanioğlu, "Blueprints."

91. Şemseddin Sami, "Semerat-ı İlm," *Hafta* (1 Safer 1299/December 23, 1881): 291–92.

92. Samipaşazade Abdülbaki, "Maarif," *Tercüman-ı Hakikat* (2 Zilhicce 1298/October 26, 1881): 2.

93. Samipaşazade Abdülbaki, "Maarif," *Tercüman-ı Hakikat* (3 Zilhicce 1298/October 27, 1881): 1–2.

94. Ibid., 2.

95. Samipaşazade Abdülbaki, "Maarif," *Tercüman-ı Hakikat* (4 Zilhicce 1298/October 28, 1881): 1.

96. "Redd-i Bâtıl ve Isbat-ı Hak," *Tercüman-ı Hakikat* (14 Zilhicce 1298/November 7, 1881): 2.

97. Ibid., 3.

98. Ibid.

99. Şükrü Hanioğlu, *A Brief History of the Late Ottoman Empire* (Princeton, NJ: Princeton University Press, 2008), 140. Other relevant works by Hanioğlu are "Blueprints" (n. 76 above) and *The Young Turks in Opposition* (New York: Oxford University Press, 1995). Rather than constituting a coherent and consistent philosophy, the outlooks of these young men primarily involved a strong belief in scientific certainty, an interest in using physiological concepts to explain social phenomena, versions of physico-chemical reductionism, and a skeptical if not antagonistic attitude toward religion in the Ottoman context.

100. In the words of Hanioğlu, the Young Turks were "an extremely marginal group at the time." See Şükrü Hanioğlu, "The Historical Roots of Kemalism," in *Democracy, Islam, and Secularism in Turkey*, ed. Ahmet Kuru and Alfred Stepan (New York: Columbia University Press, 2012), 38.

1. See, e.g., Berrak Burçak, "Science, A Remedy For All Ills: Healing 'The Sick Man of Europe': A Case for Ottoman Scientism" (PhD diss., Princeton University, 2005); Said Özervarlı, "Alternative Approaches to Modernization in the Late Ottoman Period," *International Journal of Middle East Studies* 39 (2007): 102–25; and Amit Bein, "The Istanbul Earthquake of 1894 and Science in the Late Ottoman Empire," *Middle Eastern Studies* 44 (2008): 909–24. Most such studies base their analysis on Hanioğlu's pioneering works.

2. This is primarily due to the influence of the graduates of the Imperial Military Academy in these cadres. Hanioğlu's most recent work examines the impact of scientism and materialism on the views of Kemal Atatürk, the founder of the Turkish Republic. See Şükrü Hanioğlu, *Atatürk: An Intellectual Biography* (Princeton, NJ: Princeton University Press, 2011).

3. Nabizade Nazım, "Elektrikiyyetin Sevdaya Tatbiki," *Manzara* (1 Mart 1303/ March 13, 1887): 8. This representation is also an example of the materialistic approaches popular among students of the military and medical schools in this period.

4. Sadullah Paşa, "Ondokuzuncu Asır," *Мecmua-ı Ebuzziya* no 46 (15 Cemaziyelevvel 1302/April 1, 1885): 1453–55.

5. Mahmud Es'ad, *Tarih-i Sanayi* (No publisher, 1307/1890). In the brief reference to Darwin, the hypothesis regarding "progress from the simple to the complex" is presented as enabling a synthesis of hitherto unconnected findings in the biological sciences. Ibid., 472.

6. Safvet, "Müstakbelde Fen," *Maarif* (14 Cemaziyelahir 1309/January 15, 1892): 327–28.

7. Ali Muzaffer, "Fennin Mahiyeti," *Hazine-i Fünûn* (21 Muharrem 1311/August 4, 1893): 29.

8. Shumayyil was from a Christian family, and his materialism and antireligious arguments were closely related to his advocacy for the construction of a new Ottoman identity that would not be based on religion. On Shumayyil, see Marwa Elshakry, *Reading Darwin in Arabic, 1860–1950* (Chicago: University of Chicago Press, 2014).

9. For a succinct discussion on these policies, see Şükrü Hanioğlu, *A Brief History of the Late Ottoman Empire* (Princeton, NJ: Princeton University Press, 2008), 125–29.

10. Abdülhamid II was vilified by much literature in the twentieth century for the censorship that was identified with his reign. It is indeed crucial to note that Abdülhamid's institution of censorship got particularly determined, effective, and comprehensive after the 1880s. But we should remember that the general framework of the system had already been established by the Tanzimat bureaucrats in the late 1860s. Similarly, just as he was a staunch supporter of a Westernized educational system if it was organized in a way

that would serve his purposes, Abdülhamid was also too aware of the power of the press to completely prohibit it. Instead, Hamid's censors were to make sure to render the press very effective when it came to bolstering the image of the sultan. Moreover, many journalists took advantage of the financial implications of being favored by the sultan. For a study that demonstrates the complexity of the relations between the Palace and the press in this period, see Ebru Boyar, "The Press and the Palace: The Two-Way Relationship between Abdülhamid II and the Press, 1876–1908," *Bulletin of the School of Oriental and African Studies* 69 (2006): 417–32.

11. Most texts were written in the form of didactic dialogues—a format used commonly in Islamic judicial documents. The contents of the texts were also based exclusively on Islamic principles and justifications. See Benjamin Fortna, "Islamic Morality in Late Ottoman 'Secular' Schools," *International Journal of Middle Eastern Studies* 32 (2000).

12. Roy MacLeod, "The 'Bankruptcy of Science' Debate: The Creed of Science and Its Critics, 1885–1900," *Science, Technology, and Human Values* 7 (1982): 2–15.

13. Phyllis Stock-Morton, *Moral Education for a Secular Society: The Development of Morale Laïque in Nineteenth-Century France* (Albany: State University of New York Press, 1988); Daniela S. Barberis, "Moral Education for the Elite of Democracy: The *Classe de Philosophie* between Sociology and Philosophy," *Journal of the History of the Behavioral Sciences* 38 (2002): 355–69.

14. Jeffrey Brooks, *When Russia Learned to Read: Literacy and Popular Literature, 1861–1917* (Princeton, NJ: Princeton University Press, 1985); Paul Bailey, *Reform the People: Changing Attitudes towards Popular Education in Early 20th-Century China* (Edinburgh: Edinburgh University Press, 1990). As Fortna suggests in "Islamic Morality," these examples indicate the importance of studying these policies within "world time."

15. "Mekteb-i Tıbbiye-i Şahane talebesinden birkaç zat imzalarıyle matbaamıza tebliğ olunan varakanın aynıdır," *Tercüman-ı Hakikat* (13 Ramazan 1298/August 9, 1881): 2.

16. On Remzi, see Ekrem Kadri Unat, "Muallim Miralay Dr. Hüseyin Remzi Bey ve Türkçe Tıp Dilimiz," in *IV. Türk Tıp Tarihi Kongresi Kitabı* (Ankara: TTK, 2003): 239–52.

17. Hüseyin Remzi, *İlmihal-i Tıbb* (Istanbul: Mahmud Bey Matbaası, 1305/1887–88), 5.

18. Ibid., 9.

19. Compare this to the arguments of Münif Pasha, who determinedly used the concept *ilm* to refer to all of the new sciences (chap. 3).

20. Mehmed Fahri, *Hıfzıssıhhat-ı Siyam ve Erkân-ı Sâire-i Din-i İslam* (Istanbul: Mahmud Bey Matbaası, 1308/1890–91), 5.

21. Mehmed Fahri *Yâdigâr-ı Hekîm* (Istanbul: Mahmut Bey, 1317/1899).

22. Marwa Elshakry, "Exegesis of Science in 20th-Century Arabic Interpretations," in *Nature and Scripture in the Abrahamic Religions: 1700–Present*, ed.

Jitse van der Meer and Scott Mandelbrote, vol. 2 (Leiden: Brill, 2008): 491–523. See Elshakry, *Reading Darwin in Arabic* for more on these authors.

23. On the representatives of the "scientific exegesis" approach, see Muzaffar Iqbal, *Science and Islam* (Westport, CT: Greenwood, 2007): 151–53.

24. Hüseyin Hulki, *Siyam* (Istanbul: Mihran, 1893 [1310]), 4–5.

25. Hüseyin Hulki also translated Nurican Efendi's *Aperçu historique sur la médicine arabe*, and wrote for this work on Arab medicine a passionate introduction about Arabs' contributions to science. For Ottomans, Hulki wrote, it was the greatest honor to be coreligionists with the Arabs, and that no effort to praise the Arabs could be considered sufficient. See Nurican Efendi, *Müslümanların Tababete Ettikleri Hizmet, yahud Sevabık-ı Maarifimizden Bir Nebze*, trans. Hüseyin Hulki (Istanbul: Mihran, 1883 [1300]).

26. Veli Behçet Kurdoğlu, ed., *Şair Tabibler* (Istanbul: Baha, 1967), 273.

27. Mustafa Şevket, *Kitab-ı Burhan-ı Hakikat* (Istanbul: Mihran, 1299/1882–83), 9.

28. Ibid., 38–39.

29. Ibnülemin M. K. Inal, *Son Asır Türk Şairleri* (Istanbul: Devlet Matbaası, 1939) 8: 1532.

30. Ibnülemin M. K. Inal, *Son Asır Türk Şairleri* (Istanbul: Devlet Matbaası, 1938) 6: 1150.

31. Ibnülemin M. K. Inal, *Son Asır Türk Şairleri* (Istanbul: Devlet Matbaası, 1932) 3: 507.

32. Bir Mektebli, "Müteahhirinin Mütekaddimin Üzerine Olan Fazl ve Meziyeti," *Tercüman-ı Hakikat* (21 Ramazan 1299/August 6, 1882): 2–3. It is important to note that *mektebli* is the word used for the students of the new schools, not the *medrese*s.

33. Mahmud Esad and Ali Sedad, "El-fazl-ı-li'l-mütekaddim," *Vakit* (2 Şevval 1299/August 17, 1882): 3. Mahmud Esad (1857–1917), the son of a prominent ulema, was a former *medrese* student who was at the time a student of the School of Law (Mekteb-i Hukuk). He later became a leading educator, bureaucrat, and a member of the Ottoman parliament after 1908. Thanks to his familiarity with both the religious and the new sciences, he wrote many texts on both and became a proponent of the "Islam as a pro-science religion" discourse. Ali Sedad (1857–1900) was the son of Cevdet Pasha, one of the most influential statesmen of the Tanzimat Era and the author of a book on classical logic dedicated to this son, among many other works (see chap. 2). Sedad was educated mostly by private tutors, and in 1886 he published the first book on logic in Ottoman Turkish that discussed contemporary debates.

34. Abdülhekim Hikmet, "Mevadd-ı Fenniye: Reddiye," *Tercüman-ı Hakikat* (12 Şevval 1299/August 27, 1882): 3.

35. Mahmud Esad and Ali Sedad, "Mevadd-ı Fenniye: Hikmet Efendi'nin Reddiyesine Cevabdır," *Tercüman-ı Hakikat* (18 Şevval 1299/September 2, 1882): 3.

36. Abdülhekim Hikmet, "Mevadd-ı Fenniye," *Tercüman-ı Hakikat* (22 Şevval 1299/September 6, 1882): 3.

37. Cornelius van Dyck (1818–95) was a missionary in Syria most of his life. He taught at the Syrian Protestant College, translated the New Testament into Arabic, and published many essays on the sciences. This particular essay, written in Arabic, was published in 1852 in the *Transactions of the Syrian Society of Arts and Sciences*. On van Dyck, also see Elshakry, *Reading Darwin in Arabic*.

38. Mahmud Esad and Ali Sedad, "Mevadd-ı Fenniye," *Tercüman-ı Hakikat* (21 Zilkade 1299/October 4, 1882): 2.

39. Abdülhekim Hikmet, "Mevadd-ı Fenniye," *Tercüman-ı Hakikat* (9 Zilhicce 1299/October 22, 1882): 2–3.

40. Two conditions referred to in traditional Arabic and Persian medicine that are commonly associated with leprosy.

41. Refet Hüsameddin, "Mevadd-ı Fenniye," *Tercüman-ı Hakikat* (29 Zilhicce 1299/November 11, 1882): 3.

42. Mehmed Fahri, "Isbat-ı Hakikat," *Tercüman-ı Hakikat* (4 Muharrem 1300/November 15, 1882): 3.

43. Mahmud Esad and Ali Sedad, "Mevadd-ı Fenniye," *Tercüman-ı Hakikat* (13 Muharrem 1300/November 24, 1882): 3.

44. On this theme, see David Livingstone, *Putting Science in Its Place: Geographies of Scientific Knowledge* (Chicago: University of Chicago Press, 2003).

45. Quoted in Elshakry, *Reading Darwin in Arabic*, 56. More examples can be found in Elshakry, chap. 1.

46. See M. Alper Yalçınkaya, "Science as an Ally of Religion: A Muslim Appropriation of the 'Conflict Thesis,'" *British Journal for the History of Science* 44 (2011): 161–81.

47. Şemseddin Sami, "Terakki ve Maarif," *Hafta* (22 Ramazan 1298/August 18, 1881): 4.

48. Şemseddin Sami, "Hikmet-i Tabiiyye ve Kimyanın Mahiyet ve Ehemmiyeti," *Hafta* (22 Ramazan 1298/August 18, 1881): 9.

49. Şemseddin Sami, *Gök* (Istanbul: Mihran, 1296/1879), 9.

50. Ibid., 12–13.

51. Ibid., 34–35. On this theory, see Blair Nelson, "'Men Before Adam!': American Debates over the Unity and Antiquity of Humanity," in *When Science and Christianity Meet*, ed. David C. Lindberg and Ronald Numbers (Chicago: University of Chicago Press, 2003): 161–81.

52. Şemseddin Sami, *Yer* (Istanbul: Mihran, 1296/1879).

53. Şemseddin Sami, "Ahlak-ı umumiye," *Hafta* (12 Zilkade 1298/October 6, 1881): 113–17.

54. Mehmed Nadir (1856–1927), a graduate of the Imperial Military Academy, was a prominent mathematician and the founder of one of the most successful and prestigious private schools in the Ottoman Empire (the Numune-i Terakki). Nadir's involvement with the Young Turks cost him his school, however.

55. Mehmed Nadir, "Fennin Asar-ı Celilesi," *Mirat-ı Alem* 16 (1299/1882): 254.

56. "Mukaddime," *Hadika-i Maarif* 1 (1299/1882): 5.

57. See, e.g., the first issues of *Manzara* (1 Mart 1303/March 13, 1887): 3, and *Umran* (21 Eylül 1303/October 3, 1887): 2.

58. Ahmed Hamdi Efendi, *Nasayihü'ş-şübbân* (Istanbul: Mehmed Esad, 1298/1880–81), 4.

59. Mahmud Esad, *Tarih-i Sanayi* (Izmir: Hidmet, 1307/1889–90): 504.

60. Mustafa Zihni, trans., *Mikyasü'l-Ahlak* (Istanbul: Alem, 1315/1897–98): 6–8.

61. Sadreddin Şükrü, *Nuhbetü'l-Fezail* (Istanbul: Alem, 1313/1895–96), 6.

62. Rifat, *Fezail-i Ahlak* (Istanbul: Mekteb i Sanayi, 1311/1893–94), 9.

63. Ali İhsan Eğribozi, *Çocuklara Talim-i Fezail-i Ahlak* (Istanbul: Kütübhane-i Cihan, 1312/1894–95), 76–77.

64. Ibid., 28.

65. [Faik] Reşad and Ibrahim [Aşkî], *Kıraat—Dördüncü Seneye Mahsus* (Istanbul: Karabet, 1313/1895–96), 12–13.

66. Muallim Naci, *Mekteb-i Edeb, Birinci Kısım* (Istanbul: Kitapçı Arakel, (1310/1892–93), 63.

67. M. R., "Mukaddime," *Etfal* 7 (23 Mayıs 1291/June 4, 1875): 1.

68. "Ulviyyet," *Vasıta-i Terakki* 1 (16 Nisan 1298/April 28, 1881): 2.

69. "Çocuklara Mütalaa," *Vasıta-i Terakki* 1 (16 Nisan 1298/April 28, 1881): 3.

70. "Tahsil-i İlm ü Marifet," *Çocuklara Mahsus Gazete* (9 Ramazan 1314/February 11, 1897): 1.

71. Ahmed Rasim, *Kıraat Kitabı* (Istanbul: Karabet, 1306/1888–89).

72. *Etfal* (8 Rebiyülahir 1303/January 14, 1886).

73. "Kâğıdhane karyesi mektebi," *Tercüman-ı Hakikat* (1 Zilkade 1306/June 29, 1889): 3. For very similar remarks made by students or educators at such ceremonies, see *Vakit* (14 Zilkade 1297/October 18, 1880): 3; *Tarik* (11 Receb 1301/May 7, 1884): 3; *Tarik* (22 Şaban 1301/June 16, 1884): 3; *Tarik* (15 Ramazan 1302/June 28, 1885): 1.

74. Ahmed Rasim, trans., *Fonograf*, i.

75. Necib Asım, *Kıraat-ı Fenniye* (Istanbul: Karabet, 1305/1887–88): 1.

76. İsmail Cenabî, trans., *İlm-i Eşya* (Istanbul, 1302/1884–85), [2].

77. Ömer Subhi, trans., *Coğrafya-yı Hikemî* (Istanbul: Ceride-i Askeriye, 1301/1883–84): 2–3.

78. M. Münşi, *Bedraka-i Mühendisîn* (Istanbul: Mihran, 1302/1884–85), 1.

79. Besim Ömer, *Dişlerin Hıfzıssıhhati* (Istanbul: Mihran, 1301/1883–84), 2. For further examples, see, among others, Mehmed Rakım and Mustafa Nail, *Hayat-ı Düvel* (Istanbul: Maviyan, 1306/1888–89); Haydar Daniş, trans., *Medhal-i İlm-i Heyet yahud Kozmografya* (Istanbul: Matbaa-i Osmaniye, 1307/1892–93); and Nişan Berberyan, trans., *Kimya* (Istanbul: Nişan Berberyan Matbaası, 1309/1891–92).

80. Mehmed Izzet, "Mekteb," *Gülşen* 3 (13 Şubat 1301/February 25, 1886): 10.

81. (Mizancı) Mehmed Murad, *Turfanda mı Turfa mı?* (Istanbul: Mahmud Bey, 1308/1890–91). The title is usually translated as "First Fruits or Forbidden Fruits?" or "The Early or the Spoiled Seed?"

82. Ibid., 124.

83. Ibid., 130–31.
84. Hüseyin Rahmi, *İffet* (Istanbul: Ikdam, 1314/1896–97), 12.
85. Ibid., 13–14.
86. Ahmed Midhat, *Acaib-i Alem* (Ankara: TDK., 2000 [1299/1882–83]), 19.
87. Ibid., 192.
88. Ibid., 319.

CONCLUSION

1. The Ottoman Empire had relied on ad hoc envoys for diplomatic contacts until this period. While permanent Ottoman embassies would be opened in Europe only at the end of the eighteenth century, Yirmisekiz Mehmed's long visit to France can be regarded as the first step toward this development. For Mehmed Çelebi's report, see G. Veinstein, ed., *Le Paradis des Infidèles: Relation de Yirmisekiz Çelebi Mehmed Efendi, Ambassadeur Ottoman en France sous la Régence* (Paris: La Découvert, 2004.) Also see Fatma Müge Göçek, *East Encounters West: France and the Ottoman Empire in the Eighteenth Century* (Oxford: Oxford University Press, 1995).
2. For an examination of Mehmed Çelebi's observations from a science studies perspective, see Berna Kılınç, "Yirmisekiz Mehmed Çelebi's Travelogue and the Wonders That Make a Scientific Centre," in *Travels of Learning: A Geography of Science in Europe*, ed. Ana Simões et al. (Dordrecht: Kluwer, 2003), 77–100. Also see Feza Günergun, "The Ottoman Ambassador's Curiosity Coffer: Eclipse Prediction with De La Hire's "Machine" Crafted by Bion of Paris," *Science between Europe and Asia, Boston Studies in the Philosophy of Science* 275 (2011): 103–23.
3. On the link between knowledge (*ilm*) and power in Islamic political thought and practice, see Anthony Black, *The History of Islamic Political Thought from the Prophet to the Present* (Edinburgh: Edinburgh University Press, 2011).
4. Pierre Bourdieu, "Rethinking the State: Genesis and Structure of the Bureaucratic Field," in *State/Culture: State-Formation after the Cultural Turn*, ed. George Steinmetz (Ithaca, NY: Cornell University Press, 1999), 71.
5. The wording here is *ulûm-ı lâzıme* and *fünûn-ı nâfia*.
6. Marwa Elshakry, "When Science Became Western: Historiographical Reflections," *Isis* 101 (2010): 98–109. Unlike the cases Elshakry describes, however, in the Ottoman Empire of the nineteenth century, the traditional institutional setting of "prestigious knowledge," the *medrese* system, was no longer as effective as it used to be. Consequently, its products had a more limited role to play in the Ottoman debate on science.
7. See Cemil Aydın, *The Politics of Anti-Westernism in Asia: Visions of World Order in Pan-Islamic and Pan-Asian Thought* (New York: Columbia University Press, 2007). Elshakry and Aydın also note similar attitudes in China and Japan, respectively.
8. This is why Şemseddin Sami's work on the civilization of Islam was based

on European sources, just like Ziya Pasha's popular book on the history of Andalusia.

9. For a positive approach to the discourse that separates the material from the spiritual that arose in comparable colonial settings, see Partha Chatterjee, *The Nation and Its Fragments: Colonial and Postcolonial Histories* (Princeton, NJ: Princeton University Press, 1993).

10. In his memoirs the journalist Ahmet Ihsan refers to the speech that Abdurrahman Şeref, historian and the principal of the School of Administration, gave to the incoming students on the first day of school, September 7, 1882. Standing near a map of the former Ottoman territories in Europe, Şeref asserted: "Gentlemen! We have lost all these territories due to ignorance. I am so happy to see at least fifty young men in our country who can . . . understand such a map. Study, work, and become the knowledgeable officials our nation needs. This is the purpose of this school." See Ahmet İhsan Tokgöz, *Matbuat Hatıralarım 1888–1923* (Istanbul: Ahmet İhsan, 1930), 1:22.

11. See Benjamin Fortna, "Islamic Morality in Late Ottoman 'Secular' Schools," *International Journal of Middle Eastern Studies* 32 (2000): 369–93.

12. See Benjamin Fortna, "Education and Autobiography at the End of the Ottoman Empire," *Die Welt des Islams* 41 (2001): 1–31.

13. Hanioğlu's works illustrate this effectively.

14. See in particular John H. Evans, "Epistemological and Moral Conflict between Religion and Science," *Journal for the Scientific Study of Religion* 50 (2011): 707–27, and Michael S. Evans and John H. Evans, "Arguing against Darwinism: Religion, Science, and Public Morality," in *The New Blackwell Companion to the Sociology of Religion*, ed. Bryan S. Turner (Oxford: Blackwell): 286–308.

15. Science studies have now revived interest in such questions. Recent work in the sociology and history of science, and in particular, the works of Steven Shapin, have demonstrated the importance of virtue in the form of trustworthiness in the world of contemporary science. For an important study that highlights the importance of the "good life" for the philosophers of the so-called scientific revolution, see Matthew Jones, *The Good Life in the Scientific Revolution: Descartes, Pascal, Leibniz, and the Cultivation of Virtue* (Chicago: University of Chicago Press, 2008).

16. See Kristina Petkova and Pepka Boyadjieva, "The Image of the Scientist and Its Functions," *Public Understanding of Science* 3 (1994): 215–24; Roslynn Haynes, *From Faust to Strangelove: Representations of the Scientist in Western Literature* (Baltimore: Johns Hopkins University Press, 1994); and Roslynn Haynes, "From Alchemy to Artificial Intelligence: Stereotypes of the Scientist in Western Literature," *Public Understanding of Science* 12 (2003): 243–53.

17. "Erdoğan: Batının Ahlaksızlıklarını Aldık," *Milliyet*, January 24, 2008. http://www.milliyet.com.tr/2008/01/24/son/sonsiy18.asp (accessed September 22, 2013).

18. Not a founder of the movement, Kemal Atatürk is considered a member of

the second generation of the Young Turks. See Şükrü Hanioğlu, *Atatürk: An Intellectual Biography* (Princeton, NJ: Princeton University Press, 2011), 52.

19. *Atatürk'ün Söylev ve Demeçleri* (Ankara: ATAM, 1997) 2:194–99.

20. Kemal Inal, *Eğitim ve Iktidar* (Istanbul: Utopya, 2008), 23.

21. *Atatürk'ün Söylev ve Demeçleri* 2:194–99.

22. Ibid., 42–46.

23. Mehmet Saray, "Atatürk, Milli Eğitim Davamız ve Komünizm Tehlikesi," *Sosyoloji Konferansları Dergisi* 19 (1981): 87–95.

24. Richard Olson, *Science and Scientism in Nineteenth-Century Europe* (Urbana: University of Illinois Press, 2008); Tom Sorell, *Scientism: Philosophy and the Infatuation with Science* (London: Routledge, 1991); Harry Collins and Robert Evans, "King Canute Meets the Beach Boys: Responses to 'The Third Wave,'" *Social Studies of Science* 33 (2003): 435–52.

25. On the leader and his followers, see Şerif Mardin, *Religion and Social Change in Modern Turkey: The Case of Bediuzzaman Said Nursi* (Albany: State University of New York Press, 1989); Ibrahim Abu-Rabi', ed., *Islam at the Crossroads: On the Life and Thought of Bediuzzaman Said Nursi* (Albany: State University of New York Press, 2003); Hakan Yavuz, *Toward an Islamic Enlightenment: The Gülen Movement* (New York: Oxford University Press, 2013).

Bibliography

Primary Sources

JOURNALS AND NEWSPAPERS

Âfak
Âfitab-ı Maarif
Basiret
Çocuklara Mahsus Gazete
Dağarcık
Etfal
Gayret
Gülşen
Güneş
Hadika
Hadika-i Maarif
Hafta
Hazine-i Fünûn
Hürriyet
İbret
Kasa
Maarif
Malûmat
Manzara
Mecmua-i Ebuzziya
Mecmua-i Fünûn
Mir'at
Mirat-ı Alem
Mümeyyiz
Revnak
Ruzname-i Ceride-i Havadis
Saadet

Sandık
Sirac
Takvim-i Vekayi
Tarik
Tasvir-i Efkâr
Terakki
Tercüman-ı Ahval
Tercüman-ı Hakikat
Ulûm
Umran
Vakit
Vasıta-i Terakki

PUBLISHED SOURCES

Abdurrahman Şeref. *Tarih Musahebeleri.* Istanbul: Matbaa-i Amire, 1921 [1340].

Ahmed Cevdet Paşa. *Tezakir.* Edited by Cavid Baysun. 4 vols. Ankara: Türk Tarih Kurumu, 1953–63.

Ahmed Hamdi Efendi. *Nasayihü'ş-şübbân.* Istanbul: Mehmed Esad, 1880–81 [1298].

Ahmed Lütfi Efendi *Vak'anüvis Ahmed Lütfi Efendi Tarihi.* Vols. 1–8. Istanbul: YKY, 1999.

———. *Vak'anüvis Ahmed Lütfi Efendi Tarihi.* Vols. 9–15, edited by Münir Aktepe. Istanbul: Edebiyat Fakültesi, 1984–93.

Ahmed Midhat. *Acaib-i Alem.* Ankara: TDK, 2000 [1882–83 (1299)].

———. *Beşir Fuad.* Istanbul: Oğlak, 1996 [1888 (1305)].

———. *Felsefe Metinleri.* Edited by Erdoğan Erbay and Ali Utku. Erzurum: Babil, 2002.

Ahmed Rasim, trans. *Fonoğraf: Sadayı Tahrir ve Iade Eden Alet.* Istanbul: K. Bağdadlıyan, 1884–85 [1302].

Ahmed Rasim. *Kıraat Kitabı.* Istanbul: Karabet, 1888–89 [1306].

Ahmed Remzi. *Müfredat-ı Tıb ve İlm-i Tedavi.* Istanbul, 1871–72 [1288].

Ali İhsan Eğribozi. *Çocuklara Talim-i Fezail-i Ahlak.* Istanbul: Kütübhane-i Cihan, 1894–95 [1312].

Ali Münif. *Ali Münif Efendi'nin Hatıraları.* Edited by Taha Toros. Istanbul: Isis, 1996.

Al-Tahtawi, Rifaa. *An Imam in Paris: Al-Tahtawi's Visit to France 1826–31.* Translated and edited by David Newman. London: Saqi Books, 2002.

Atatürk'ün Söylev ve Demeçleri. Ankara: ATAM, 1997.

Basiretçi Ali Efendi. *Istanbul Mektupları.* Edited by Nuri Sağlam. Istanbul: Kitabevi, 2001.

Berberyan, Nişan, trans. *Kimya.* Istanbul: Nişan Berberyan Matbaası, 1891–92 [1309].

Besim Ömer. *Dişlerin Hıfzıssıhhati.* Istanbul: Mihran, 1883–84 [1301].

Beşir Fuad. *Şiir ve Hakikat.* Edited by Handan Inci. Istanbul: YKY, 1999.

Bianchi, Thomas X. "Miroir des corps. Écrit en turc par Chani-zadéh." *Revue Encyclopédique* 10 (1821): 294–99.

———. *Vocabulaire Français-Turc*. Paris: Everat, 1831.

Çetin, Atilla. "Maarif Nazırı Ahmed Zühdü Paşa"nın Osmanlı Imparatorluğundaki Yabancı Okullar Hakkındaki Raporu." *Güney-doğu Avrupa Araştırmaları Dergisi* 10 (1981): 189–219.

Clarke, Hyde. "Public Instruction in Turkey." *Journal of the Statistical Society of London* 30 (1867): 502–34.

"Commercial Relations with Turkey." *The British and Foreign Review; or, European Quarterly Journal* 5 (1837): 468–69.

Coumbary, Aristide. "Fall of a Meteorite." *Nature* (March 24, 1870): 538.

———. "Lettre de M. Aristide Coumbary." *Comptes Rendus* (May 29, 1865): 1114–15.

Daubrée, G. A. "Note sur un metéorite tombée le 20 Mai, 1874, en Turquie á Virba près Vidin." *Comptes Rendus* 79 (1874): 276–77.

De Kay, James Ellsworth. *Sketches of Turkey in 1831 and 1832 by an American*. New York: J. and J. Harper, 1833.

De Leon, Edwin. "The Old Ottoman and the Young Turk." *Harper's Magazine* 44 (1872): 606–12.

Derviş Mehmed Emin Paşa. *Usûl-i Kimya*. Istanbul: Darü't-tıbaatü'l-Amire, 1847–48 [1264].

Emin. *Menakıb-ı Kethüdazade el-Hac Mehmed Arif Efendi*. Istanbul, 1877.

Ersoy, Mehmet Akif. *Safahat*. Ankara: TC Kültür Bakanlığı, 1989.

Faik Hilmi. *Cep Defteri*. Istanbul: Kitapçı Tahir Efendi, 1887–88 [1305].

Fatma Aliye. *Ahmet Cevdet Paşa ve Zamanı*. Istanbul: Bedir, 1995 [1913].

Halid Eyüb Yenişehirlizade. *Islam ve Fünun*. Istanbul: Cemal Efendi Matbaası, 1897–98 [1315].

Haydar Daniş, trans. *Medhal-i İlm-i Heyet yahud Kozmografya*. Istanbul: Matbaa-i Osmaniye, 1892–93 [1307].

Hertslet, Lewis, ed. *A Complete Collection of the Treaties and Conventions and Reciprocal Regulations at Present Subsisting between Britain and Foreign Powers*. London: Henry Butterworth, 1840.

Heuschling, Xavier. *L'Empire de Turquie*. Bruxelles: H. Tarlier; Paris: Guillemin, 1860.

Hüseyin Hulki. *Siyam*. Istanbul: Mihran, 1892–93 [1310].

Hüseyin Rahmi. *İffet*. Istanbul: Ikdam, 1896–97 [1314].

Hüseyin Remzi. *İlmihal-i Tıbb*. Istanbul: Mahmud Bey Matbaası, 1887–88 [1305].

Inal, Ibnülemin M. K. *Son Asır Türk Şairleri*. 12 vols. Istanbul: Devlet Matbaası, 1930–42.

Inalcık, Halil. "The Hatt-i Sherif of Gülhane. 3 November 1839." In *The Middle East and North Africa in World Politics: A Documentary Record*, edited by J. C. Hurewitz. New Haven, CT: Yale University Press, 1975.

Ishak Efendi. *Mecmua-i Ulum-ı Riyaziye*. Cairo: Bulak Matbaası, 1841–45 [1831–34].

İsmail Cenabî, trans. *İlm-i Eşya*. Istanbul, 1884–85 [1302].

Kaplan, Mehmet et al., eds. *Yeni Türk Edebiyatı Antolojisi*. Istanbul: Istanbul Üniversitesi Edebiyat Fakültesi Yayınları, 1978.

Kurdoğlu, Veli Behçet, ed. *Şair Tabibler*. Istanbul: Baha, 1967.

Kurzman, Charles, ed. *Modernist Islam, 1840–1940*. New York: Oxford University Press, 2002.

Langlès, L.-M. "Lettre de M. L. Langlès à M. Millin, sur un ouvrage imprimé en français en 1803 dans l'imprimerie de scutari." *Magasin Encyclopedique* 5 (1809): 5–11.

MacFarlane, Charles. *Reminiscences of a Literary Life*. London: John Murray, 1917.

———. *Turkey and Its Destiny*. 2 vols. London: John Murray, 1850.

Mahmud Es'ad. *Tarih-i Sanayi*. Izmir: Hizmet, 1890 [1307].

Mahmud Raif. *Journal du Voyage de Mahmoud Raif Efendi en Angleterre. écrit par lui meme*. Reproduced in Vahdettin Engin, "Mahmud Raif Efendi Tarafından Kaleme Alınmış Ingiltere Seyahati Gözlemleri" in *Prof. Dr. Ismail Aka Armağanı*. İzmir: Beta, 1999.

———. *Osmanlı İmparatorlıuğu'nda Yeni Nizamların Cedveli*. Translated by Arslan Terzioğlu and Hüsrev Hatemi. Istanbul: Turing, 1988.

Meclis-i Mebusan Zabıt Ceridesi. Vol. 1. Edited by Hakkı Tarık Us. Istanbul: Vakit, 1939.

———. Vol. 2. Edited by Hakkı Tarık Us. Istanbul: Vakit, 1954.

Mehmed Ali Fethi. *İlm-i Tabakât-ı Arz*. Istanbul: Darü't-Tibaatü'l-Amire, 1852–53 [1269].

Mehmed Fahri. *Hıfzıssıhhat-ı Siyam ve Erkân-ı Sâire-i Din-i İslam*. Istanbul: Mahmud Bey Matbaası, 1890–91 [1308].

———. *Yâdigâr-ı Hekîm*. Istanbul: Mahmut Bey, 1899–1900 [1317].

Mehmed Mansur. *Meşhur İskenderiye Kütübhanesine Dair Risaledir*. Istanbul: Ceride-i Askeriye Matbaası, 1882–83 [1300].

Mehmed Murad (Mizancı). *Turfanda mı Turfa mı?* Istanbul: Mahmud Bey Matbaası, 1890–91 [1308].

Mehmed Rakım and Mustafa Nail. *Hayat-ı Düvel*. Istanbul: Maviyan, 1888–89 [1306].

Mehmed Süreyya. *Sicill-i Osmanî*. 6 vols. Istanbul: Tarih Vakfı Yurt Yayınları, 1996.

Mehmed Tahir. *Türklerin Ulum ve Fününa Hizmetleri*. Istanbul: Ikdam, 1896–97 [1314].

Menemenlizade Tahir. *Yad-i Mazi*. Istanbul: Cemal Efendi, 1888.

Millingen, Frederick. *La Turquie sous le Règne d'Abdul-Aziz, 1862–1867*. Paris: Librarie Internationale, 1868.

"Moltke's Campaign against the Egyptians." *Macmillan's Magazine* 46 (1882): 473–81.

Mordtmann, Andreas David. *Istanbul ve Yeni Osmanlılar. Stambul und das Moderne Türkentum*. Translated by Gertraude Habermann Songu. Istanbul: Pera, 1999 [1877].

Muallim Naci. *Mekteb-i Edeb. Birinci Kısım*. Istanbul: Kitapçı Arakel, 1892–93 [1310].

———. *Musa bin Ebu'l Gazan, yahud Hamiyyet*. Istanbul: Matbaa-i Ebuzziya, 1881–82 [1299].

———. *Terkib-i Bend-i Muallim Naci*. [Ruse: Bulgaria]: Tuna Vilayeti Matbaası, [1890?].

Münşi. *Bedraka-i Mühendisîn*. Istanbul: Mihran, 1884–85 [1302].

Musa Akyiğitzade. *Avrupa Medeniyetinin Esasına Bir Nazar*. Istanbul: Cemal Efendi Matbaası, 1897–98 [1315].

Mustafa Behçet. *Hezar Esrar*. Istanbul: Muhib, 1868–69 [1285].

Mustafa Sami. "Avrupa Risalesi," in *Bir Osmanlı Bürokratının Avrupa İzlenimleri: Mustafa Sami Efendi ve Avrupa Risalesi*, edited by Fatih Andı. Istanbul: Kitabevi, 1996.

Mustafa Şevket. *Kitab-ı Burhan-ı Hakikat*. Istanbul: Mihran, 1882–83 [1299].

Mustafa Zihni, trans. *Mikyas-ü'l Ahlak*. Istanbul: Alem, 1897–98 [1315].

Namık Kemal. *Namık Kemal'in Hususi Mektupları*. Edited by Fevziye A. Tansel. Ankara: Türk Tarih Kurumu, 1969.

———. *Namık Kemal'in Türk Dili ve Edebiyatı Üzerine Görüşleri ve Yazıları*. Edited by Kazım Yetiş. Istanbul: İÜ Edebiyat Fakültesi, 1989.

Necib Asım. *Kıraat-ı Fenniye*. Istanbul: Karabet, 1887–88 [1305].

Nurican Efendi. *Müslümanların Tababete Ettikleri Hizmet, yahud Sevabık-ı Maarifimizden Bir Nebze*. Translated by Hüseyin Hulki. Istanbul: Mihran, 1882–83 [1300].

Ömer Subhi, trans. *Coğrafya-yı Hikemî*. Istanbul: Ceride-i Askeriye, 1883–84 [1301].

Osmanoğlu, Ayşe. *Babam Abdülhamid*. Istanbul: Güven, 1960.

Pardoe, Julia. *The City of the Sultan and Domestic Manners of the Turks in 1836*. London: Henry Colburn, 1838.

Planat, Jules. *Histoire de la régénération de l'Egypte*. Paris: J. Barbezat, 1830.

Porter, James. *Turkey: Its History and Progress*. 2 vols. London: Hurst and Blackett, 1854.

Prince de Joinville. *Memoirs Vieux Souvenirs of Prince de Joinville*. Translated by Mary Loyd. New York: Macmillan, 1895.

"Proceedings at Boston, May 7th, 1884." *Journal of the American Oriental Society* 11 (1885): clxxvi–ccii.

"Public Instruction in Turkey." *American Journal of Education* 4 (1870): 17–31.

Reşad and Ibrahim. *Kıraat—Beşinci Seneye Mahsus*. Istanbul: Karabet, 1895–96 [1313].

———. *Kıraat—Dördüncü Seneye Mahsus*. Istanbul: Karabet, 1895–96 [1313].

Rifat. *Fezail-i Ahlak*. Istanbul: Mekteb-i Sanayi, 1893–94 [1311].

Royer, Alphonse. "Sultan Mahmoud II." *Revue de Paris* 53 (1837): 205–40.

Sadık Rifat Pasha. *Müntehabat-ı Asar* Istanbul: Divitciyan, 1873–76 [1290–93].

———. *Zeyl-i Risale-i Ahlak*. Istanbul: Darü't-Tabaatü'l-Amire, 1856–57 [1273].

Sadreddin Şükrü. *Nuhbet-ü'l Fezail*. Istanbul: Alem, 1895–96 [1313].

Şanizade Mehmed Ataullah. *Miyarü'l-Etıbbâ*. Istanbul: Tabhane-i Şahane, 1820.

Şemseddin Sami. *Gök*. Istanbul: Mihran, 1879 [1296].

————. *Medeniyyet-i İslamiyye*. Istanbul: Mihran, 1879 [1296].

————. *Yer*. Istanbul: Mihran, 1879 [1296].

Şerafeddin Mağmumi. *Vücud-ı Beşer*. Istanbul: Nişan Berberyan, 1892–93 [1310].

Seyyid Mustafa. *Diatribe de l'Ingénieur Séïd Moustapha sur l'état actuel de l'art militaire. du génie et des sciences à Constantinople*, edited by L. Langles. Paris: Ferra, 1810 [1803].

Smithsonian Institution. *List of Foreign Correspondents of the Smithsonian Institution*. Washington, DC: Smithsonian, 1872.

"State of the Turkish Press at Constantinople." *The Oriental Herald and Journal of General Literature* 17 (1828): 309–15.

Süslü, Zeynep and Ismail Kara. "Şemseddin Sami'nin 'Medeniyet'e Dair Dört Makalesi." *Kutadgu Bilig* 4 (2003): 259–83.

Tarhan, Abdülhak Hamid. *Abdülhak Hamid'in Tiyatroları*. Vol. 1. Edited by İnci Enginün. Istanbul: Dergah, 1998.

Taşköprülüzade Ahmed. *Osmanlı Bilginleri (eş-Şakâiku'n-nu'mâ-niyye fî ulemâi'd-devleti'l-Osmâniyye)*. Translated by Muharrem Tan. Istanbul: Iz, 2007.

Thomason, Edward. *Sir Edward Thomason's Memoirs During Half a Century*. London: Longman, Brown, Green and Longmans, 1845.

Tokgöz, Ahmet İhsan. *Matbuat Hatıralarım, 1888–1923*. Istanbul: Ahmet İhsan, 1930.

Veinstein, Gilles, ed. *Le Paradis des Infidèles: Relation de Yirmisekiz Çelebi Mehmed Efendi, Ambassadeur Ottoman en France sous la Régence*. Paris: La Découvert, 2004.

Von Moltke, Helmuth. *Türkiye'deki Durum ve Olaylar Üzerine Mektuplar*. Translated by Hayrullah Örs. Ankara: Türk Tarih Kurumu, 1960 [1841].

Wanda. *Souvenirs Anecdotiques sur la Turquie 1820–1870*. Paris: Firmin-Didot, 1884.

Secondary Sources

Abrams, Philip. "Notes on the Difficulty of Studying the State." *Journal of Historical Sociology* 1 (1977): 58–89.

Abu-Manneh, Butrus. "The Islamic Roots of the Gülhane Rescript." *Die Welt des Islams* 34 (1994): 173–203.

Abu-Rabi', Ibrahim, ed. *Islam at the Crossroads: On the Life and Thought of Bediuzzaman Said Nursi*. Albany: State University of New York Press, 2003.

Adas, Michael. *Machines as the Measure of Men: Science, Technology, and Ideologies of Western Dominance*. Ithaca, NY: Cornell University Press, 1990.

Adıvar, Adnan. *Osmanlı Türklerinde İlim*. 4th ed. Istanbul: Remzi, 1982. Revised Turkish translation of *La Science chez les Turcs Ottomans* Paris: G. P. Maisonneuve, 1939.

Ahmed Emin. *The Development of Modern Turkey as Measured by Its Press*. New York: AMS Press, 1968 [1914].

Akgündüz, Hasan. *Klasik Dönem Osmanlı Medrese Sistemi: Amaç, Yapı, İşleyiş.* Istanbul: Ulusal, 1997.

Akı, Niyazi. *Türk Tiyatro Edebiyatı Tarihi.* Vol. 1. Istanbul: Dergah, 1989.

———. *XIX. Yüzyıl Türk Tiyatrosunda Devrin Hayat ve İnsanı:Sosyopsikolojik Deneme.* Erzurum: Atatürk Üniversitesi, 1974.

Aksan, Virginia. *An Ottoman Statesman in War and Peace: Ahmed Resmi Efendi 1700–1783.* Leiden: Brill, 1995.

Aksoy, Musa. *Moderniteye Karşı Geleneğin Direnişi: Hacı İbrahim Efendi.* Ankara: Akçağ, 2005.

Akyüz, Kenan. *Batı Te'sirinde Türk Şiiri Antolojisi.* 3rd ed. Ankara: Doğuş, 1970.

———. *Encümen-i Dâniş.* Ankara: Ankara Üniversitesi Basımevi, 1975.

Akyüz, Yahya. *Türk Eğitim Tarihi, Başlangıçtan 2001'e.* Istanbul: Alfa, 2001.

Algar, Hamid. *Mīrzā Malkum Khān: A Study in the History of Iranian Modernism.* Berkeley: University of California Press, 1973.

Arjomand, Kamran. "The Emergence of Scientific Modernity in Iran: Controversies Surrounding Astrology and Modern Astronomy in the Mid-Nineteenth Century." *Iranian Studies* 30 (1997): 5–24.

Asad, Talal. "The Concept of Cultural Translation in British Social Anthropology." In *Writing Culture: The Poetics and Politics of Ethnography*, edited by James Clifford and George Marcus, 141–64. Berkeley: University of California Press, 1986.

Atay, Hüseyin. *Osmanlılarda Yüksek Din Eğitimi: Medrese Programları, İcazetnameler, Islahat Hareketleri.* Istanbul: Dergah, 1983.

Aydın, Cemil. *The Politics of Anti-Westernism in Asia: Visions of World Order in Pan-Islamic and Pan-Asian Thought.* New York: Columbia University Press, 2007.

Aydüz, Selim. "Osmanlı İmparatorluğunda Müneccimbaşılık." In *Osmanlı Bilimi Araştırmaları*, edited by Feza Günergun, 159–207. Istanbul: IU Edebiyat Fakültesi, 1995.

Aynî, Mehmed Ali. *Darülfünûn Tarihi*, edited by Aykut Kazancıgil. Istanbul: Kitabevi, 2007 [1927].

Ayvazoğlu, Beşir. "Edebiyatımızda Endülüs," in *Endülüs'ten İspanya'ya*, 79–85. Ankara: Türkiye Diyanet Vakfı, 1996.

Baber, Zaheer. *The Science of Empire: Scientific Knowledge, Civilization, and Colonial Rule in India.* Albany: State University of New York Press, 1996.

Babich, Babette. "Nietzsche's Critique of Scientific Reason and Scientific Culture: On 'Science as a Problem' and 'Nature as Chaos.'" In *Nietzsche and Science*, edited by Gregory M. Moore and Thomas Brobjer, 133–53. Aldershot: Ashgate, 2004.

Bailey, Paul. *Reform the People: Changing Attitudes towards Popular Education in Early 20th Century China.* Edinburgh: Edinburgh University Press, 1990.

Balcı, Sezai. *Osmanlı Devleti'nde Tercümanlık ve Bab-ı Âli Tercüme Odası.* PhD diss., Ankara University, 2006.

Barberis, Daniela S. "Moral Education for the Elite of Democracy: The *Classe de philosophie* between Sociology and Philosophy." *Journal of the History of the Behavioral Sciences* 38 (2002): 355–69.

Barkey, Karen. *Empire of Difference: The Ottomans in Comparative Perspective.* Cambridge: Cambridge University Press, 2008.

Barnes, Barry, and David Bloor. "Relativism, Rationalism and the Sociology of Knowledge." In *Rationality and Relativism*, edited by Martin Hollis and Steven Lukes, 21–47. Oxford: Blackwell. 1982.

Barton, Ruth. "Just before Nature: The Purposes of Science and the Purposes of Popularization in Some English Popular Science Journals of the 1860s." *Annals of Science* 55 (1998): 1–33.

Baş, Eyüp. *Dil-Tarih İlişkisi Bağlamında Osmanlı Türklerinde Tarih Yazıcılığı: XVI. ve XVII. Yüzyıl Örnekleriyle.* Ankara: Türkiye Diyanet Vakfı, 2006.

Bates, Don. "Why Not Call Modern Medicine 'Alternative'?" *Perspectives in Biology and Medicine* 43 (2000): 502–18.

Bauman, Zygmunt. *Modernity and the Holocaust.* Cambridge: Polity, 1989.

———. *Postmodern Ethics.* Cambridge: Polity, 1993.

Bein, Amit. "The Istanbul Earthquake of 1894 and Science in the Late Ottoman Empire." *Middle Eastern Studies* 44 (2008): 909–24.

Belhoste, Bruno, Amy Dahan-Dalmédico, and Antoine Picon, eds. *La Formation polytechnicienne: 1794–1994.* Paris: Dunod, 1994.

Berkes, Niyazi. *The Development of Secularism in Turkey.* New York: Routledge, 1988 [1964].

Beydilli, Kemal. *İlk Mühendislerimizden Seyyid Mustafa ve Nizâm-i Cedîd'e dair risâlesi.* Istanbul: Edebiyat Fakültesi, 1983–87.

———. *Mahmud Râif Efendi ve Nizâm-ı Cedîd'e dâir eseri.* Ankara: Türk Tarih Kurumu, 2001.

———. *Türk Bilim ve Matbaacılık Tarihinde Mühendishâne. Mühendishâne Matbaası ve Kütüphânesi, 1776–1826.* Istanbul: Eren, 1995.

Bilim, Cahit. "İlk Türk Bilim Akademisi: Encümen-i Daniş." *Hacettepe Üniversitesi Edebiyat Fakültesi Dergisi* 3 (1985): 81–104.

Black, Anthony. *The History of Islamic Political Thought from the Prophet to the Present.* Edinburgh: Edinburgh University Press, 2011.

Bourdieu, Pierre. *Distinction: A Social Critique of the Judgment of Taste.* Cambridge. MA: Harvard University Press, 1979.

———. "The Forms of Capital." In *Handbook of Theory and Research for the Sociology of Education*, edited by John Richardson, 241–58. New York: Greenwood, 1986.

———. *The Logic of Practice.* Cambridge: Cambridge University Press, 1990.

———. "Rethinking the State: Genesis and Structure of the Bureaucratic Field." In *State/Culture: State-Formation after the Cultural Turn*, edited by George Steinmetz, 53–75. Ithaca, NY: Cornell University Press. 1999.

———. "Social Space and Symbolic Power." *Sociological Theory* 7 (1989): 14–25.

———. *The State Nobility: Elite Schools in the Field of Power.* Stanford, CA: Stanford University Press, 1996.

Bourdieu, Pierre, and Jean-Claude Passeron. *Reproduction in Education, Society and Culture*. London: Sage, 1977.

Boyar, Ebru. "The Press and the Palace: The Two-Way Relationship between Abdülhamid II and the Press, 1876–1908." *Bulletin of the School of Oriental and African Studies* 69 (2006): 417–32.

Brooks, Jeffrey. *When Russia Learned to Read: Literacy and Popular Literature, 1861–1917*. Princeton, NJ: Princeton University Press, 1985.

Budak, Ali. *Batılılaşma Sürecinde Çok Yönlü Bir Osmanlı Aydını: Münif Paşa*. Istanbul: Kitabevi, 2004.

Burçak, Berrak. "Modernization, Science and Engineering in the Early Nineteenth-Century Ottoman Empire." *Middle Eastern Studies* 44 (2008): 69–83.

———. "Science, A Remedy For All Ills. Healing 'The Sick Man of Europe': A Case for Ottoman Scientism." PhD diss., Princeton University, 2005.

Çankaya, Ali. *Yeni Mülkiye Tarihi ve Mülkiyeliler*. Vol. 2. Ankara: SBF, 1969.

Carroll, Patrick. *Science, Culture, and Modern State Formation*. Berkeley: University of California Press, 2006.

Çeçen, M. Korkut. "Encümen-i Şuara'nın Tanzimat Birinci Dönem Sanatçılarına Etkisi." *Çukurova Üniversitesi Sosyal Bilimler Enstitüsü Dergisi* 15 (2006): 133–52.

Çelik, Hüseyin. *Ali Süavi ve Dönemi*. Istanbul: İletişim, 1994.

Çelik, Zeynep. *Displaying the Orient: Architecture of Islam at Nineteenth-Century World's Fairs*. Berkeley: University of California Press, 1992.

Collins, Harry. "Public Experiments and Displays of Virtuosity: The Core-Set Revisited." *Social Studies of Science* 18 (1988): 725–48.

Collins, Harry, and Robert Evans. "King Canute Meets the Beach Boys: Responses to 'The Third Wave.'" *Social Studies of Science* 33 (2003): 435–52.

Cunningham, Andrew. "Getting the Game Right: Some Plain Words on the Identity and Invention of Science." *Studies in the History and Philosophy of Science* 19 (1988): 365–89.

Darnton, Robert. *Mesmerism and the End of the Enlightenment in France*. Cambridge, MA: Harvard University Press, 1968.

Daston, Lorraine, and Peter Galison. *Objectivity*. Cambridge, MA: MIT Press, 2007.

Davison, Roderic. "The Advent of the Electric Telegraph in the Ottoman Empire." In *Essays in Ottoman and Turkish History. 1774–1923: The Impact of the West*, edited by Roderic Davison, 133–65. Austin: University of Texas Press, 1990.

———. "Halil Şerif Paşa: The Influence of Paris and the West on an Ottoman Diplomat," in *Nineteenth-Century Ottoman Diplomacy and Reforms*, 81–94. Istanbul: Isis, 1999 [1986].

Dear, Peter. "Religion, Science and Natural Philosophy: Thoughts on Cunningham's Thesis." *Studies in the History and Philosophy of Science* 32 (2001): 377–86.

Deringil, Selim. "Legitimacy Structures in the Ottoman State: The Reign of Abdulhamid II, 1876–1909." *International Journal of Middle Eastern Studies* 23 (1991): 345–59.

———. *The Well-Protected Domains: Ideology and the Legitimation of Power in the Ottoman Empire, 1876–1909*. London: Tauris, 1998.

DeWitt, Anne. *Moral Authority, Men of Science, and the Victorian Novel*. Cambridge: Cambridge University Press, 2013.

Dizer, Muammer. "Rasathane-i Amire." *Bilim Tarihi* 2 (1993): 3–10.

Eickelman, Dale. "Mass Higher Education and the Religious Imagination in Contemporary Arab Societies." *American Ethnologist* 19 (1992): 643–55.

Ege, Nezahat Nureddin. *Prens Sabahaddin: Hayatı ve İlmî Müdafaaları*. Istanbul: Güneş, 1977.

Ekhtiar, Maryam. "Nasir al-Din Shah and the Dar al-Funun: The Evolution of an Institution." *Iranian Studies* 34 (2001): 153–63.

El-Abbadi, Mostafa, and Omnia Mounir Fathallah, eds. *What Happened to the Ancient Library of Alexandria?* Leiden: Brill, 2008.

Eldem, Edhem. "Capitulations and Western Trade." In *The Cambridge History of Turkey*, vol. 3: *The Later Ottoman Empire, 1603–1839*, edited by Suraiya Faroqhi, 283–335. New York: Cambridge University Press. 2006.

———. "Fransa'ya Eğitime Gönderilen Sadrazam Ibrahim Edhem." *Popüler Tarih* 2 (2006): 50–53.

Elman, Benjamin A. "'Universal Science' Versus 'Chinese Science': The Changing Identity of Natural Studies in China, 1850–1930." *Historiography East and West* 1 (2003): 68–116.

Elshakry, Marwa. "Exegesis of Science in 20th Century Arabic Interpretations." In *Nature and Scripture in the Abrahamic Religions: 1700–Present*, vol. 2, edited by Jitse van der Meer and Scott Mandelbrote, 491–524. Leiden: Brill, 2008.

———. *Reading Darwin in Arabic*. Chicago: University of Chicago Press, 2014.

———. "When Science Became Western: Historiographical Reflections." *Isis* 101 (2010): 98–109.

Enginün, İnci. "Edebiyatımızda Endülüs" in *Araştırmalar ve Belgeler*, 32–41. Istanbul: Dergah, 2000.

Epstein, Steven. *Impure Science: AIDS, Activism, and the Politics of Knowledge*. Berkeley: University of California Press, 1996.

———. *Inclusion: The Politics of Difference in Medical Research*. Chicago: University of Chicago Press, 2007.

Ergin, Osman Nuri. *İstanbul Mektepleri ve İlim. Terbiye ve San'at Müesseseleri Dolayisiyle Türkiye Maarif Tarihi*. 2nd ed. Istanbul: Eser, 1977.

Evans, John H. "Epistemological and Moral Conflict between Religion and Science." *Journal for the Scientific Study of Religion* 50 (2011): 707–27.

Evans, Michael S. "Defining the Public, Defining Sociology: Hybrid Science, Public Relations, and Boundary-Work in Early American Sociology." *Public Understanding of Science* 18 (2009): 5–22.

Evans, Michael S., and John H. Evans. "Arguing against Darwinism: Religion, Science, and Public Morality." In *The New Blackwell Companion to the Sociology of Religion*, edited by Bryan S. Turner, 286–308. Oxford: Blackwell, 2010.

Ezrahi, Yaron. *The Descent of Icarus: Science and the Transformation of Contemporary Democracy*. Cambridge, MA: Harvard University Press, 1992.

Fara, Patricia. "An Attractive Therapy: Animal Magnetism in Eighteenth-Century England." *History of Science* 33 (1995):127–77.

Findley, Carter. *Bureaucratic Reform in the Ottoman Empire: The Sublime Porte, 1789–1922*. Princeton, NJ: Princeton University Press, 1980.

———. *Ottoman Civil Officialdom: A Social History*. Princeton, NJ: Princeton University Press, 1989.

———. "An Ottoman Occidentalist in Europe: Ahmed Midhat Meets Madame Gulnar, 1889." *American Historical Review* 103 (1998): 15–49.

———. *Turkey, Islam, Nationalism, and Modernity*. New Haven, CT: Yale University Press, 2010.

Fortna, Benjamin. "Education and Autobiography at the End of the Ottoman Empire." *Die Welt des Islams* 41 (2001): 1–31.

———. *Imperial Classroom: Islam, the State, and Education in the Late Ottoman Empire*. Oxford: Oxford University Press, 2002.

———. "Islamic Morality in Late Ottoman 'Secular' Schools." *International Journal of Middle Eastern Studies* 32 (2000): 369–93.

Foucault, Michel. "What Is An Author?" In *Aesthetics, Method, and Epistemology*, edited by James D. Faubion, 205–22. New York: New Press, 1998.

Fox, Robert, and George Weisz, eds. *The Organization of Science and Technology in France, 1808–1914*. Cambridge: Cambridge University Press, 1980.

Fyfe, Aileen, and Bernard Lightman, eds. *Science in the Marketplace: Nineteenth-Century Sites and Experiences*. Chicago: University of Chicago Press, 2007.

Gallagher, Nancy E. *Medicine and Power in Tunisia, 1780–1900*. Cambridge: Cambridge University Press, 1983.

Gaziano, Emanuel. "Ecological Metaphors as Scientific Boundary Work: Innovation and Authority in Interwar Sociology and Biology." *American Journal of Sociology* 101 (1996): 874–907.

Georgeon, François. "La formation des élites à la fin de l'Empire ottoman: Le cas de Galatasaray." *Revue du monde musulman et de la Méditerranée* 72 (1994): 15–25.

Gieryn, Thomas. "Boundary-Work and the Demarcation of Science from Non-science: Strains and Interests in Professional Ideologies of Scientists." *American Sociological Review* 48 (1983): 781–95.

———. *Cultural Boundaries of Science: Credibility on the Line*. Chicago: University of Chicago Press, 1999.

———. "Cultural Boundaries: Settled and Unsettled." In *Clashes of Knowledge: Orthodoxies and Heterodoxies in Science and Religion*, edited by Peter Meusburger et al., 91–100. Berlin: Springer, 2008.

Göçek, Fatma Müge. *East Encounters West: France and the Ottoman Empire in the Eighteenth Century*. Oxford: Oxford University Press, 1987.

———. *Rise of the Bourgeoisie, Demise of Empire: Ottoman Westernization and Social Change*. New York: Oxford University Press, 1996.

Göçgün. Önder, ed. *Ziya Paşa'nın Hayatı. Eserleri. Edebi Şahsiyeti ve Bütün Şiirleri*. Ankara: Kültür ve Turizm Bakanlığı, 1987.

Gregory, Frederick. *Scientific Materialism in Nineteenth-Century Germany*. Dordrecht: D. Reidel, 1977.

Günergun, Feza. "Derviş Mehmed Emin pacha 1817–1879, serviteur de la science et de l'État ottoman." In *Medecins et Ingénieurs Ottomans à l'âge des Nationalismes*, edited by Méropi Anastassiadou-Dumont, 171–83. Paris: Maisonneuve et Larose, 2002.

———. "Ondokuzuncu Yüzyıl Türkiye'sinde Kimyada Adlandırma." *Osmanlı Bilimi Araştırmaları* 1 (2003): 1–33.

———. "The Ottoman Ambassador's Curiosity Coffer: Eclipse Prediction with De La Hire's 'Machine' Crafted by Bion of Paris." *Science between Europe and Asia. Boston Studies in the Philosophy of Science* 275 (2011): 103–23.

Hamadeh, Shirine. "Ottoman Expressions of Early Modernity and the 'Inevitable' Question of Westernization." *Journal of the Society of Architectural Historians* 63 (2004): 32–51.

Hanioğlu, Şükrü. *Atatürk: An Intellectual Biography*. Princeton, NJ: Princeton University Press, 2011.

———. "Blueprints for a Future Society: Late Ottoman Materialists on Science, Religion and Art." In *Late Ottoman Society: The Intellectual Legacy*, edited by Elisabeth Özdalga, 28–116. London: Routledge, 2005.

———. *A Brief History of the Late Ottoman Empire*. Princeton, NJ: Princeton University Press, 2008.

———. "The Historical Roots of Kemalism." In *Democracy, Islam, and Secularism in Turkey*, edited by Ahmet Kuru and Alfred Stepan, 32–56. Columbia University Press, 2012.

———. *The Young Turks in Opposition*. New York: Oxford University Press, 1995.

Haraway, Donna. *Modest_Witness@Second_Millennium. FemaleMan©_Meets_Onco-Mouse™: Feminism and Technoscience*. New York: Routledge, 1997.

Harding, Sandra. *Science and Social Inequality: Feminist and Postcolonial Issues*. Urbana: University of Illinois Press, 2006.

Harrison, Peter. "'Science' and 'Religion': Constructing the Boundaries." *Journal of Religion* 86 (2006): 81–106.

Haynes, Roslynn. "From Alchemy to Artificial Intelligence: Stereotypes of the Scientist in Western Literature." *Public Understanding of Science* 12 (2003): 243–53.

———. *From Faust to Strangelove: Representations of the Scientist in Western Literature*. Baltimore: Johns Hopkins University Press, 1994.

Heper, Metin. "Center and Periphery in the Ottoman Empire: With Special Reference to the Nineteenth Century." *International Political Science Revie/Revue internationale de science politique* 1 (1980): 81–105.

Hess, David. *Science in the New Age: The Paranormal, Its Defenders and Debunkers, and American Culture*. Madison: University of Wisconsin Press, 1993.

Heyd, Uriel. "The Ottoman Ulema and Westernization in the Time of Selim III and Mahmud II." In *The Modern Middle East: A Reader*, edited by Albert Hourani et al., 29–59. Berkeley: University of California Press, 1993.

Huff, Toby. *The Rise of Early Modern Science: Islam, China, and the West*. Cambridge: Cambridge University Press, 1993.

İhsanoğlu, Ekmeleddin. *Başhoca İshak Efendi: Türkiye'de Modern Bilimin Öncüsü*. Ankara: Kültür Bakanlığı, 1989.

———. "Cemiyet-i Ilmiye-i Osmaniye'nin Kuruluş ve Faaliyetleri." In *Osmanlı İlmî ve Meslekî Cemiyetleri*, edited by Ekmeleddin İhsanoğlu, 197–220. Istanbul: İÜ Edebiyat Fakültesi, 1987.

———. *Darülfünun: Osmanlı'da Kültürel Modernleşmenin Odağı*. Istanbul: IRCICA, 2010.

———. "Institutionalisation of Science in the Medreses of Pre-Ottoman and Ottoman Turkey." In *Turkish Studies in the Philosophy and History of Science*, edited by Gürol Irzık and G. Güzeldere, 265–84. Dordrecht: Springer, 2005.

———. "19. yy. Başında Kültür Hayatı ve Beşiktaş Cemiyet-i İlmiyesi." *Belleten* 51 (1987): 801–20.

———. "Ottoman Science: The Last Episode in the Islamic Scientific Tradition and the Beginning of European Scientific Tradition." In *Science, Technology and Industry in the Ottoman World. Proceedings of the XXth International Congress of History of Science*, vol. 6, edited by E. İhsanoğlu et al., 11–48. Turnhout: Brepols, 2000.

———. *Science, Technology, and Learning in the Ottoman Empire: Western Influence, Local Institutions, and the Transfer of Knowledge*. Aldershot: Ashgate, 2004.

———. "Tanzimat Döneminde İstanbul'da Darülfünun Kurma Teşebbüsleri." In *150. Yılında Tanzimat*, edited by Hakkı D. Yıldız, 397–439. Ankara: Türk Tarih Kurumu, 1992.

Inal, Kemal. *Eğitim ve İktidar*. Istanbul: Utopya, 2008.

Inalcık, Halil and Donald Quaetert, eds. *An Economic and Social History of the Ottoman Empire, 1300–1914*. Cambridge: Cambridge University Press, 1994.

Iqbal, Muzaffar. *Science and Islam*. Westport, CT: Greenwood, 2007.

Irwin, Alan, and Brian Wynne, eds. *Misunderstanding Science? The Public Reconstruction of Science and Technology*. Cambridge: Cambridge University Press, 1996.

Izgi, Cevat. *Osmanlı Medreselerinde İlim*. 2 vols. Istanbul: İz, 1997.

Jamison, Andrew. "National Styles of Science and Technology: A Comparative Model." *Sociological Inquiry* 57 (1987): 144–58.

Jones, Matthew. *The Good Life in the Scientific Revolution: Descartes, Pascal, Leibniz, and the Cultivation of Virtue*. Chicago: University of Chicago Press, 2008.

Kara, Ismail. "Modernleşme dönemi Türkiyesi'nde 'ulûm,' 'fünûn' ve 'sanat' kavramlarının algılanışı üzerine birkaç not." In *Din ve Modernleşme Arasında: Çağdaş Türk Düşüncesinin Meseleleri*, 126–97. Istanbul: Dergah, 2003.

———. "Tarih ve Hurafe." In *Din ve Modernleşme Arasında: Çağdaş Türk Düşüncesinin Meseleleri*, 75–109. Istanbul: Dergah, 2003.

———. *Türkiye'de İslâmcılık Düşüncesi: Metinler, Kişiler*. Istanbul: Risale, 1986.

Karaçavuş, Ahmet. *Tanzimat Dönemi Osmanlı Bilim Cemiyetleri*. PhD diss., Ankara University, 2006.

Karal, Enver Ziya. "Osmanlı Tarihinde Türk Dili Sorunu." In *Bilim, Kültür ve Öğretim Dili Olarak Türkçe*, edited by Aydın Sayılı, 7–96. Ankara: Türk Tarih Kurumu, 1978.

Karpat, Kemal. *Ottoman Population, 1830–1914: Demographic and Social Characteristics*. Madison: University of Wisconsin Press, 1985.

———. *The Politicization of Islam: Reconstructing Identity, State, Faith, and Community in the Late Ottoman State*. New York: Oxford University Press, 2001.

Kasaba, Reşat. *The Ottoman Empire and the World-Economy: The Nineteenth Century*. Albany: State University of New York Press, 1988.

Kayalı, Hasan. *Arabs and Young Turks: Ottomanism, Arabism, and Islamism in the Ottoman Empire, 1908–1918*. Berkeley: University of California Press, 1997.

———. "Islam in the Thought and Politics of Two Late Ottoman Intellectuals: Mehmed Akif and Said Halim." *Archivum Ottomanicum* 19 (2001): 307–33.

Kaynar, Reşat. *Mustafa Reşit Paşa ve Tanzimat*. Ankara: TTK, 1954.

Kazancıgil, Aykut. *XIX. Yuzyılda Osmanlı İmparatorluğunda Anatomi*. Istanbul: Özel, 1991.

Keddie, Nikki R. *An Islamic Response to Imperialism: Political and Religious Writings of Sayyid Jamal ad-Din "al-Afghani."* Berkeley: University of California Press, 1968.

Keyder, Çağlar. "Europe and the Ottoman Empire in Mid-Nineteenth Century: Development of a Bourgeoisie in the European Mirror." In *East Meets West: Banking and Commerce in the Ottoman Empire*, edited by Philip Cottrell, 41–58. Aldershot: Ashgate, 2008.

———. *State and Class in Turkey*. London: Verso, 1987.

Kılınç, Berna. "Yirmisekiz Mehmed Çelebi's Travelogue and the Wonders that Make a Scientific Centre." In *Travels of Learning: A Geography of Science in Europe*, edited by Ana Simões et al., 77–100. Dordrecht: Kluwer, 2003.

Koloğlu, Orhan. *Takvim-i Vekâyi: Türk Basınında 150 Yıl*. Ankara: ABS, 1982.

Küçük, Harun. "Early Enlightenment in Istanbul." PhD diss., University of California, San Diego, 2012.

Kuntay, Mithat Cemal. *Namık Kemal, Devrinin İnsanları ve Olayları Arasında*. Istanbul: Milli Eğitim Basımevi, 1949.

Kuran, Timur. "The Economic Ascent of the Middle East's Religious Minorities: The Role of Islamic Legal Pluralism." *Journal of Legal Studies* 33 (2004): 475–515.

Kutay, Cemal. *Sultan Abdülaziz'in Avrupa Seyahati*. Istanbul: Boğaziçi, 1991.

Lamont, Michèle. *Money, Morals, and Manners: The Culture of the French and American Upper-Middle Class*. London: University of Chicago Press, 1992.

Lamont, Michèle, and Annette Lareau. "Cultural Capital: Allusions, Gaps and Glissandos in Recent Theoretical Developments." *Sociological Theory* 6 (1988): 153–68.

Lamont, Michèle, and Virag Molnar. "The Study of Boundaries in the Social Sciences." *Annual Review of Sociology 28* (2002): 167–95.

Lewis, Bernard. *What Went Wrong? The Clash between Islam and Modernity in the Middle East.* New York: Oxford University Press, 2002.

Livingstone, David. *Putting Science in Its Place: Geographies of Scientific Knowledge.* Chicago: University of Chicago Press, 2003.

Lucier, Paul. "The Professional and the Scientist in Nineteenth Century America." *Isis* 100 (2009): 699–732.

Ma, Hongming. *The Images of Science Through Cultural Lenses: A Chinese Study on the Nature of Science.* Rotterdam, The Netherlands: Sense, 2012.

MacLeod, Roy. "The 'Bankruptcy of Science' Debate: The Creed of Science and Its Critics, 1885–1900." *Science, Technology. and Human Values* 7 (1982): 2–15.

———, ed. *Nature and Empire: Science and the Colonial Enterprise. Osiris*, 2nd ser., vol. 15. Chicago: University of Chicago Press, 2001.

Mahmud Cevad İbnü'ş-şeyh Nâfi. *Maarif-i Umumiye Nezareti Tarihçe-i Teşkilat ve İcraatı: XIX. Asır Osmanlı Maarif Tarihi.* Edited by Taceddin Kayaoğlu. Ankara: Yeni Türkiye, 2001.

Mardin, Şerif. *The Genesis of Young Ottoman Thought: A Study in the Modernization of Turkish Political Ideas.* Princeton, NJ: Princeton University Press, 1962.

———. *Religion and Social Change in Modern Turkey: The Case of Bediuzzaman Said Nursi.* Albany: State University of New York Press, 1989.

———. "Super-Westernization in Urban Life in the Last Quarter of the Nineteenth Century." In *Turkey: Geographical and Social Perspectives*, edited by Peter Benedict et al., 403–46. Leiden: Brill, 1974.

Marlow, Louise. *Hierarchy and Egalitarianism in Islamic Thought.* Cambridge: Cambridge University Press, 1997.

Martin, Emily. *Flexible Bodies: Tracking Immunity in American Culture: From the Days of Polio to the Age of AIDS.* Boston: Beacon Press, 1994.

Mauss, Marcel. *The Gift.* Oxon: Routledge, 2002.

Messick, Brinkley. *The Calligraphic State: Textual Domination and History in a Muslim Society.* Berkeley: University of California Press, 1993.

Mizrachi, N., T. Shuval, and S. Gross. "Boundary at Work: Alternative Medicine in Biomedical Settings." *Sociology of Health and Illness* 27 (2005): 20–43.

Moore, Kelly. *Disrupting Science: Social Movements, American Scientists, and the Politics of the Military, 1945–1975.* Princeton, NJ: Princeton University Press, 2011.

Morus, Iwan Rhys. *Frankenstein's Children: Electricity, Exhibition, and Experiment in Early Nineteenth-Century London.* Princeton, NJ: Princeton University Press, 1998.

Nelson, Blair. "'Men Before Adam!': American Debates over the Unity and Antiquity of Humanity." In *When Science and Christianity Meet*, edited by David C. Lindberg and Ronald Numbers, 161–81. Chicago: University of Chicago Press, 2003.

Neumann, Christoph K. "Political and Diplomatic Developments." In *The

Cambridge History of Turkey: The Later Ottoman Empire, 1603–1839, edited by Suraiya Faroqhi, 44–62. Cambridge: Cambridge University Press, 2006.

———. "Whom Did Ahmed Cevdet Represent?" In *Late Ottoman Society: The Intellectual Legacy*, edited by Elisabeth Özdalga, 117–34. New York: Routledge, 2005.

Okay, Orhan. *Beşir Fuad: İlk Türk Pozitivist ve Natüralisti*. Istanbul: Hareket, 1969.

Olson, Richard. *Science and Scientism in Nineteenth-Century Europe*. Urbana: University of Illinois Press, 2008.

Özervarlı, Said. "Alternative Approaches to Modernization in the Late Ottoman Period." *International Journal of Middle East Studies* 39 (2007): 102–25.

Pakalın, Mehmed Zeki. *Safvet Paşa*. Istanbul: Ahmet Sait, 1943.

Palladino, Paolo, and Michael Worboys. "Science and Imperialism." *Isis* 84 (1993): 91–102.

Paul, Harry W. *The Sorcerer's Apprentice: The French Scientist's Image of German Science 1840–1919*. Gainesville: University of Florida Press, 1972.

Petitjean, Patrick. "Science and the 'Civilizing Mission': France and the Colonial Enterprise." In *Science across the European Empires, 1800–1950*, edited by Benedikt Stutchey, 107–28. Oxford: Oxford University Press, 2005.

Petkova, Kristina, and Pepka Boyadjieva. "The Image of the Scientist and Its Functions." *Public Understanding of Science* 3 (1994): 215–24.

Pistor-Hatam, Anja. "Iran and the Reform Movement in the Ottoman Empire: Persian Travellers. Exiles and Newsmen under the Impact of the Tanzimat." In *Proceedings of the Second European Conference of Iranian Studies*, 561–78. Rome: Is.M.E.O, 1995.

Piterberg, Gabriel. *An Ottoman Tragedy: History and Historiography at Play*. Berkeley: University of California Press, 2003.

Portes, Alejandro. "Social Capital: Its Origins and Applications in Modern Sociology." *Annual Review of Sociology* 24 (1998): 1–24.

Prakash, Gyan. *Another Reason: Science and the Imagination of Modern India*. Princeton, NJ: Princeton University Press, 1999.

Prasad, Monica, et al. "The Undeserving Rich: 'Moral Values' and the White Working Class." *Sociological Forum* 24 (2009): 225–53.

Quataert, Donald. "Ottoman Workers and the State." In *Workers and Working Classes in the Middle East: Struggles, Histories, Historiographies*, edited by Zachary Lockman, 21–40. Albany: State University of New York Press, 1994.

Radcliffe, David Hill. "Charles Macfarlane: Reminiscences of a Literary Life" http://lordbyron.cath.lib.vt.edu/contents.php?doc=ChMacfa.1917. Accessed August 22, 2013.

Reingold, Nathan, and Marc Rothenberg. *Scientific Colonialism: A Cross-Cultural Comparison*. New York: Smithsonian, 1987.

Rodrigue, Aron. *French Jews, Turkish Jews: The Alliance Israélite Universelle and the Politics of Jewish Schooling in Turkey, 1860–1925*. Bloomington: Indiana University Press, 1990.

Rojo, L. M., and T. van Dijk. "'There Was a Problem, and It Was Solved!': Legitimating the Expulsion of 'Illegal' Migrants in Spanish Parliamentary Discourse." *Discourse and Society* 8 (1997): 523–66.

Rosenthal, Franz. *Knowledge Triumphant: The Concept of Knowledge in Medieval Islam.* Leiden: Brill, 1970.

Ross, Sydney. "'Scientist': The Story of a Word." *Annals of Science* 18 (1962): 65–85.

Roth, A. L., J. Dunsby, and L. A. Bero. "Framing Processes in Public Commentary on US Federal Tobacco Control Regulation." *Social Studies of Science* 33 (2003): 7–44.

Şakul, Kahraman. "Nizam-ı Cedid Düşüncesinde Batılılaşma ve İslami Modernleşme." *Divan İlmi Araştırmalar Dergisi* 19 (2005): 117–50.

Saray, Mehmet. "Atatürk, Milli Eğitim Davamız ve Komünizm Tehlikesi." *Sosyoloji Konferansları Dergisi* 19 (1981): 87–95.

Sarı, Nil. "Cemiyet-i Tıbbiyye-i Osmaniyye ve Tıp Dilinin Türkçeleşmesi Akımı," in *Osmanlı İlmî ve Meslekî Cemiyetleri,* edited by Ekmeleddin İhsanoğlu, 121–42. Istanbul: Edebiyat Fakültesi, 1987.

Schaffer, Simon. "The Consuming Flame: Electrical Showmen and Tory Mystics in the World of Goods." In *Consumption and the World of Goods,* edited by J. Brewer and R. Porter, 489–526. London: Routledge, 1993.

———. "Natural Philosophy and Public Spectacle in the Eighteenth Century." *History of Science* 21 (1983): 1–41.

Schayegh, Cyrus. *Who Is Knowledgeable Is Strong: Science, Class, and the Formation of Modern Iranian Society, 1900–1950.* Berkeley: University of California Press, 2009.

Sewell, William, Jr. "The Concepts of Culture." In *Beyond the Cultural Turn: New Directions in the Study of Society and Culture,* edited by Victoria E. Bonnell and Lynn Hunt, 35–61. Berkeley: University of California Press, 1999

Shapin, Steven. "Cordelia's Love: Credibility and the Social Studies of Science." *Perspectives on Science* 3 (1995): 255–75.

———. "Of Gods and Kings: Natural Philosophy and Politics in the Leibniz–Clarke Disputes." *Isis* 72 (1981): 187–215.

———. *The Scientific Life: A Moral History of a Late Modern Vocation.* Chicago: University of Chicago Press, 2008.

———. *A Social History of Truth: Civility and Science in Seventeenth-Century England.* Chicago: University of Chicago Press, 1994.

Shapin, Steven, and Barry Barnes. "Science, Nature and Control: Interpreting Mechanics' Institutes." *Social Studies of Science* 7 (1977): 31–74.

Shapin, Steven, and Simon Schaffer. *Leviathan and the Air-Pump: Hobbes, Boyle and the Experimental Life.* Princeton, NJ: Princeton University Press, 1985.

Shaw, Wendy M. *Possessors and Possessed: Museums, Archaeology, and the Visualization of History in the Late Ottoman Empire.* Berkeley: University of California Press, 2003.

Sinaplı, Ahmet Nuri. *Devlete Millete Beş Padişah Devrinde Kıymetli Hizmetlerde Bulunan Şeyhülvüzera Serasker Mehmed Namık Paşa.* Istanbul: Yenilik, 1987.

Şişman, Adnan. *Tanzimat Döneminde Fransa'ya Gönderilen Osmanlı Öğrencileri 1839–1876*. Ankara: Türk Tarih Kurumu, 2004.

Somel, Selçuk Akşin. "Kırım Savaşı. Islahat Fermanı ve Eğitim." http://research .sabanciuniv.edu/5529/1/Kirim_Savasi._Islahat_Fermani_ve_Egitim.pdf Accessed September 9, 2013.

———. *The Modernization of Public Education in the Ottoman Empire*. Leiden: Brill, 2001.

Sorell, Tom. *Scientism: Philosophy and the Infatuation with Science*. London: Routledge, 1991.

Southerton, Dale. "Boundaries of 'Us' and 'Them': Class, Mobility and Identification in a New Town." *Sociology* 36 (2002): 171–93.

Star, Susan Leigh, and J. R. Griesemer. "Institutional Ecology: 'Translations' and Boundary Objects: Amateurs and Professionals in Berkeley's Museum of Vertebrate Zoology, 1907–39." *Social Studies of Science* 19 (1989): 387–420.

Stöckelová, Tereza. "Immutable Mobiles Derailed: STS, Geopolitics, and Research Assessment." *Science, Technology, and Human Values* 37 (2012): 286–311.

Stock-Morton, Phyllis. *Moral Education for a Secular Society: The Development of Morale Laïque in Nineteenth-Century France*. Albany: State University of New York Press, 1988.

Strauss, Johann. "The Greek Connection in Nineteenth Century Ottoman Intellectual History." In *Greece and the Balkans: Identities, Perceptions and Cultural Encounters since the Enlightenment*, edited by Dimitris Tziovas, 47–67. Aldershot: Ashgate, 2003.

———. "The Millets and the Ottoman Language: The Contribution of Ottoman Greeks to Ottoman Letters, 19th–20th Centuries." *Die Welt des Islams* 35 (1995): 189–249.

Sungu, Ihsan. "Tanzimat ve Yeni Osmanlılar." In *Tanzimat I*, 777–857. Istanbul: Maarif, 1940.

Swartz, David. "Bridging the Study of Culture and Religion: Pierre Bourdieu's Political Economy of Symbolic Power." *Sociology of Religion* 57 (1996): 71–85.

Tampakis, Konstantinos. "Onwards Facing Backwards: The Rhetoric of Science in Nineteenth-century Greece." *British Journal for the History of Science* (2013), doi:10.1017/S000708741300040X. Accessed September 23, 2013.

Tanpınar, Ahmet Hamdi. *XIX. Asır Türk Edebiyatı Tarihi*. Istanbul: İbrahim Horoz, 1956.

Teber, Serol. *Paris Komünü'nde Üç Yurtsever Türk: Mehmet. Reşat ve Nuri Beyler*. Istanbul: De, 1986.

Teich, Mikulas. "Circulation, Transformation: Conservation of Matter and the Balancing of the Biological World in the Eighteenth Century." *Ambix* 29 (1982): 17–28.

Thiem, Jon. "The Great Library of Alexandria Burnt: Towards the History of a Symbol." *Journal of the History of Ideas* 40 (1979): 507–26.

Thorpe, Charles. *Oppenheimer: The Tragic Intellect*. Chicago: University of Chicago Press, 2006.

Thurs, Daniel P. *Science Talk: Changing Notions of Science in American Popular Culture.* New Brunswick, NJ: Rutgers University Press, 2007.

Timmermans, Stefan, and Marc Berg. "Standardization in Action: Achieving Local Universality through Medical Protocols." *Social Studies of Science* 27 (1997): 273–305.

Topham, Jonathan. "Science and Popular Education in the 1830s: The Role of the Bridgewater Treatises." *British Journal for the History of Science* 25 (1992): 397–430.

Topuz, Hıfzı. *II. Mahmut'tan holdinglere Türk basın tarihi.* Istanbul: Remzi, 2003.

Türker, Deniz. *The Oriental Flâneur: Khalil Bey and the Cosmopolitan Experience.* MA thesis, MIT, 2007.

Türköne, Mümtaz'er. *Siyasi Düşünce Olarak İslamcılığın Doğuşu.* Istanbul: İletişim, 1991.

Turner, Frank M. "The Victorian Conflict between Science and Religion: A Professional Dimension." *Isis* 69 (1978): 356–76.

Unan, Fahri. *Kuruluşundan Günümüze Fatih Külliyesi.* Ankara:Türk Tarih Kurumu, 2003.

———. "Taşköprülü-zâde'nin Kaleminden XVI. Yüzyılın 'İlim ve Âlim' Anlayışı." *Osmanlı Araştırmaları* 17 (1997): 149–264.

Unat, Ekrem Kadri. "Muallim Miralay Dr. Hüseyin Remzi Bey ve Türkçe Tıp Dilimiz." In *IV. Türk Tıp Tarihi Kongresi Kitabı,* 239–52. Ankara: TTK, 2003.

Uzluk, Feridun Nafiz. *Hekimbaşı Mustafa Behçet: Zâtı, Eserleri Üstüne Bir Araştırma.* Ankara: Ankara Üniversitesi Tıp Tarihi Enstitüsü, 1954.

Vassiadis, George A. *The Syllogos Movement of Constantinople and Ottoman Greek Education, 1861–1923.* Athens: Centre for Asia Minor Studies, 2007.

Willis, Martin. "George Eliot's *The Lifted Veil* and the Cultural Politics of Clairvoyance." In *Victorian Literary Mesmerism,* edited by Martin Willis and Catherine Wynne, 145–62. Amsterdam: Rodopi, 2006.

Worringer, Renée. "'Sick Man of Europe' or 'Japan of the Near East'?: Constructing Ottoman Modernity in the Hamidian and Young Turk Eras." *International Journal of Middle East Studies* 36 (2004): 207–30.

Yalçınkaya, Alaeddin. "Mahmud Raif Efendi as the Chief Secretary of Yusuf Agah Efendi, The First Permanent Ottoman-Turkish Ambassador to London, 1793–1797." *OTAM* 5 (1994): 385–434.

Yalçınkaya, M. Alper. "Science as an Ally of Religion: A Muslim Appropriation of the 'Conflict Thesis.'" *British Journal for the History of Science* 44 (2011): 161–81.

Yavuz, Hakan. *Toward an Islamic Enlightenment: The Gülen Movement.* New York: Oxford University Press, 2013.

Yeo, Richard. *Defining Science: William Whewell, Natural Knowledge and Public Debate in Early Victorian Britain.* Cambridge: Cambridge University Press, 1993.

Yetiş, Kazım. "Cevdet Paşa'nın Belâgât-ı Osmaniyyesi ve Uyandırdığı Akisler." In *Belagatten Retoriğe,* edited by Kazım Yetiş, 246–99. Istanbul: Kitabevi, 2006.

———. "Yeni bir Edebiyat Anlayışı." In *Dönemler ve Problemler Aynasında Türk Edebiyatı,* edited by Kazım Yetiş, 35. Istanbul: Kitabevi, 2007.

Yinanç, Mükrimin Halil. "Tanzimattan sonra bizde tarihçilik." In *Tanzimat I: Yüzüncü Yılı Münasebetile*. Ankara: Türk Tarih Kurumu, 1940.

Yücel, Yaşar. *Osmanlı devlet teşkilâtına dair kaynaklar*. Ankara: Türk Tarih Kurumu Basımevi, 1988.

Zaman, Muhammad Q. *The Ulama in Contemporary Islam: Custodians of Change*. Princeton, NJ: Princeton University Press, 2002.

Ziadat, Adel. *Western Science in the Arab World: The Impact of Darwinism, 1860–1930*. London: Macmillan, 1986.

Index

Cevdet Pasha (Ahmed), 62–64, 95, 136, 191, 252n93, 276n33
chemistry, 35–36, 49–50, 65–66, 68, 70–71, 73, 76, 87, 92–93, 126, 130, 132, 137, 142, 159, 163, 168, 190, 195, 199, 212, 270n57; organic, 115, 181–82
China, 50, 86, 93, 164, 184, 279n7
Christianity, 50, 109, 155, 160, 205, 272n78
circulation of matter, 114–15
citizenship, 3, 92, 100, 165, 179, 184, 218, 245n7. *See also* Ottoman Citizenship Law; Ottomanism
civilization, 55, 60, 62, 79–81, 86, 88–89, 91, 93–94, 101–3, 105, 107, 130, 139, 145, 147, 155–56, 162, 186, 196, 198–99, 204, 208, 212–13, 222, 249n54; Arab, 138, 147, 158, 163, 165; Eurocentric notions criticized, 105–7, 116, 129; of Europe, 2, 33, 54, 84–85, 127, 147, 158–61, 163, 207; Islamic, 108–12, 118, 158–59
civilizing mission, 49, 87, 93
Clarke, Hyde, 89
class, 24, 73, 198; of knowers, 13, 16–17, 49, 56, 82, 211; traditional Ottoman ruling, 10–11, 28, 32; working, 8, 94, 233n26. *See also* bourgeoisie; bureaucrats; ulema
classification: of knowledge, 30, 35, 41; of philosophy, 61; processes, 5; of sciences in Islam, 13, 86, 93; struggles, 7, 10
Çocuklara Mahsus Gazete, 200
Collins, Harry, 74
colonialism, 4, 33, 230n9, 259n4
community: identity of, 18, 67, 81, 97, 138, 156–57, 161, 164, 178; loyalty/disloyalty to, 18, 55, 129, 179, 193, 219; moral, 2, 118, 221; Muslim, 8, 17, 43, 45, 48, 67, 80–81, 96–100, 108–10, 120–21, 136, 140, 150–51, 177–78, 216–17; non-Muslim, 8, 45, 79–80, 121, 125, 220, 245n6; service to, 213. *See also* Armenian Ottomans; Greek: community of the Ottoman Empire; Jewish Ottomans; Turks
Comte, Auguste, 193
Constantinidis, Alexander, 92
cosmography, 70, 168
cosmopolitanism: criticized, 223; of *Journal of Sciences*, 84, 92, 108; of Ottoman Learned Academy, 63; of Ottoman society, 82, 99; of Ottoman Society of Science, 82
Coumbary, Aristide, 150–51

Council for Military Affairs, 70
Council of Education, 71, 94, 102, 143; Provisional, 59–60
Council of Poets, 248n43
Council of Public Works, 56, 59

Dağarcık, 114, 116, 118, 126
Dar-al Funun, 78–79
Darülfünûn: first attempt, 72–75, 77–78; professors for, 71; recommended by Council of Education, 59–60; second attempt, 141–44; textbooks for, 61; third attempt, 144–45
Darwin, Charles, 87, 116, 181, 187, 189, 274n5
Day-Age Theory, 196
demonstration, 73–78, 144. *See also* experiment
Deringil, Selim, 152
Derviş Paşa, 65–66, 73–74, 76, 252n99
diplomacy, 11, 70
diplomats, 16, 26, 36, 37, 42, 46–47, 98, 122, 209–11. *See also* ambassadors
disciplining, 18–19, 56, 58, 102, 131, 155, 172, 179, 184, 205, 256n68, 271n74
discourse, 3, 6–7, 11, 15, 19, 40–41, 50, 77, 82, 88–89, 95, 97, 100, 131, 134, 136–37, 158, 167–68, 182–86, 193, 198, 208, 222–23, 273n99; alternative, 17–18; alternative (Young Ottoman), on science, 34, 67, 97, 99–104, 107, 110–12, 117, 125, 141, 143, 146, 149–51, 164, 217; authoritative, 233n20; official, 3, 9, 16–18, 22; official, on science, 42–44, 48, 58, 60, 67, 72, 79–80, 86, 94, 96–97, 105, 113, 119, 122–23, 152, 161, 179, 205–6, 209–12, 215, 217, 244n1; Ottomanist, 81, 84
disobedience. *See* obedience
Diyojen, 150
doctrine, Islamic. See *akaid*
Donizetti, Giuseppe, 28
Draper, John William, 155, 268n9
due reciprocity, principle of, 24, 210–11
Duhamel, Jean-Marie, 174
Duruy, Victor, 79–80
Dyck, Cornelius van, 191–93, 276n37

Earth, 86–87, 195–96, 201, 205; age of, 87, 181; emergence of humans on, 114, 116; formation of, 114, 126
Ebuzziya Tevfik, 108, 121, 129, 137, 163
École des mines, 22, 65–66, 251n84